Mitochondrial Inhibitors and Neurodegenerative Disorders

Contemporary Neuroscience

Cerebral Signal Transduction: From First to Fourth Messengers, edited by **Maarten E. A. Reith,** 2000

Central Nervous System Diseases: Innovative Animal Models from Lab to Clinic, edited by **Dwaine F. Emerich, Reginald L. Dean, III, and Paul R. Sanberg,** 2000

Mitochondrial Inhibitors and Neurodegenerative Disorders, edited by **Paul R. Sanberg, Hitoo Nishino, and Cesario V. Borlongan,** 2000

Neurotransmitter Transporters: Structure, Function, and Regulation, edited by **Maarten E. A. Reith,** 1997

Motor Activity and Movement Disorders: Research Issues and Applications, edited by **Paul R. Sanberg, Klaus-Peter Ossenkopp,** and **Martin Kavaliers,** 1996

Neurotherapeutics: Emerging Strategies, edited by **Linda M. Pullan** and **Jitendra Patel,** 1996

Neuron–Glia Interrelations During Phylogeny: II. Plasticity and Regeneration, edited by **Antonia Vernadakis** and **Betty I. Roots,** 1995

Neuron–Glia Interrelations During Phylogeny: I. Phylogeny and Ontogeny of Glial Cells, edited by **Antonia Vernadakis** and **Betty I. Roots,** 1995

The Biology of Neuropeptide Y and Related Peptides, edited by **William F. Colmers** and **Claes Wahlestedt,** 1993

Psychoactive Drugs: Tolerance and Sensitization, edited by **A. J. Goudie** and **M. W. Emmett-Oglesby,** 1989

Experimental Psychopharmacology, edited by **Andrew J. Greenshaw** and **Colin T. Dourish,** 1987

Developmental Neurobiology of the Autonomic Nervous System, edited by **Phyllis M. Gootman,** 1986

The Auditory Midbrain, edited by **Lindsay Aitkin,** 1985

Neurobiology of the Trace Elements, edited by **Ivor E. Dreosti** and **Richard M. Smith**

Vol. 1: *Trace Element Neurobiology and Deficiencies,* 1983

Vol. 2: *Neurotoxicology and Neuropharmacology,* 1983

Mitochondrial Inhibitors and Neurodegenerative Disorders

Edited by

Paul R. Sanberg, PhD, DSc

University of South Florida College of Medicine, Tampa, FL

Hitoo Nishino, MD, PhD

Nagoya City University Medical School, Nagoya, Japan

Cesario V. Borlongan, PhD

National Institutes of Health, Baltimore, MD

Foreword by

Joseph T. Coyle, MD

Eben S. Draper Professor of Psychiatry

Harvard University Medical School, Boston, MA

Humana Press

Totowa, New Jersey

999 Riverview Drive, Suite 208
Totowa, New Jersey 07512

This publication is printed on acid-free paper. ∞
ANSI Z39.48-1984 (American Standards Institute) Permanence of Paper for Printed Library Materials.

Cover illustration:

For additional copies, pricing for bulk purchases, and/or information about other Humana titles, contact Humana at the above address or at any of the following numbers: Tel.: 973-256-1699; Fax: 973-256-8341; E-mail: humana@humanapr.com or visit our website at http://www.humanapress.com

Printed in the United States of America. 10 9 8 7 6 5 4 3 2 1

Library of Congress Cataloging in Publication Data

Main entry under title:
Mitochondial inhibitors and neurodegenerative disorders / edited by Paul R. Sanberg, Hitoo Nishino, Cesario V. Borlongan.
p. cm.—(Contemporary neuroscience)
Includes index.
ISBN 0-89603-805-X (alk. paper)
1. Nervous system—Degeneration—Pathophysiology. 2. Nervous system—Degeneration—animal models. 3. Mitochondrial pathology. 4. Neurotoxic aganets. I. Sanberg , Paul R. II. Nishino, Hitoo. III. Borlongan, Cesario V. IV. Series.
[DNLM: 1. Neurodegenerative Diseases—chemically induced. 2. Propionic Acids—toxicity. 3. Mitochondria—metabolism. 4. Neurotoxins—toxicity. WL 359 M684 1999]
RC394.D35M56 1999
616.8'047—dc21
DNLM/DLC 98-55467
for Library of Congress CIP

Dedications

To my father and best friend, Bernard Sanberg, in memorium—Paul

To my wife, Akiko, and loving mother and father—Hitoo

To my inspirations, Christine Stahl and Mia Borlongan—Cesar

Foreword

Mitochondria have long been the Rodney Dangerfield of cellular organelles. Believed to be the remnants of bacterial infection of eukaryotic cells eons ago, the mitochondrion evolved a symbiotic relationship in which it dutifully served as the efficient source of ATP for cell function. The extraordinary dependence of cells on the energy provided by mitochondrial oxidative metabolism of glucose, especially through critical organs such as the heart and brain, is underlined by the fatal consequences of toxins that interfere with the mitochondrial electron transport system.

Consistent with their ancestry, the mitochondria have their own DNA that encodes many but not all of their proteins. The mitochondria and their genes come from the mother via the ovum since sperm do not possess mitochondria. This extranuclear form of inheritance derived exclusively from the female side has proven to be a powerful tool for tracing the evolution by the number of base substitutions in mtDNA.

That mitochondrial gene mutations might be a source of human disease became evident a decade ago with the characterization of a group of multisystem disorders, typically involving the nervous system, which are transmitted from mother to child. Specific point mutations in mtDNA have been associated with the different syndromes.

The central role of mitochondria in neurodegenerative disorders has become apparent over the last decade as the molecular mechanisms causing cell death have come under scientific scrutiny. Reactive oxygen species were shown to be mediators of delayed neuronal degeneration caused by activation of ionotropic glutamate receptors. Oxidative stress was also shown to precipitate programmed cell death or apoptosis. The linkage between these two phenomena related to the facts that the mitochondria are the source of 80% or more of the oxyradicals generated in the neuron and that Ca^{2+} dysregulation causing excessive activation of glutamate ionotropic receptors disrupts the mitochondrial electron.

In this context, *Mitochondrial Inhibitors and Neurodegenerative Disorders* provides a timely, in-depth review of the effects of mitochondrial toxins on the nervous system. What is particularly interesting about the clinical manifestations of the mitochondrial poisons is the uneven vulnerability of neurons, with neurons of the extrapyramidal system exhibiting particular susceptibility. This selective vulnerability mimics that of hereditary neurodegenerative disorders such as Huntington's and Parkinson's Disease. Furthermore, experimental studies indicate that activation of the receptor mediates this selective vulnerability. The insights derived from this line of research suggest novel therapeutic approaches that could prevent the onset of these disorders in individuals at risk.

Joseph T. Coyle, MD

Preface

Mitochondrial Inhibitors and Neurodegenerative Disorders critically surveys all the recent work on the utilization of mitochondrial inhibitors to deepen understanding of the various mechanisms involved in neurodegenerative disorders. The many facets of advances in this field can be divided into the three major areas that we have included here. The first section is concerned with the role of mitochondrial inhibitors in neurodegenerative disorders, a topic that has been the subject of much research this past decade; many neurotoxins that disrupt normal mitochondrial energy metabolism have been identified. The chapters tackled in this first section deal largely with discovery of environmental mitochondrial toxins. A short historical background of these neurotoxins is presented to provide the reader with an understanding of the basic neurochemistry and mode of action of these drugs as they relate to mitochondrial dysfunction.

The second section deals with the development of animal models of those human diseases that in recent years have been suggested to be caused by abnormal mitochondrial function. At the forefront of these mitochondrial deficiency-related disorders is Huntington's disease, and the chapters in this section have thus been written by investigators who have examined these neurotoxic models [specifically 3-nitropropionic acid (3-NP)] into replicating the cellular and anatomical, as well as the behavioral, alterations seen in this disorder. Because of our own keen interest and the significant increase in the recent literature validating the utility of 3-NP in modeling many of the symptoms of Huntington's disease, we have chosen to review the many studies on this neurotoxin. The bulk of information on 3-NP is the concentration of this book and should provide "proof of principle" that mitochondrial inhibitors, in general, play an important role in the etiology of central nervous system disorders.

Finally, any validation of the usefulness of a drug for modeling specific human disease leads to the development of treatment strategies. The third section of *Mitochondrial Inhibitors and Neurodegenerative Disorders* thus discusses recent therapeutic modalities directed toward rescuing the central nervous system from abnormal mitochondrial functioning.

We very much hope that *Mitochondrial Inhibitors and Neurodegenerative Disorders* will guide students and researchers alike in further establishing the neurobehavioral foundations of the human disorders that are mimicked by administration of mitochondrial inhibitors.

Paul R. Sanberg, PhD, DSc
Hitoo Nishino, MD
Cesario V. Borlongan, PhD

Contents

Contributors

ROGER ALBIN • *Department of Neurology, University of Michigan & Geriatrics Research Education and Clinical Center, Ann Arbor VAMC, Ann Arbor, MI*

TAJRENA ALEXI • *Department of Pediatrics, RCDMB Starship Hospital, Garfton, Auckland, New Zealand*

SAFIA BAGGIA • *Portland State University, Portland, Oregon*

TERRENCE J. BAZZETT • *Department of Psychology, SUNY Geneseo, Geneseo, NY*

JILL B. BECKER • *Department of Psychology, Reproductive Sciences Program, Neuroscience Program, University of Michigan, Ann Arbor, MI*

MARÍA ISABEL BEHRENS • *Las Condes, Santiago, Chile*

ZBIGNIEW BINIENDA • *Division of Neurotoxicology, National Center for Toxicological Research, Jefferson, AR*

VIMALA BONDADA • *Sanders-Brown Center on Aging and Department of Anatomy and Neurobiology, University of Kentucky, Lexington, KY*

CESARIO V. BORLONGAN • *Cellular Neurophysiology, Intramural Research Program, National Institute on Drug Abuse, National Institutes of Health, Baltimore, MD*

D. ALLAN BUTTERFIELD • *Director, Center of Membrane Sciences, and Faculty Associate, Sanders-Brown Center on Aging, Lexington, KY*

JOHN M. CARNEY • *Centaur Pharmaceuticals, Sunnyvale, CA*

CHUANG C. CHIUEH • *Laboratory Chief, Unit on Neurodegeneration and Neuroprotection, LCS, NIMH, NIH, Bethesda, MD*

SHRIPAD B. DESHPANDE • *Department of Physiology, Nagoya City University Medical School, Mizuho-cho, Mizuho-ku, Nagoya, Japan*

JIE DONG • *Brain Research Laboratory, Psychology Department, Central Michigan University, Mt. Pleasant, MI*

GARY L. DUNBAR • *Department of Psychology, Director, Brain Research Laboratory, Central Michigan University, Mount Pleasant, MI*

STEPHEN B. DUNNETT • *MRC Cambridge Centre for Brain Repair and Department of Experimental Psychology, University of Cambridge, Cambridge, UK*

BARRY J. EVERITT • *MRC Cambridge Centre for Brain Repair and Department of Experimental Psychology, University of Cambridge, Cambridge, UK*

RICHARD L. M. FAULL • *Department of Anatomy with Radiology, School of Medicine, University of Auckland, Auckland, New Zealand*

THOMAS B. FREEMAN • *Department of Neurosurgery, University of South Florida College of Medicine, Tampa, FL*

ATSUO FUKUDA • *Department of Physiology, Nagoya City University Medical School, Mizuho-cho, Mizuho-ku, Nagoya, Japan*

S. PRASAD GABBITA • *Sanders-Brown Center on Aging, Department of Chemistry and Center of Membrane Sciences, University of Kentucky, Lexington, KY*

JAMES W. GEDDES • *University of Kentucky Medical Center, Lexington, KY*

DANIEL H. GOULD • *Department of Pathology, College of Veterinary Medicine and Biomedical Sciences, Colorado State University, Fort Collins, CO*

DAVID L. GUSTINE • *USDA-ARS PSWMRL, USDA Pasture Laboratory, University Park, PA*

KRISTI L. HAIK-CREGUER • *Brain Research Laboratory, Psychology Department, Central Michigan University, Mt. Pleasant, MI*

BRADLEY F. HAMILTON • *Bayer Corporation, Agriculture Division, Stilwell, KS*

ROBERT A. HAUSER • *Deparment of Neurology, University of South Florida College of Medicine, Tampa, FL*

PAUL E. HUGHES • *Department of Pharmacology and Clinical Pharmacology, School of Medicine, University of Auckland, Auckland, New Zealand*

GOPAL KRISHNA • *Unit on Neurodegeneration and Neuroprotection, Laboratory of Clinical Science, National Institute of Mental Health, Bethesda, MD*

MICHIKO KUMAZAKI • *Department of Physiology, Nagoya City University Medical School, Mizuho-Cho, Mizuho-ku, Nagoya, Japan*

WEN LIN • *Department of Neurology, University of Wisconsin School of Medicine and Veterans Administration Medical Center, Milwaukee, WI*

ALBERT C. LUDOLPH • *Department of Neurology, University of Ulm, Ulm, Germany*

ALICIA MELDRUM • *MRC Cambridge Centre for Brain Repair, Cambridge, UK*

KEIYA NAKAJIMA • *Department of Physiology, Nagoya City University Medical School, Mizuho-cho, Mizuho-ku, Nagoya, Japan*

HITOO NISHINO • *Department of Physiology, Nagoya City University Medical School, Mizuho-cho, Mizuho-ku, Nagoya, Japan*

KEITH J. PAGE • *MRC Cambridge Centre for Brain Repair and Department of Experimental Psychology, University of Cambridge, Cambridge, UK*

ZHEN PANG • *Sanders-Brown Center on Aging and Department of Anatomy and Neurobiology, University of Kentucky, Lexington, KY*

NORMAN C. REYNOLDS, JR. • *Department of Neurology, University of Wisconsin School of Medicine and Veterans Administration Medical Center, Milwaukee, WI*

MATTHIAS RIEPE • *Department of Neurology, University of Ulm, Ulm, Germany*

MOHAMMAD SABRI • *Oregon Health Sciences University, Portland, OR*

PAUL R. SANBERG • *Department of Neurosurgery, and Program in Neuroscience, University of South Florida School of Medicine, Tampa, FL*

ANDREW C. SCALLET • *Division of Neurotoxicology, National Center for Toxicological Research/FDA, Jefferson, AK*

DEBORAH A. SHEAR • *Brain Research Laboratory, Psychology Department, Central Michigan University, Mt. Pleasant, MI*

YASUNOBU SHIMANO • *Department of Physiology, Nagoya City University Medical School, Mizuho-cho, Mizuho-ku, Nagoya, Japan*

TORU SHIMIZU • *Department of Psychology and the Neuroscience Program, University of South Florida, Tampa, FL*

PETER S. SPENCER • *Center for Research on Occupational and Environmental Toxicology and Department of Neurology, Oregon Health Sciences University, Portland, OR*

CHRISTINE E. STAHL • *Uniformed Services University of Health Sciences, Bethesda, MD 20814*

KUNIO TORII • *Department of Physiology, Nagoya City University Medical School, Mizuho-cho, Mizuho-ku, Nagoya, Japan*

CHUCHARIN UNGSUPARKORN • *Department of Physiology, Nagoya City University Medical School, Mizuho-ku, Nagoya, Japan*

YUN WANG • *Cellular Neurophysiology, Intramural Research Program, National Institute on Drug Abuse, National Institutes of Health, Baltimore, MD*

MOUSSA B. H. YOUDIM • *NIH Fogarty International Center for Advance Studies in Human Health, Unit on Neurodegeneration and Neuroprotection, Laboratory of Clinical Science, National Institute of Mental Health, Bethesda, MD*

GAIL D. ZEEVALK • *Department of Neurology, UMDNJ-Robert Wood Johnson Medical School, Piscataway, NJ*

I

Mitochondrial Toxins

Symptomatology, Origin, and Chemistry

1

Clinical Manifestations and Mechanisms of Action of Environmental Mitochondrial Toxins

Mohammad I. Sabri, Peter S. Spencer, Safia Baggia, and Albert C. Ludolph

INTRODUCTION

There is increasing evidence that defects in mitochondrial energy metabolism play an important role in the pathogenesis of major human neurodegenerative diseases such as Alzheimer's disease (AD), Parkinson's disease (PD), Huntington's disease (HD), amyotrophic lateral sclerosis (ALS), and dystonia. AD is the most common form of dementia that occurs in the elderly and may result from various genetic and environmental influences *(1)*. A genetic defect arising from mitochondrial DNA (mt DNA) that is inherited solely from the mother could account for defects in the electron transport chain and contributes to deficits in energy levels in AD *(2–4)*. Recent work has shown that mutations in cytochrome *c* oxidase may impair energy metabolism that may lead to a cascade of events resulting in AD. PD is characterized clinically by bradykinesia, rigidity, and tremors and pathologically by the damage of dopaminergic neurons in the substantia nigra. Some forms of PD are inherited and other forms may be triggered by environmental agents. A significant decrease of mitochondrial complex I activity has been observed in the substantia nigra of Parkinson's patients *(5–8)*.

HD is a rare genetic degenerative disorder of the brain characterized by irregular, spasmodic, involuntary movement of the limbs or facial muscles and severe mental deterioration *(9)*. HD is caused by a mutation that leads to unstable CAG trinucleotide repeats in the coding sequence of a gene on chromosome 4 that codes huntingtin, a low-molecular-weight protein of 340 kDa. Neither the function of the huntingtin protein nor the biochemical basis of the pathogenesis of HD is understood. However, several lines of evidence suggest that the expanded polyglutamine segment in the huntingtin

From: *Mitochondrial Inhibitors and Neurodegenerative Disorders*
Edited by: P. R. Sanberg, H. Nishino, and C. V. Borlongan © Humana Press Inc., Totowa, NJ

protein causes either a primary or a secondary impairment of energy metabolism leading to neuronal degeneration in the striatum *(10)*.

ALS is a progressive, degenerative disease of the motor neurons of the brain and spinal cord *(11)*. ALS, commonly known as Lou Gehrig's disease, after the famous baseball player who succumbed to this disease, is characterized by a general weakening and wasting of the voluntary muscles that leads to complete paralysis. The etiology of ALS is unknown, although a number of causal factors, namely, aluminum, selenium, heavy metals, and viruses have been suggested *(12)*. Identification of mutations in copper/zinc superoxide dismutase (SOD-1) in a subset of cases of familial ALS *(13)*, as well as mutations in neurofilament heavy chains in some cases of sprodic ALS *(14)*, has led to substantial advances in our understanding of the pathogenesis of this disease. Free radicals have been suggested as key mediators in ALS *(16)*. When an agent interferes with oxidative phosphorylation, ATP synthesis falls, and electrons that move along the transport chain "leak" onto oxygen to form the superoxide anion. The superoxide anion, if not sequestered in time, can damage nerve cells in the brain and spinal cord *(17)*.

Dystonia or dystonic symptoms are a consequence of an abnormality in the basal ganglia. The etiopathogenesis of dystonic syndromes is unknown and may have genetic and environmental components *(18,19)*. Ingestion of sugarcane contaminated with a 3-nitropropionic acid (3-NPA)-producing fungus has been reported to cause irreversible generalized dystonia in humans *(20)*. 3-NPA, an inhibitor of succinic dehydrogenase, a component of mitochondrial complex II, is believed to be the agent that produces encephalopathy and its tardive effects *(19)*.

MITOCHONDRIAL TOXINS AND NEURODEGENERATIVE DISEASES

More than 70,000 chemicals are currently used in industry, most of which have not been tested for their neurotoxic properties *(21)*. The Health Care Financing Administration of the U.S. Department of Health and Human Services reported that $23 billion were spent in 1980 alone for the care of people with neurological diseases *(22)*. The first convincing evidence that a chemical agent, i.e., 1-methyl-4-phenyl-1,2,3,6-tetrahydropyridine (MPTP), causes Parkinsonism in humans was discovered accidently *(23)*. Another human neurodegenerative disease, amyotrophic lateral sclerosis–Parkinsonism-dementia complex (ALS–PDC) may also be triggered by environmental agents *(15,24)*. The occurrence of tropical ataxic neuropathy and konzo in Africa has been attributed to dietary cyanide, a potent inhibitor of mitochondrial enzyme cytochrome *c* oxidase *(25)*. Consumption of *Lathyrus*

Table 1
Selected Neurological Incidents and Exposure to Mitochondrial Toxins

Year	Neurotoxin	Neurologic Effects
1950	Mercury	Hundreds poisoned and died after eating shellfish contaminated with mercury in Japan. Alkyl mercury inhibits citric acid cycle and mitochondrial electron transport chain.
1950	Manganese	150 ore miners suffered chronic manganese (Mn^{2+}) intoxication involving severe neurobehavioral problems in Morocco. Mn^{2+} accumulates preferentially in mitochondria and inhibits oxidative phosphorylation.
1971	Mercury	Several thousand poisoned and hundreds died after consuming bread made from seed grains treated with mercury as fungicide.
1983	MPTP	MPTP contamination in illicit drug found to cause symptoms identical to those of Parkinson's disease in California. MPTP is oxidized to MPP^+, a potent inhibitor of mitochondrial complex I.
1991	3-NPA	3-NPA was responsible for several deaths in China from mildewed sugarcane contaminated with the fungus *Arthrinium* spp. 3-NPA is an irreversible inhibitor of the mitochondrial enzyme succinic dehydrogenase (complex II).
1993	L-BOAA	L-BOAA, the toxic component of *Lathyrus sativus*, causing lathyrism in hundreds of people in Ethiopia and India. Some investigators have suggested that L-BOAA is a potent inhibitor of mitochondrial complex I *(28)*.
1994	Cyanide	The occurrence of neurodegenerative diseases such as tropical ataxic neuropathy and konzo in Africa is attributed to dietary cyanide exposure. Cyanide is a potent inhibitor of mitochondrial enzyme cytochrome *c* oxidase.

Abbreviation: MPTP, 1-methyl-4-phenyl-1,2,3,6-tetrahydropyridine; MPP^+, 1-methyl-4-phenyl-pyridinium ion; 3-NPA, 3-nitropropionic acid; L-BOAA, β-*N*-oxalylamino-L-alanine.

sativus, a protein-rich legume that harbors neurotoxic β-*N*-oxalylamino-L-alanine (L-BOAA), causes a neurological disorder, lathyrism *(26)*. Although the results are not yet confirmed *(27)*, L-BOAA is claimed to be a potent mitochondrial complex I inhibitor *(28)*.

The evidence is mounting that disruption of mitochondrial energy metabolism may be a common biochemical mechanism linking exposure to certain environmental toxins and the onset of neurodegenerative diseases. Table 1 lists a few neurotoxic incidents caused by environmental toxins, some of which are potent mitochondrial toxins. Evidence is mounting that exposure to mitochondrial toxins underlies neuronal degeneration in a number of human neurological diseases including PD, AD, ALS, HD, and dystonia *(4,18,24,29–34)*.

SELECTED MITOCHONDRIAL TOXINS

A number of environmental toxins compromise mitochondrial integrity and inhibit ATP synthesis. The purpose of this chapter is to review clinical manifestations of 3-NPA, MPTP, and cyanide and discuss mechanisms by which they cause neurodegeneration.

3-Nitropropionic Acid

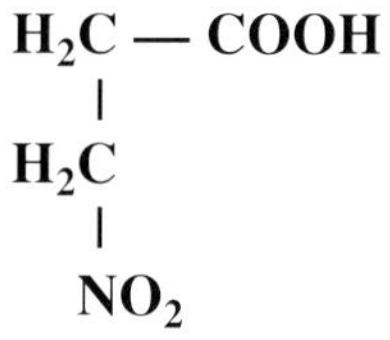

3-NPA and its derivatives are widely distributed aliphatic nitrocompounds in toxic plants such as *Astragalus* spp. *(35)*. 3-NPA was identified in 1954 as the component of *Indigofera endecaphylla (36)*. 3-NPA is also produced by the fungus *Arthrinium* spp., which was responsible for the development of an acute encephalopathy in humans *(18,20)*. Sugarcane is a favorite fruit of Chinese children. It is grown in the southern part of China and normally harvested in October. Each year a large amount of sugarcane is transported to northern China to be stored over the winter for selling during the Chinese New Year in early spring. Owing to improper storage conditions, the sugarcane becomes mildewed and causes acute intoxication, preferentially in children. Outbreaks of acute mildewed sugarcane poisoning usually occur between January and March in northern China *(18,20,37,38)*. Adults may also develop gastrointestinal symptoms after consuming mildewed sugarcane, but they rarely develop central nervous system (CNS) disorders. The high susceptibility of acute mildewed sugarcane poisoning in children may be due to (1) high toxin intake due to higher consumption of mildewed sugarcane or (2) immature blood–brain barrier that may be less resistant to the toxin. Analysis of the mildewed sugarcane collected from patient's families revealed that *Arthrinium* was the predominant fungus, accounting for 50–70% of the total 92 strains of fungi in the samples *(37)*. Subsequent studies revealed that the toxic agent isolated from the *Arthrinium* cultures was 3-NPA *(39)*. Further studies showed that only the *Arthrinium* culture induces paralysis in mice and cats and causes convulsions in dogs showing neurotoxicity from *Arthrinium* cultures; pathological examination of these animals revealed cerebral edema similar to that observed in patients poisoned with mildewed sugarcane *(37)*.

The prime clinical feature of all cases of acute mildewed sugarcane poisoning is gastrointestinal irritation with an abrupt onset. Involvement of the CNS is usually manifested as loss of consciousness, frequent convulsions, extensor plantar reflexes, or diffuse EEG abnormalities indicating a diffuse encephalopathy *(37)*. In moderately affected patients, forced upward gaze, conjugated deviation of the eyes, and horizontal or vertical nystagmus are frequent findings. Patients with coma persisting for less than 3 d usually recover fully. In severe cases, development of delayed dystonia is a common feature. The characteristics of the dystonic syndrome following acute encephalopathy induced by mildewed sugarcane (3-NPA) are: (1) appearance of coma in severe cases usually persisting for more than 3 d; (2) delayed occurrence of dystonia usually at 11–60 d after onset or 7–40 d after regaining consciousness; (3) dramatic involuntary movements, facial grimacing, sustained athetosis, spasmodic torticollis, torsion spasm, jerk-like movement resembling chorea or paroxysmal painful spasms of the extremities; and (4) motor aphasia or (5) nonprogressive and (6) irreversible dysarthria. The CT scans in dystonic patients show bilateral hypodensities in lenticular nuclei that likely explain the extrapyramidal symptoms *(20,40)*. Since 1972, there have been 217 outbreaks and more than 884 patients, 88 of whom died, in China involving mainly children.

3-NPA produces basal ganglia degeneration and extrapyramidal signs in humans and in experimental animals *(18,20,41–44)*. A number of investigators found age-dependent vulnerability of striatal neurons following intrastriatal, subacute, or chronic administration of 3-NPA in rats *(41,45)*. Some laboratories have reported neurochemical and histologic changes following intrastriatal injection of 3-NPA *(41,46)*. Noninvasive spectroscopic imaging has been used to detect 3-NPA-induced neurochemical alterations in brain *(47)*. Locomotor changes, vacuous chewing movements, a putative analogue of tardive dyskinesia, and dysfunction of the blood–brain barrier have been studied in rats systemically treated with 3-NPA *(42,48–50)*. Axonal degeneration has been reported in the caudate–putamen region of rats treated with multiple doses of 3-NPA *(51)*. Recent work has shown that chronic exposure to 3-NPA replicates the cognitive and motor deficits *(52)* and behavioral pathology of HD in baboons and rats, respectively *(53)*. Rodents and primates appear to be good animal models of HD *(54)*.

The chemical structure of 3-NPA is isoelectronic with that of succinate *(36)*. 3-NPA inhibits the activity of succinic dehydrogenase, an enzyme of the citric acid cycle and a component of mitochondrial complex II

(18,55–57). 3-NPA reduces energy supplies (ATP) of cultured cortical explants and causes neuronal degeneration by an excitotoxic mechanism *(34,46,57,58)*. Exposure of cultured striatal or cortical neurons to 3-NPA results in neuronal cell death by an apoptotic mechanism *(59)*.

3-NPA decreases synaptosomal respiration in a concentration-dependent manner *(60)*. The earliest sign of impairment of energy metabolism is a fall in the phosphocreatine/creatine ratio *(60)*. In the initial phase of intoxication, 3-NPA selectively inhibits the citric acid cycle of γ-aminobutyric acidergic (GABAergic) neurons; glial metabolic activity remains unaffected during this time *(61)*. These studies may explain why the caudate/putamen neurons, which are GABAergic, are selectively damaged by 3-NPA. Some investigators have suggested that an impairment of energy metabolism by 3-NPA may underlie neuronal death by an excitotoxic mechanism and formation of free radicals *(18,46,62,63)*. 3-NPA toxicity is significantly attenuated in SOD-1 transgenic mice *(64)*. Impaired energy metabolism and oxidative stress appear to play an important role in causing neurodegenerative diseases *(29)*.

There is no effective treatment for acute mildewed sugarcane poisoning. Pretreatment of animals with nerve growth factor *(65,66)*, prior decortication *(46)*, treatment with glutamate antagonists *(46,57)*, nitric oxide synthase inhibitors *(67)*, or oral supplementation with creatine and cyclocreatine *(68)* are reported to protect neurons against 3-NPA neurotoxicity. The combination of NMDA receptor antagonist, MK-801, with coenzyme Q_{10} has proven to be a more effective treatment for 3-NPA neurotoxicity *(69)*. Treatment with Q_{10} and nicotinamide and free radical scavengers has been shown to ameliorate striatal lesions produced by mitochondrial toxin *(70)*.

MPTP and MPP⁺

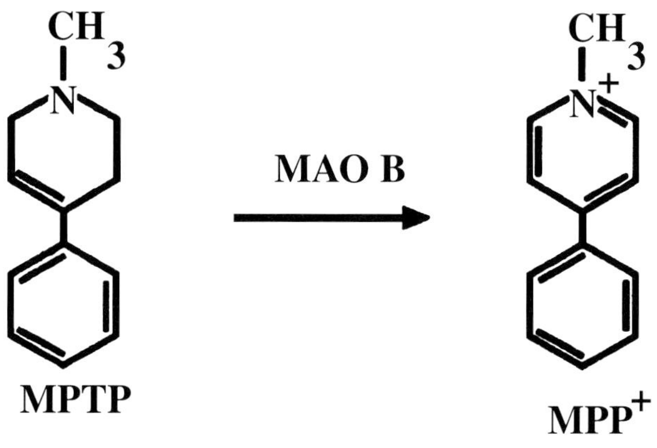

MPTP is a piperidine derivative that causes irreversible symptoms of Parkinsonism in humans. In 1982, some young drug addicts developed severe Parkinsonism after injecting a newly synthesized heroin contaminated with MPTP. Administration of pure MPTP to monkeys produces symptoms similar to those seen in humans with Parkinson's disease.

The acute effects of MPTP in rhesus monkeys include abnormal movements, decreased spontaneous activity, loss of facial expression, postural tremor, extension of the head, rigidity of the upper and lower extremities, twitching of the facial muscles, and facial grimacing *(71)*. Increased bradykinesia, frequent "nodding off," and usually sitting hunched over in a tightly flexed posture are observed in squirrel monkeys *(72)*. MPTP causes degeneration of the pars compacta of the substantia nigra, a hallmark of Parkinson's disease *(71,72)*. Thus, MPTP is a valuable tool for creating an animal model of PD and studying the mechanism of degeneration of pars compacta dopaminergic neurons. It was soon discovered that MPTP itself is not toxic; MPTP must be oxidized to MPP^+ by monoamine oxidase B of astrocytes to cause neurotoxicity.

Further studies showed that although primates are sensitive to MPTP, rodents, particularly rats, are refractory to its neurotoxic effects. Subsequent studies showed that mice are quite sensitive to MPTP toxicity, and the mouse became a useful animal model for studying the pharmacology of MPTP *(73,74)*. MPP^+ is selectively and efficiently taken up into dopaminergic nerve terminals by a high-affinity dopamine transporter *(75)*. MPP^+ binds to mitochondria, where it blocks NADH-coenzyme Q reductase (complex 1) activity *(8)*. MPP^+ selectively compromises cellular energy (ATP) generation in dopaminergic neurons *(76)* and causes neurodegeneration, which may be mediated by oxidative stress *(74,77,78)*.

Some investigators have proposed that energy deficit is the primary cause of MPTP/MPP^+ neurotoxicity *(79,80)*. MPP^+ interacts with mitochondrial complex I, irreversibly inhibits complex I enzyme activity, and causes the generation of increased free radicals *(63,78,81)*. Free radical scavengers have been shown to attenuate MPTP neurotoxicity *(82)*. MPTP depletes glutathione (GSH) levels both in vitro and in vivo *(78)*; inhibitors of GSH synthesis potentiate MPTP neurotoxicity *(83)*. *N*-Methyl-D-aspartate (NMDA) receptors appear to play a crucial role in MPTP/MPP^+ neurotoxicity because this effect is blocked by NMDA receptor antagonists *(84–86)*. Some investigators, however, report no protection of NMDA receptor antagonists against MPTP/MPP^+ neurotoxicity *(87–89)*. This discrepancy may be due to MPTP/MPP^+ dosage used and time point of intervention with neuroprotective compounds. Recent studies have shown that NMDA receptor antagonists that are effective in primates do not protect mice from MPTP toxicity *(74)*. Another line of study suggests that nitric oxide may be a key mediator of MPTP neurotoxicity that can be blocked by 7-nitroindazole or *S*-methylthiocitrulline, potent inhibitors of neuronal nitric oxide synthase (NOS) *(88,90,91)*. Mice lacking the NOS gene are reported to be refractory to MPTP neurotoxicity *(92)*.

The discovery of MPTP and its neurotoxic effects on human and experimental animals raises the possibility that some forms of Parkinson's disease may be caused by an environmental agent *(23)*. Pyridines related to MPTP are found in the environment both as industrial pollutants and in foods. It is conceivable that low-level exposure over a lifetime causes a slow and steady loss of dopaminergic cells that becomes critical late in life when only few cells are left *(93)*.

Cyanide

Cyanide is a highly toxic occupational and environmental chemical; victims may die within minutes of exposure *(94)*. Humans are exposed to cyanide from smoking, alkylcyanides used as solvents, cyanide salts used for polishing and metal cleaning, the antihypertensive drug sodium nitroprusside, and from consumption of cyanophoric plants (e.g., cassava roots), lima beans, and almonds *(94,95)*. Cyanide intoxication is the result of a complex series of effects, with primary sites of action in the cardiovascular and central nervous systems *(96–98)*. After absorption, cyanide reacts readily with the trivalent iron of cytochrome *c* oxidase in mitochondria. Cellular respiration is inhibited, resulting in lactic acidosis and cytotoxic hypoxia. Respiration is stimulated because chemoreceptive cells respond to decreased oxygen. A transient stage of CNS stimulation with hypernea and headache is observed. Hypoxic convulsions occur, leading to death due to respiratory failure. Most people with acute cyanide exposure die quickly, but some recover. Sequelae include extrapyramidal syndromes, personality changes, and memory defects *(99)*.

Cyanide inhibits cytochrome *c* oxidase activity, lowers energy supplies, causes neuronal degeneration, and produces neurological dysfunction including Parkinsonism and dystonia *(100,101)*. Chronic cyanide exposure has also been implicated in motor neuron disease *(102)*. Magnetic resonance imaging (MRI) shows bilateral lesions of the basal ganglia, and positron emission tomography (PET) with 6-fluoro-L-dopa displays marked dysfunction of dopaminergic transmission similar to that observed in Parkinsonism *(103)*. Cyanide depletes GABA and elevates glutamate concentrations in brain *(104)*. The dopaminergic system of rodents is highly susceptible to cyanide neurotoxicity *(105,106)*. Some investigators have suggested that cyanide selectively affects basal ganglia by an excitotoxic mechanism following disruption of energy metabolism *(107)*. Whether cyanide-induced cytochrome *c* oxidase inhibition is the primary biochemical lesion in cyanide toxicity remains unresolved, as cyanide has been shown to depress synaptic transmission without inhibiting cytochrome *c* oxidase activity *(108)*.

Cyanide increases cytosolic free Ca^{+2} in energy-compromised neurons by the activation of NMDA receptors and initiates a series of intracellular cascades that culminate in cell death *(109–111)*. In PC12 cells, cyanide activates phospholipase A2, stimulates inositol triphosphate generation through an interaction with the glutamate/metabotropic receptors *(112)*, and induces an apoptotic cell death *(113)*. The toxic effect of cyanide can be partially blocked with NMDA receptor antagonists *(111)*. Cyanide inhibits brain catalase, superoxide dismutase, and glutathione peroxidase and increases lipid peroxidation in the striatum *(114)*. These studies suggest that oxidative stress plays an important role in the expression of cyanide neurotoxicity. Nitric oxide has also been proposed as a mediator of convulsions associated with cyanide toxicity *(115)*.

In parts of Africa, where cyanogenic cassava consumption is high and protein intake is low, cyanide exposure is implicated in causing neurodegenerative diseases, namely tropical ataxic neuropathy and konzo, a paralytic disorder characterized by spastic paraparesis *(25,116)*. Cassava-consuming populations subsisting on a low-protein diet on a chronic basis are candidates for neurological diseases *(117)*. Cassava harbors a cyanogenic glucoside, linamarin, that liberates cyanide in the body. Free cyanide is rapidly, but reversibly, trapped by methemoglobin to form cyano-methemoglobin.

Cyanide is detoxified to thiocyanate (SCN^-) by the enzyme rhodanese, which requires sulfane sulfur derived from dietary sulfur amino acids, cysteine and methionine. In protein-deficient individuals, in whom sulfur amino acid concentrations are low, detoxification of cyanide to SCN^- may be impaired and cyanide may be converted to neurotoxic cyanate (OCN^-) *(118)*. High concentration of OCN^- inhibits cytochrome *c* oxidase activity in vitro *(119)*, uncouples oxidative phosphorylation *(120)*, blocks the activity of glutathione reductase, and reduces glutathione levels both in vitro and in vivo *(121)*.

A better understanding of chemical mechanisms linked to outbreaks of neurological disease is needed to design preventive measures for cyanide neurotoxicity. The hypothesis that sulfur amino acid deficiency in protein malnutrition plays an important role in cyanide detoxification can be tested in cassava-consuming populations *(122)*. Detection of high levels of cyanate in the blood of malnourished individuals may explain, in part, neurological deficits in cassava-consuming populations.

ACKNOWLEDGMENTS

The author wishes to thank Jerry Schnell, Ph.D., for critically reading the manuscript and making useful suggestions. The able secretarial assistance of Ms. Emily McKinzie in the production of the manuscript is gratefully acknowledged. This work was partly supported by the Oregon Health Sciences Foundation.

REFERENCES

1. Davis RE, Miller S, Herrnstadt C, et al. Mutations in mitochondrial cytochrome *c* oxidase genes segregate with late-onset Alzheimer disease. Proc Natl Acad Sci USA 1997;94:4526–4531.
2. Hoyer S. Intermediary metabolism disturbance on AD/SADT and its relation to molecular events. Neuropsychopharmacology 1993;17:628–632.
3. Wallace DC. Mitochondrial genetics: a paradigm for aging and degenerative diseases? Science 1992;256:628–632.
4. Wallace DC. Diseases of the mitochondrial DNA. Annu Rev Biochem 1992;61:1175–1212.
5. Schapira AHV, Mann VM, Cooper JM, et al. Anatomic and disease specificity of NADH CoQ1 reductase (complex I) deficiency in Parkinson's disease. J Neurochem 1990;55:2142–2145.
6. Schapira AHV, Cooper JM, Dexter D, et al. Mitochondrial complex I deficiency in Parkinson's disease. J Neurochem 1990;54:823–827.
7. Langston JW. Mechanism of MPTP toxicity: more answers, more questions. Trends Pharmacol Sci 1985;6:375–378.
8. Nicklas WJ, Vyas I, Heikkila RE. Inhibition of NADH-linked oxidation in brain mitochondria by 1-methyl–4-phenyl-pyridine, a metabolite of the neurotoxin, 1-methyl–4-phenyl–1,2,5,6-tetrahydropyridine. Life Sci 1985;36: 2503–2508.
9. Koroshetz WJ, Jenkins BG, Rosen BR, et al. Energy metabolism defects in Huntington's disease and effects of coenzyme Q10. Ann Neurol 1997; 41:160–165.
10. Gu M, Gash MT, Mann VM, et al. Mitochondrial defect in Huntington's disease caudate nucleus. Ann Neurol 1996;39:385–389.
11. Mitsumoto H, Chad DA, Pioro EP. Amyotrophic Lateral Sclerosis. FA Davis, Philadelphia; 1998.
12. De Belleroche J, Orrell RW, Virgo L. Amyotrophic lateral sclerosis: recent advances in understanding disease mechanisms. J Neuropathol Exp Neurol 1996;55:747–757.
13. Rosen DR, Siddique T, Patterson D, et al. Mutations in Cu/Zn superoxide dismutase gene are associated with familial amyotrophic lateral sclerosis. Nature 1993;362:59–63.

14. Figlewicz DA, Krizus A, Martinole M. Variants of the human heavy neurofilament subunit are associated with the development of amyotrophic lateral sclerosis. Hum Mol Genet 1995;3:1757–1761.
15. Spencer PS, Nunn PB, Hugon J, et al. Guam amyotrophic lateral sclerosis—parkinsonism-dementia linked to a plant excitant neurotoxin. Science 1987;237:517–522.
16. Halliwell B. Reactive oxygen species and the central nervous system. J Neurochem 1992;59:1609–1623.
17. Patterson D, Warner HR, Fox LM. Superoxide dismutase, oxygen radical metabolism, and amyotrophic lateral sclerosis. Mol Gene Med 1994;4:79-119.
18. Ludolph AC, He F, Spencer PS, et al. 3-Nitropropionic acid—exogenous animal neurotoxic and possible human striatal toxin. Can J Neurol Sci 1991; 18:492-498.
19. Ludolph AC, Seelig M, Ludolph A, et al. 3-Nitropropionic acid decreases cellular energy levels and causes neuronal degeneration in cortical explants. Neurodegeneration 1992;1:155–161.
20. He F, Zhang S, Zhang C, et al. Mycotoxic induced encephalopathy and dystonia in children. In: Volans GN, Sims J, Sullivan FM, Turner P, eds. Basic Science in Toxicology. Taylor and Francis, London, 1990, pp. 596–604.
21. National Research Council. Environmental Neurotoxicology. National Academy of Sciences Press, Washington, DC, 1992.
22. Rice ME, Lee EJK, Choy Y. High levels of ascorbic acid, not glutathione, in the CNS of anoxia-tolerant reptiles contrasted with levels in anoxia-intolerant species. J Neurochem 1995;64:1790–1799.
23. Langston JW, Ballard P, Tetrud JW, et al. Chronic Parkinsonism in humans due to a product of meperidine analog synthesis. Science 1983;219:979–980.
24. Spencer PS, Butterfield PG. Environmental agents and Parkinson's disease. In: Ellenberg JH, Koller WC, Langston JW. eds. Etiology of Parkinson's Disease. Marcel Dekker, New York, 1995, pp. 319–365.
25. Tylleskar T. The causation of konzo. Ph.D. Thesis, Uppsala University, Sweden, Faculty of Medicine, 1994.
26. Spencer PS. Lathyrism. In: de Wolff FA ed. Handbook of Clinical Neurology, Vol. 21: Intoxications of the Nervous System, Part II. Elsevier, Amsterdam, 1995, pp. 1–20.
27. Sabri MI, Lystrup B, Roy DN, et al. Action of β-*N*-oxalylamino-L-alanine on mouse brain NADH-dehydrogenase activity. J Neurochem 1995;65:1842–1848.
28. Pai KS, Ravindranath VL. BOAA induces selective inhibition of brain mitochondrial enzyme NADH-dehydrogenese. Brain Res 1993;621:215–221.
29. Bowling AC, Beal MF. Bioenergetic and oxidative stress in neurodegenerative diseases. Life Sci 1995;56:1151–1171.
30. Beal MF, Hyman BT, Koroshetz W. Do defects in mitochondrial energy metabolism underlie the pathology of neurodegenerative diseases? Trends Neurosci 1993;16:125–161.
31. Beal MF. Does impairment of energy metabolism result in excitotoxic neuronal death in neurodegenerative illnesses? Ann Neurol 1992;31:119–130.

32. Lees GJ. Contributory mechanisms in the causation of neurodegenerative disorders. Neuroscience 1993;54:287–322.
33. Di Monte DA. Mitochondrial DNA and Parkinson's disease. Neurology 1991;41:38–42.
34. Dawson Jr R, Beal MF, Bondy SC, et al. Excitotoxins, aging, and environmental neurotoxins: implications for understanding human neurodegenerative diseases. Toxicol Appl Pharmacol 1995;134:1–17.
35. James LF, Hartley WJ, Williams MC, et al. Field and experimental studies in cattle and sheep poisoned by nitro-bearing Astragalus or their toxins. Am J Vet Res 1980;41:377–382.
36. Alston TA, Mela L, Bright HJ. 3-Nitropropionate, the toxic substances of Indigofera, is a suicide inactivator of succinate dehydrogenase. Proc Natl Acad Sci USA1977;74: 3767–3771.
37. Liu XJ. Investigations on the etiology of mildewed sugarcane poisoning. A review. Chin J Prevent Med 1986;20:306–308.
38. Ming L. Moldy sugarcane poisoning—a case report with a brief review. Clin Toxicol 1995;33:363–367.
39. Hu W. The isolation and structure identification of a toxic substance, 3-nitropropionic acid, produced by Arthrinium from mildewed sugarcanes. Chin J Prevent Med 1986;20:321–323.
40. He F. Extrapyramidal lesions induced by mildewed sugarcane poisoning. Three cases report. Chin Med J 1987;67:395–396.
41. Brouillet E, Jenkins BG, Hyman BT, et al. Age-dependent vulnerability of the striatum to the mitochondrial toxin 3-nitropropionic acid. J Neurochem 1993;60:356–359.
42. Borlongan CV, Koutouzis TK, Randall TS, et al. Systemic 3-nitropropionic acid: behavioral deficits and striatal damage in adult rats. Brain Res Bull 1995;36:549–556.
43. Wullner U, Young AB, Penney JB, et al. 3-Nitropropionic acid toxicity in the striatum. J Neurochem 1994;63:1772–1781.
44. Hamilton BF, Gould DH. Nature and distribution of brain lesions in rats intoxicated with 3-nitropropionic acid: a type of hypoxic (energy deficient) brain damage. Acta Neuropathol 1987;72:286–297.
45. Beal MF, Brouillet E, Jenkins B, et al. Age-dependent striatal excitotoxic lesions produced by the endogenous mitochondrial inhibitor malonate. J Neurochem 1993;61:1147–1150.
46. Beal MF, Brouillet E, Jenkins BG, et al. Neurochemical and histologic characterization of striatal excitotoxic lesions produced by the mitochondrial toxin 3-nitropropionic acid. J Neurosci 1993;13:4181–4192.
47. Jenkins BG, Brouillet E, Chen Y-CI, et al. Non-invasive neurochemical analysis of focal excitotoxic lesions in models of neurodegenerative illness using spectroscopic imaging. J Cereb Blood Flow Metab 1996;16: 450–461.
48. Andreassen OA, Jørgensen HA. The mitochondrial toxin 3-nitropropionic acid induces vacuous chewing movements in rats. Implications for tardive dyskinesia? Psychopharmacology 1995;119:474–476.

49. Nishino H, Shimano Y, Kumazaki M, et al. Chronically administered 3-nitropropionic acid induces striatal lesions attributed to dysfunction of the blood–brain barrier. Neurobiology 1994;2:223–234.
50. Koutouzis TK, Borlongan CV, Sorcia T, et al. Systemic 3-nitropropionic acid: long-term effects on locomotor behavior. Brain Res 1994;646:242–246.
51. Allen CN, Marino JL, Meshul CK. Development of 3-nitropropionic acid (3-NPA)-induced neuronal degeneration: a silver impregnation study. Neurodegeneration 1994;3:225–233.
52. Palfi S, Ferrante RJ, Brouillet E, et al. Chronic 3-nitropropionic acid treatment in baboons replicates the cognitive and motor deficits of Huntington's disease. J Neurosci 1996;16:3019–3025.
53. Borlongan CV, Koutouzis TK, Freeman TB, et al. Behavioral pathology induced by repeated systemic injections of 3-nitropropionic acid mimics the motoric symptoms of Huntington's disease. Brain Res 1995;697:254–257.
54. Beal MF. Neurochemistry and toxin models in Huntington's disease. Curr Opin Neurol 1994;7:542–547.
55. Gould DH, Gustine DL. Basal ganglia degeneration, myelin alterations, and enzyme inhibition induced in mice by the plant toxin 3-nitropropionic acid. Neuropathol Appl Neurobiol 1982;8:377–393.
56. Coles CJ, Edmondson DE, Singer TP. Inactivation of succinate dehydrogenase by 3-nitropropionate. J Bio Chem 1979;254:5161–5167.
57. Ludolph AC, Seelig M, Ludolph A, et al. 3-Nitropropionic acid decreases cellular energy levels and causes neuronal degeneration in cortical explants. Neurodegeneration 1992;1:155–161.
58. Zeevalk GD, Derr-Yellin E, Nicklas WJ. NMDA receptor involvement in toxicity to dopamine neurons in vitro caused by the succinate dehydrogenase inhibitor 3-nitropropionic acid. J Neurochem 1995;64:455–458.
59. Behrens MI, Koh J, Canzoniero LMT, et al. 3-Nitropropionic acid induces apoptosis in cultured striatal and cortical neurons. NeuroReport 1995;6:545-548.
60. Erecinska M, Nelson D. Effects of 3-nitropropionic acid on synaptosomal energy and transmitter metabolism: relevance to neurodegenerative brain diseases. J Neurochem 1994;63:1033–1041.
61. Hassel B, Sonnewald U. Selective inhibition of the tricarboxylic acid cycle of GABAergic neurons with 3-nitropropionic acid in vivo. J Neurochem 1995;65:1184–1191.
62. Simpson JR, Isacson O. Mitochondrial impairment reduces the threshold for in vivo NMDA-mediated neuronal death in the striatum. Exp Neurol 1993;121:57-64.
63. Schulz JB, Henshaw DR., Siwek D, et al. Involvement of free radicals in excitotoxicity in vivo. J Neurochem 1995;64:2239–2247.
64. Beal MF, Ferrante RJ, Henshaw R, et al. 3-Nitropropionic acid neurotoxicity is attenuated in copper/zinc superoxide dismutase transgenic mice. J Neurochem 1995;65:919–922.
65. Frim DM, Simpson J, Uhler TA, et al. Striatal degeneration induced by mitochondrial blockade is prevented by biologically delivered NGF. J Neurosci Res 1993;35:452–458.

66. Galpern WR, Matthews RT, Beal MF, et al. NGF attenuates 3-nitrotyrosine formation in a 3-NP model of Huntington's disease. NeuroReport 1996;7:2639–2642.
67. Schulz JB, Matthews RT, Jenkins BG, et al. Blockade of neuronal nitric oxide synthase protects against excitotoxicity in vivo. J Neurosci 1995; 15:8419–8429.
68. Matthews RT, Yang L, Jenkins, BG, et al. Neuroprotective effects of creatine and cyclocreatine in animal models of Huntington's disease. J Neurosci 1998;18:156–163.
69. Schulz JB, Matthews RT, Henshaw, DR, et al. Neuroprotective strategies for treatment of lesions produced by mitochondrial toxins: implications for neurodegenerative diseases. Neuroscience 1996;71:1043–1048.
70. Beal MF, Henshaw DR, Jenkins BG, et al. Coenzyme Q10 and nicotinamide blocked striatal lesions produced by the mitochondrial toxin malonate. Ann Neurol 1994;36:882–888.
71. Burns RS, Chiueh CC, Markey SP, et al. A primate model of parkinsonism: selective destruction of dopaminergic neurons in the pars compacta of the substantia nigra by *N*-methyl-4-phenyl-1,2,3,6-tetrahydropyridine. Proc Natl Acad Sci USA 1983;80:4546–4550.
72. Langston JW, Forno LS, Rebert CS, et al. Selective nigral toxicity after systemic administration of 1-methyl-4-phenyl-1,2,5,6-tetrahydropyrine (MPTP) in the squirrel monkey. Brain Res 1984;292:390–394.
73. Heikkila RE, Hess A, Duvoisin RC. Dopaminergic neurotoxicity of 1-methy-4-phenyl-1,2,5,6-tetrahydropyridine (MPTP) in the mouse: relationships between monoamine oxidase, MPTP metabolism and neurotoxicity. Life Sci 1985;36:231–236.
74. Fornai F, Vaglini F, Maggio R, et al. Excitatory amino acids and MPTP toxicity. Adv Neurol 1996;69:167–176.
75. Javitch JA, D'Amato RJ, Strittmatter SM, et al. Parkinsonism-inducing neurotoxin, *N*-methyl-4-phenyl-1,2,3,6-tetrahydropyridine: uptake of the metabolite *N*-methyl-4-phenylpyridine by dopamine neurons explains selective toxicity. Proc Natl Acad Sci USA 1985;82:2173–2177.
76. Chan P, DeLanney LE, Irwin I, et al. Rapid ATP loss caused by 1-methyl-4-phenyl-1,2,3,6-tetrahydropyridine in mouse brain. J Neurochem 1991; 57:348-351.
77. Fallon J, Matthews RT, Hyman BT, et al. MPP^+ produces progressive neuronal degeneration which is mediated by oxidative stress. Exp Neurol 1997;144:193–198.
78. Sriram K, Pai KS, Boyd MR, et al. Evidence for generation of oxidative stress in brain by MPTP: in vitro and in vivo studies in mice. Brain Res 1997; 749:44–52.
79. Bates TE, Heales SJR, Davies SEC, et al. Effects of 1-methyl-4-phenylpyridinium on isolated rat brain mitochondria: evidence for a primary involvement of energy depletion. J Neurochem 1994;63:640–648.
80. Gluck MR, Youngster SK, Rasmay RR, et al. Studies on the characterization of the inhibitory mechanism of 4'-alkylated 1-methyl-4-phenylpyridinium and

phenylpyridine analogues in mitochondria and electron transport particles. J Neurochem 1994;63:655–661.
81. Cleeter MWJ, Cooper JM, Schapira AHV. Irreversible inhibition of mitochondrial complex I by 1-methyl-4-phenylpyridinium: evidence for free radical involvement. J Neurochem 1992;58:786–789.
82. Schulz JB, Henshaw DR, Matthews RT, et al. Coenzyme Q10 and nicotinamide and a free radical spin trap protect against MPTP neurotoxicity. Exp Neurol 1995;132:279–283.
83. Wullner U, Loschmann PA, Schulz JB, et al. Glutathione depletion potentiates MPTP and MPP^+ toxicity in nigral dopaminergic neurons. NeuroReport 1996;7:921–923.
84. Turski L, Turski WA. Towards an understanding of the role of glutamate in neurodegenerative disorders: energy metabolism and neuropathology. Experientia 1993;49:1064–1072.
85. Turski L, Bressler K, Rettig K-J, et al. Protection of substantia nigra from MPP^+ neurotoxicity by *N*-methyl-D-aspartate antagonists. Nature 1991; 349:414–418.
86. Storey E, Hyman BT, Jenkins B, et al. 1-Methyl-4-phenylpyridinium produces excitotoxic lesions in rat striatum as a result of impairment of oxidative metabolism. J Neurochem 1992;58:1975–1978.
87. Sonsalla PK, Zeevalk GD, Manzino L, et al. MK–801 Fails to protect against the dopaminergic neuropathology produced by systemic 1-methyl-4-phenyl-1,2,3,6-tetrahydropyridine in mice or intranigral 1-methyl-4-phenylpyridinium in rats. J Neurochem 1992;58:1979–1982.
88. Finiels-Marlier F, Marini AM, Williams P, et al. The *N*-methyl-D-aspartate antagonist MK-801 fails to protect dopaminergic neurons from 1-methyl-4-phenylpyridinium toxicity in vitro. J Neurochem 1993;60:1968–1971.
89. Chan P, Di Monte DA, Langston JW, et al. (+) MK–801 does not prevent MPTP-induced loss of nigral neurons in mice. J Pharmacol Exp Ther 1997;280:439-446.
90. Schulz JB, Matthews RT, Muqit MMK, et al. Inhibition of neuronal nitric oxide synthase by 7-nitroindazole protects against MPTP-induced neurotoxicity in mice. J Neurochem 1995;64:936–939.
91. Matthews RT, Yang L, Beal MF. *S*-Methylthiocitrulline, a neuronal nitric oxide synthase inhibitor, protects against malonate and MPTP neurotoxicity. Exp Neurol 1997;143:282–286.
92. Przedborski S, Jackson-Lewis V, Yokoyama R, et al. Role of neuronal nitric oxide in 1-methyl-4-phenyl-1,2,3,6-tetrahydropyridine (MPTP)-induced dopaminergic neurotoxicity. Proc Natl Acad Sci USA 1996;93:4565–4571.
93. Hall ZW. An Introduction to Molecular Neurobiology. Sinauer, Sunderland, MA, 1992.
94. U.S. Department of Health and Human Services. Toxicological profile for cyanide. Agency for Toxic Substances and Disease Registry, Atlanta, GA, 1997.
95. Borowitz JL, Kanthasamy AG, Isom GE. Toxicodynamics of cyanide. In: Somani SM, ed. Chemical Warfare Agents. Academic Press, San Diego, 1992, pp. 209-236.

96. Way JL. Cyanide intoxication and its mechanism of antagonism. Annu Rev Pharmacol Toxicol 1984;24:451–481.
97. Brierley JB, Prior PF, Calverley J. Cyanide intoxication in macaca mulata: physiological and neuropathological aspects. J Neurol Sci 1977;31:133–157.
98. Hall A, Rumack B. Clinical toxicology of cyanide. Ann Emerg Med 1986;16:1067–1074
99. Marrs TC, Maynard RA. Neurotoxicity of chemical warfare agents. In: de Wolff FA, ed. Handbook of Clinical Neurology, Vol. 21 (64): Intoxications of the Nervous System, Part II. Elsevier, Amsterdam, 1995, pp. 1–20.
100. Carella F, Grassi MP, Savoiardo M, et al. Dystonic-Parkinsonian syndrome after cyanide poisoning: clinical and MRI findings. J Neurol Neurosurg Psychiatry 1988;51:1345–1348.
101. Valenzuela R, Court J, Godoy J. Delayed cyanide-induced dystonia. J Neurol Neurosurg Psychiatry 1992;55:198–199.
102. Kato T, Kameyama M, Nakamura S, et al. Cyanide metabolism in motor neuron disease. Acta Neurol Scand 1985;72:151–156.
103. Rosenberg NL, Myers JA, Martin WRW. Cyanide induced Parkinsonism: clinical, MRI and 6-fluorodopa PET studies. Neurology 1989;38:142–144.
104. Persson S-A, Cassal G, Sellstrom A. Acute cyanide intoxication and central transmitter systems. Fund Appl Toxicol 1985;5:5150–5159.
105. Kanthasamy AG, Borowitz JL, Pavlakovic G, et al. Dopaminergic neurotoxicity of cyanide: neurochemical, histological, and behavioral characterization. Toxicol Appl Pharmacol 1994;126:156–163.
106. Cassal G, Persson S-A. Effects of acute lethal cyanide intoxication on central dopaminergic pathways. Pharmacol Toxicol 1992;70:148–151.
107. Spencer PS, Ludolph AC, Kisby GE. Are human neurodegenerative disorders linked to environmental chemicals with excitotoxic properties? Ann NY Acad Sci 1992;648:154–160.
108. Aitken P, Brartmen D. The effects of cyanide on neural and synaptic function in hippocampal slices. Neurotoxicology 1989;10:239–248.
109. Novelli A, Reilly JA, Lysko PG, et al. Glutamate becomes neurotoxic via the *N*-methyl-D-aspartate receptor when intracellular energy levels are reduced. Brain Res 1988;451:205–212.
110. Maduh EV, Johnson JD, Ardelt BK, et al. Cyanide-induced neurotoxicity: mechanism of attenuation by chlorpromazine. Toxicol Appl Pharmacol 1988;96:60–67.
111. Patel MN, Yim GKW, Isom GE. *N*-Methyl-D-aspartate receptors mediate cyanide-induced cytotoxicity in hippocampal cultures. Neurotoxicology 1993;14:35–40.
112. Yang C-W, Isom GE. Metabotropic receptor mediated inositol triphosphate generation: stimulation by cyanide. Toxicologist 1994;14:351.
113. Mills EM, Isom GE. The induction of apoptosis in differentiated PC12 cells by cyanide. Toxicologist 1994;14:362.
114. Ardelt BK, Borowitz JL, Isom GE. Brain lipid peroxidation and antioxidant protectant mechanisms following acute cyanide intoxication. Toxicology 1989;56:147–154.

115. Gunasekar PG, Sun PW, Kanthasamy AG, et al. Cyanide-induced neurotoxicity involves nitric oxide and reactive oxygen species generation after *N*-methyl-D-aspartate receptor activation. J Pharmacol Exp Ther 1996; 277:150–155.
116. Tylleskar T, Banea M, Bikangi N, et al. *Cassava cyanogens* and Konzo, and upper motoneuron disease found in Africa. Lancet 1992;339:208–211.
117. Rosling H. Molecular anthropology of cassava cyanogenesis. In: Sobral BWS, ed. The Impact of Plant Molecular Genesis. Birkhauser, Boston, 1996, p. 315.
118. Swenne I, Eriksson U, Christoffersson R. Cyanide detoxification in rats exposed to acetonitrile and fed a low protein diet. Fund Appl Toxicol 1996;31:66–71.
119. Tor-Agbidye J, Agoston T, Lystrup B, et al.Selective inhibition of brain mitochondrial cytochrome c oxidase (complex IV) by sodium cyanate. J Neurochem 1995;46:S96.
120. Cammer W. Release of mitochondrial respiratory control by cyanate salts. Biochim Biophys Acta 1982;697:343–346.
121. Sabri MI, Tor-Agbidye J, Palmer VS. Glutathione and glutathione reductase activity are reduced in rodent brain by sodium cyanate. J Neurochem 1996;66:514c.
122. Tor-Agbidye J. Cyanide metabolism in sulfur amino acid deficiency: Relevance to cassava-related neurodegenerative diseases. Ph.D. Thesis, Oregon State University, School of Veterinary Medicine, Corvallis, OR, 1997.

2

History of 3-Nitropropionic Acid

Occurrence and Role in Human and Animal Disease

Bradley F. Hamilton, Daniel H. Gould, and David L. Gustine

INTRODUCTION

For many years prior to its recent discovery and exploitation as a chemical tool for investigation of various neurodegenerative disorders in humans, 3-nitropropionic acid (3-NPA) intoxication following ingestion of plants has been a substantial problem in domestic livestock. More recently, reports from China have documented the tragic consequences of 3-NPA intoxication in humans consuming moldy sugarcane. The purpose of this chapter is to provide a brief review of the occurrence of 3-NPA in nature; the basic metabolism of 3-NPA as it has been reported in veterinary intoxications; and the incidence, clinical/neurological effects, and pathology of the intoxication in domestic livestock and in humans.

OCCURRENCE AND BIOCHEMISTRY OF 3-NPA

3-Nitropropionate and 3-nitropropanol, and its glucose esters and glycoside, respectively, are rare natural products produced in just a few plant and fungal species. The isoelectronic form of 3-NPA can be converted at physiological pH to the highly reactive dianion (Fig. 1) *(1)*, which irreversibly inhibits succinate dehydrogenase (EC No. 1.3.99.1). This is the biochemical basis for 3-NPA toxicity.

Plant-produced metabolites containing oxidized nitrogen—classified as cyanogenic glycosides, glucosinolates, and nitro compounds—are derived from amino acid precursors through a common biosynthetic pathway. The initial steps common to biosynthesis of these classes of metabolites are *N*-hydroxylation of the starting amino acid, followed by oxidative decarboxylation of the hydroxyamino acid to form the corresponding aldoxime.

From: *Mitochondrial Inhibitors and Neurodegenerative Disorders*
Edited by: P. R. Sanberg, H. Nishino, and C. V. Borlongan © Humana Press Inc., Totowa, NJ

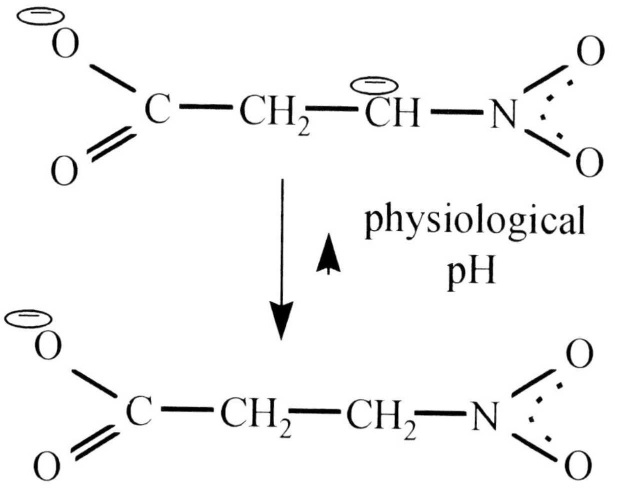

Fig. 1. 3-Nitropropionate is slowly converted to the 3-NPA dianion (pK = 9.3) at physiological conditions *(1)*.

The biosynthetic pathways then follow different routes for the synthesis of the three classes of natural products *(2)*. Cyanogenic glycosides occur in approx 2000 plant species, many of which are economic food sources, and, as their biosynthetic pathways are known, the steps for 3-NPA synthesis can be deduced. Presumably, aspartic acid is the precursor to propionaldoxime, which is converted to aci-3-nitropropionic acid, the immediate precursor of 3-NPA (Fig. 2).

3-NPA was originally isolated by Gorter in 1920 as hiptagenic acid (cited by Carter and McChesney) *(3)*. Gorter isolated the glucose ester hiptagen from the bark of *Hiptage benghalensis* (*mandoblota*) and produced a hydrolytic product he named hiptagenic acid that he mistakenly characterized as a hydroxamic acid. In 1934, Carrie *(4)* isolated karakin, a glucose ester of 3-NPA, from the karaka tree (*Corynocarpus laevigatus*) and demonstrated the hydrolytic release of a compound identical to hiptagenic acid. The toxin karakin was originally named and crystallized by Skey in 1872, who thought it was a glucoside (cited by Carrie) *(4)*. The structure of hiptagenic acid was later correctly identified as 3-NPA by Carter and McChesney *(3)*, which they recognized was the first organic nitro compound isolated from plants. Glucose esters of 3-NPA have been characterized from creeping indigo [*Indigofera spicata* (*endecaphylla*)] *(5)*, *Viola odorata* (cited by Wilson) *(6)*, various *Astragalus* species *(7,8)*, crownvetch (*Coronilla varia*) *(9)*, and *Lotus pedunculatus (10)*. 3-NPA is also produced by the fungi *Aspergillus*

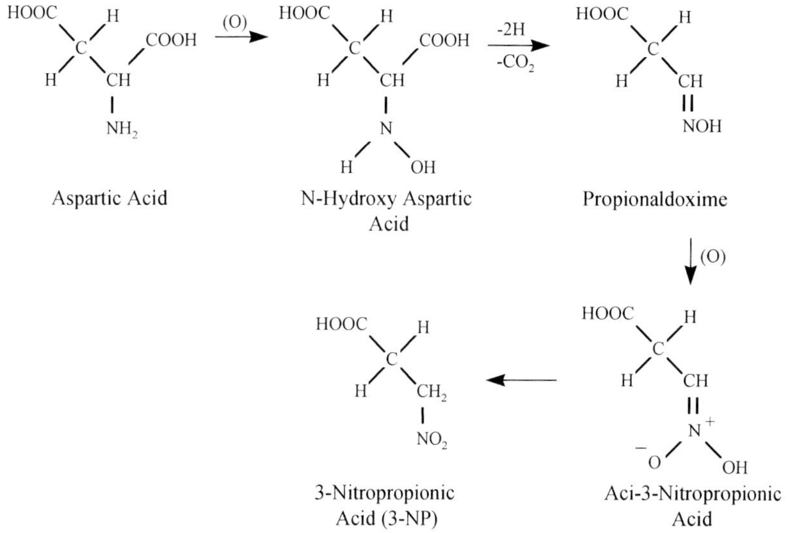

Fig. 2. Biosynthetic pathway for 3-NPA proposed by Conn *(2)*.

flavus, *Aspergillus wentii*, *Penicillium atrovenetum*, *Arthrinium sacchari*, *Arthrinium saccharicola,* and *Arthrinium phaeospermum (6,11–13)*. Its production in fungi may have evolved as an intermediate in oxidative conversion of amino acid amino nitrogen to nitrate under limiting nutrient conditions *(6,11)*. The closely related aliphatic nitro analog 3-nitropropanol is also produced in various *Astragalus* species, where it was first reported as the glycoside miserotoxin in *Astragalus miser (14)*.

Acute clinical intoxication was observed in several species treated with 3-NPA, 3-NPA esters, 3-nitropropanol, and miserotoxin. These species include chickens *(15–17)*, swine *(15)*, rabbits *(16)*, mice *(6,18)*, meadow voles *(15)*, sheep *(17,19)*, and cattle *(16,17,19)*. Poisoning by these compounds probably is through the common mechanism of 3-NPA toxicity, as 3-nitropropanol is metabolized to 3-NPA after absorption from the digestive tract. This was shown by Pass et al. *(20)* who found that inhibition of rat liver alcohol dehydrogenase prevented the toxicity of 3-nitropropanol, presumably by blocking its metabolic conversion to 3-NPA. 3-Nitropropanol was toxic if the enzyme was not inhibited.

Differences in ruminal metabolism of these related aliphatic nitro compounds dictate their toxicity and are due to differences in the chemistry of the alcohol and carboxylic acid groups. 3-Nitropropanol is rapidly released from its glycoside miserotoxin by microbial hydrolysis in the rumen. It is slowly metabolized to 3-amino-1-propanol, which cannot be utilized further by rumen microorganisms for energy *(21–24)*. 3-Nitropropanol is rapidly absorbed into the blood, transported to the liver, and converted to 3-NPA, which accounts for the toxicity of miserotoxin in ruminants. Conversely, 3-NPA esters are rapidly hydrolyzed in the rumen *(25)* and the free 3-NPA is slowly metabolized to β-alanine *(24)*, which rumen microorganisms can utilize for energy. Pass et al. *(26)* further found that 3-nitropropanol was more rapidly absorbed than 3-NPA from the digestive system of sheep. These observations account for the decreased toxicity of 3-NPA in ruminants relative to that of 3-nitropropanol. In nonruminants, 3-NPA esters can be rapidly hydrolyzed by mammalian tissue esterases to release 3-NPA, while 3-nitropropanol is oxidized by hepatic alcohol dehydrogenase to 3-NPA.

The biochemical basis for 3-NPA toxicity is its irreversible inhibition of succinate dehydrogenase (SDH) and the competitive inhibition of fumarase by the dianion form. Gustine and Moyer *(27)* predicted a toxic dose of 3-NPA should produce a physiological concentration of approx 0.02 m*M* 3-NPA dianion. This concentration would only partially inhibit the total SDH activity, but because the effect is irreversible and 3-NPA dianion is continuously formed, nearly all the 3-NPA would gradually react with SDH. A dianion concentration of 0.02 m*M* would also be sufficient to inhibit fumarase. Pass et al. *(28)* examined the effects of 3-NPA on cultured murine embryonal carcinoma cells and concluded that 3-NPA induces toxicity by inhibiting SDH and thus reducing ATP levels. This combination would cause clinically significant inhibition of respiration and, depending on 3-NPA dianion cellular concentration, would lead to cell death.

Methemoglobinemia is another biochemical effect of intoxication with 3-NPA owing to the generation of inorganic nitrite, probably as a consequence of liver metabolism *(29–31)*. However, methemoglobinemia has generally been considered a minor or even inconsequential part of 3-NPA intoxication. Mice and rats treated with sodium nitrite to induce comparable or higher levels of methemoglobinemia remain normal clinically without the characteristic neuropathological effects typical of 3-NPA *(18,29)*. In addition, treatment of rabbits with methylene blue to alleviate the methemoglobinemia does not prevent fatal 3-NPA intoxication *(16)*. These observations underscore the importance of the enzyme inhibition mentioned previously in the pathophysiology of 3-NPA toxicity.

INTOXICATION IN ANIMALS

Nearly all reports of 3-NPA poisoning in animals under natural conditions in North America involve cattle and sheep consuming plants containing miserotoxin (the glycoside of 3-nitropropanol) on grazing ranges in western Canada, the western United States, and northern Mexico *(32–39)*. A variety of *Astragalus* species are responsible for these incidents, and generally the circumstances surrounding a serious epidemic of poisoning entail overgrazing and/or periods of drought followed by rainfall wherein there is either a reduction in more desirable forage or a relative overpopulation of the toxic plant. Under these conditions, cattle and sheep are more likely to consume toxic quantities of offending *Astragalus* species. Lactating females are reportedly more susceptible *(33)*. In one incident involving cattle and sheep in New Mexico, mortality was 2–3%, and morbidity was 15–20% *(36)*. Poisoned animals may survive but then remain unthrifty even when exposure to the toxin is eliminated *(32,33,35,36)*.

The earliest reports of poisoning by *Astragalus* species containing aliphatic nitro compounds were in the 1920s and 1930s in agricultural bulletins issued in the United States and Canada (reviewed in James) *(38)*; these were followed by field studies in Texas and British Columbia *(32,33)*. Subsequent to the initial isolation and identification of miserotoxin as the toxic principle in *Astragalus miser (14)*, several studies have confirmed the basis for the natural intoxication by comparing the effects of feeding the whole plant or plant extracts with administration of the purified toxin either as 3-nitropropionic acid or 3-nitropropanol *(16,19,34,40,41)*. Poisoning due to 3-NPA is distinct from two additional syndromes of *Astragalus* poisoning in livestock: selenium toxicosis and locoism *(35,37,38)*. Although there are a few exceptions, a single poisonous species of *Astragalus* is generally associated with only one of these three toxic syndromes *(39)*.

Astragalus species are distributed worldwide *(39)* and so are undoubtedly associated with livestock intoxication in other parts of the world as well. Tarazona and Sanz *(42)* refer to poisoning of sheep in Spain with *Astragalus lusitanicus*; they isolated and identified aliphatic nitro compounds at concentrations similar to those in toxic species of North American *Astragalus*. Sager and Nieto *(43,44)* reported the presence of 3-nitropropanol in two *Astragalus* species in Argentina. Livestock intoxication is apparently not well documented in Argentina, but anecdotal evidence indicates 3-NPA toxicity due to *Astragalus* consumption occurs there in cattle, sheep, and llamas (Ricardo L. Sager, Instituto Nacional de Tecnologia Agropecuaria, Villa Mercedes, Argentina; personal communication, 1997).

Although most cases of livestock poisoning appear to be due to *Astragalus* species, reports from Hawaii described the poisonous effects of creeping indigo (*Indigofera endecaphylla*) on domestic livestock attributable to its content of 3-NPA *(45,46)*. In the northeastern United States, crownvetch (*Coronilla varia*) contains 3-NPA, and although clearly toxic to nonruminants under experimental conditions, it is not associated with a substantial occurrence of toxicity to ruminants or nonruminants under normal grazing conditions *(15,47,48)*. And in New Zealand, Bell *(49)* refers to early reports of poisoning in pigs and cattle consuming karaka fruit which contains 3-NPA esters.

Concerning *Astragalus* poisoning in North America, the clinical signs of livestock intoxication have been documented in field outbreaks as well as in experimental studies using whole plant, plant extract, or purified compound *(16,19,30,32,35,37–39,50)*. An acute and chronic syndrome has been described related to the amount and rate of toxin consumed. The chronic form is perhaps more common under range conditions where there is opportunity for exposure to a low level of toxin over an extended period of time. Either form of the intoxication induces respiratory, cardiovascular, and neurological signs.

In the acute intoxication in cattle, clinical signs include general weakness, a placid stupefied demeanor, and incoordination. Incoordination of the hindlimbs is especially notable, characterized by knuckling of the fetlock joint and interference of the hindlegs during ambulation. With further development of the intoxication, there is respiratory distress, frothy salivation, foaming at the nose, and cyanosis. If a lethal dose is consumed, recumbency, coma, and death generally occur within a period of a few hours to 1 d. Sheep by comparison generally show more respiratory distress, fewer neurological signs, and often die suddenly.

Respiratory distress and neurological signs are also prominent in the chronic intoxication. The labored respiration, which may be triggered by exertion, is characterized by a loud inspiratory rasp. Incoordination is evident, once again mainly exhibited in the hindlimbs: knuckling of the fetlock joint, interference and crossing of the hindlegs during ambulation, goose-stepping gait, and paresis. The respiratory distress and clicking sound of the dew claws due to hindleg interference have led to the common names "roaring disease" and "cracker heels" in some geographical areas. With continued exposure to the toxin, there is emaciation and death. Signs of intoxication may persist for long periods after exposure ends, and the affected animals may never fully recover.

The pathology findings in cattle and sheep intoxicated with 3-NPA have been reported in a limited number of animals *(16,19)*. Some of these animals were clinical cases from incidents of range poisoning, whereas others were given whole plant or purified compound under experimental conditions. The dose, frequency, and length of treatment varied widely encompassing a range of acute to chronic responses. Gross pathology observations included pulmonary congestion and edema, petechial hemorrhages on the surface of the heart, and liver swelling and congestion. There was pulmonary emphysema and pneumonia in animals ingesting plant material for more than a few days. Micropathology findings included alveolar emphysema, bronchiolar constriction, interlobular edema, and fibrosis in the lungs; focal parenchymal hemorrhages in the brain; and Wallerian degeneration in the spinal cord and sciatic nerves. Degenerative brain lesions were limited to focal thalamic malacia in one cow intoxicated under range conditions, and spongy vacuolation in the globus pallidus of another cow treated for approx 2 mo with whole plant. It is not known whether this paucity of brain pathology compared to laboratory rodents *(18,51)* is indicative of the true response of domestic livestock to 3-NPA or due to the relatively limited microscopic survey undertaken in this study. Maricle et al. *(52)* have reported an absence of significant diagnostic changes in routine hematological or serum biochemical parameters in cattle grazing timber milkvetch (*Astragalus miser* var. *serotinus*).

INTOXICATION IN HUMANS

Outbreaks of moldy sugarcane toxicity occurring in humans in China have provided an important comparative perspective on 3-NPA toxicity. Reviews of this subject area are available *(53,54)*. Sugarcane grown in southern China is commonly shipped to northern provinces and stored through the winter. This stored product is associated with moldy sugarcane toxicity due to fungal production of 3-NPA. During the years 1972 to 1989 there were 884 cases of moldy sugarcane toxicity, which included 88 deaths *(53)*. Because many outbreaks do not come to the attention of public health authorities, it is assumed that many cases are not reported *(55)*.

Most cases of poisoning were associated with sugarcane that had been in storage for at least 2 mo, and in most outbreaks the storage period was 3–4 mo *(13)*. Thus, cases most commonly occurred in the spring. In southern China, where sugar is rarely stored for more than 2 wk, moldy sugarcane toxicity is uncommon. No cases of poisoning have been reported in three sugarcane-producing provinces *(13)*.

Most of the victims of moldy sugarcane toxicity are children. Clinical signs characteristically develop 2–3 h after ingestion of moldy sugarcane. In some cases signs are observed less than one-half hour after ingestion. There are two clinical patterns of moldy sugarcane toxicity *(55)*. In milder cases, patients recover in a few days and the illness is characterized by gastrointestinal symptoms including abdominal pain, nausea, vomiting, diarrhea, and sometimes headache and lethargy. The severe form usually affects children, and the interval between moldy sugarcane ingestion and symptoms is usually less than 2 h. In such severe cases gastrointestinal signs may or may not become manifest. Seizures develop suddenly and occur frequently. Coma may also develop quickly and persist for 1 wk or longer. During the acute phase there may be limb weakness, abnormal eye position, nystagmus, aphasia, and difficulty swallowing. After a delay of 7–40 d, dystonia occurs in 10–50% of patients affected with moldy sugarcane toxicity *(13)*. Other symptoms disappear, but dystonia is persistent and nonprogressive. It is characterized by choreoathetosis, torsion spasms, and painful paroxysmal spasms of the extremities *(56)*. In mildly affected patients, computed tomography (CT) scans may show no alteration or, in some cases, diffuse mild hypodensity suggesting brain edema *(55)*. Dystonic patients had bilateral hypodensities in the lenticular nuclei *(56,57)*, presumed to represent focal areas of softening. Both putamen and globus pallidus were consistently involved, and there was infrequent involvement of the caudate nucleus *(56)*. At autopsy two patients with acute disease appeared to have cerebral edema, while a third had congestion of the brain *(13)*.

Fungal isolates from sugarcane samples involved in outbreaks of toxicity were dominated by *Arthrinium* species (46–70%) *(53)*. Toxigenic strains of microorganisms isolated from moldy sugarcane involved in human poisoning were identified by mouse inoculation studies. Almost all of these were *Arthrinium* species *(13)*. Toxigenic *Arthrinium* cultures and poisonous sugarcane juice produced similar clinical alterations when administered intragastrically to weanling mice *(13,58)*. There was variable distribution of toxin in the sugarcane *(58)*, which may in part explain the varying degrees of toxicity observed in a group of people consuming moldy sugarcane.

Cultures of highly toxigenic *Arthrinium* produced 3-NPA *(12)*, and 3-NPA purified from such cultures, as well as commercially obtained 3-NPA, produced toxicity in mice similar to that produced by the toxigenic culture itself *(59)*. The 3-NPA content of sugarcane involved in outbreaks of toxicity was 285 ppm to 6660 ppm *(13)*. *Arthrinium* species and their toxic metabolite, 3-NPA, appear to be the main etiological agents of moldy sugarcane toxicity. On the basis of 3-NPA content of moldy sugarcane samples consumed by two poisoning victims and estimating the amount of sugarcane consumed by each victim, the 3-NPA ingested was approx 5.7 mg/kg body wt and 2.2 mg/kg body wt for a 4-yr-old child and an 8-yr-old child respectively *(60)*.

As commented upon in the review of Ludolf et al. *(54)*, two other reports of neurological disease in Chinese children invite comparison to moldy sugarcane toxicity. One was an outbreak of an acute degenerative striatal disease following a winter of famine *(61)*. Fourteen adults and children derived from three families with close social contact were affected. Three months before the outbreak, stocks of food had failed and all the villagers subsisted on poor quality maize said to be fermented. In nine cases there was practically complete recovery. In three, the course was progressive without death; and in one, death occurred after 2 yr. In another case death occurred in 4 wk. The age range was 4 yr to 56 yr. The youngest were affected most severely; fatal cases were the youngest. Onset of symptoms was abrupt and included failures of muscles of equilibration, speech defects, and disorders of movement. Some of those affected manifested lethargy, but coma was not described. In the one case subjected to postmortem examination, there was necrosis (with gross softening) in the globus pallidus bilaterally and in the substantia nigra unilaterally.

The second report inviting comparison is by Verhaart *(62)*. Four Chinese infants, 5–8 mo of age, were affected with extensive symmetrical disintegration of the striatum. Less developed lesions were in the globus pallidus, corpus subthalamicus, red nucleus, and the corpora quadrigemina. All four were exclusively breast fed. Conventional etiologic agents were excluded.

Finally, the work of Bell *(49)* is relevant to 3-NPA toxicity of humans. This investigation concerns karaka nuts, which are the fruits of *Corynocarpus laevigatus*, a decorative tree native to New Zealand and the Chatham Islands. Karakin and other 3-NPA glucosides are present in the kernel of the fruit. Its hydrolysis derivative is 3-NPA. The Maori people consumed this as a staple vegetable after cooking and washing. In 1924, Best, as cited by Bell *(49)*, reported that when raw karaka nuts were consumed painful contractions of the limbs could occur. In 1871, Skey, as cited by Bell *(49)*, reported that the toxicity caused by the consumption of raw kernels was usually in children. Symptoms included violent convulsions in which the arms were stretched out violently and rigidly. Many cases were fatal.

ACKNOWLEDGMENT

We wish to thank Dr. Xiu Ou, Department of Pathology, College of Veterinary Medicine and Biomedical Sciences, Colorado State University, Fort Collins, CO, for translation of original papers in Chinese.

REFERENCES

1. Porter DJT, Bright HJ. 3-Carbanionic substrate analogues bind very tightly to fumarase and aspartase. J Biol Chem 1980;255:4772–4780.
2. Conn EE. Biosynthetic relationship among cyanogenic glycosides, glucosinolates, and nitro compounds. In: Cutler HG, ed. Biologically Active Natu-

ral Products: Potential Use in Agriculture, Series 380. American Chemical Society, New York, 1988, pp. 143–154.
3. Carter CL , McChesney WJ. Hiptagenic acid identified as β-nitropropionic acid. Nature 1949;164:575–576.
4. Carrie MS. Karakin, the glucoside of *Corynocarpus laevigata*. J Soc Chem Indust 1934;288T–289T.
5. Morris MP, Pagan C, Warmke HE. Hiptagenic acid, a toxic component of *Indigophera endecaphylla*. Nature 1954;164:575–576.
6. Wilson BJ. Miscellaneous *Aspergillus* toxins. In: Ciegler A, Kadis S, Ajl SJ, eds. Microbial Toxins, Vol. 6: Fungal Toxins Academic Press, New York, 1971, 207–294.
7. Finnigan RA, Stephani RA. The structure of karakin. Lloydia 1970;33:491.
8. Harlow MC, Stermitz FR, Thomas RD. Isolation of nitro compounds from *Astragalus* species. Phytochemistry 1975;14:1421–1423.
9. Gustine DL, Shenk JS, Moyer BG, et al. Isolation of β-nitropropionic acid from crownvetch. Agron J 1974;66:636–639.
10. Gnanasunderam C, Sutherland OR. Hiptagen and other aliphatic nitro esters in *Lotus pedunculatus*. Phytochemistry 1986;25:409–410.
11. Wilson BJ. Miscellaneous *Penicillium* toxins. In: Ciegler A, Kadis S, Ajl SJ, eds. Microbial Toxins, Vol. 6: Fungal Toxins. Academic Press, New York, 1971, pp. 460–517.
12. Hu WJ, Liang XT, Chen XM, et al. Isolation and structural determination of sugarcane poisoning *Arthrinium* toxicity material 3-nitropropionic acid. Chin J Prev Med 1986;20:321–323.
13. Liu X, Luo X, Hu W. Studies on the epidemiology and etiology of moldy sugarcane poisoning in China. Biomed Environ Sci 1992;5:161–177.
14. Stermitz FR, Norris FA, Williams MC. Miserotoxin, a new naturally occurring nitro compound. J Am Chem Soc 1969;91:4599–4600.
15. Shenk JS, Wangsness PJ, Leach RM, et al. Relationship between β-nitropropionic acid content of crownvetch and toxicity in nonruminant animals. J Anim Sci 1976;42:616-621.
16. Williams MC, VanKampen KR, Norris FA. Timber milkvetch poisoning in chickens, rabbits, and cattle. Am J Vet Res 1969;30:2185–2190.
17. Williams MC, James LF, Bleak AT. Toxicity of introduced nitro-containing *Astragalus* to sheep, cattle and chicks. J Range Manage 1976;29:30–33.
18. Gould DH, Gustine DL. Basal ganglia degeneration, myelin alterations, and enzyme inhibition induced in mice by the plant toxin 3-nitropropanoic acid. Neuropathol Appl Neurobiol 1982;8:377-393.
19. James LF, Hartley WJ, Williams WC, et al. Field and experimental studies in cattle and sheep poisoned by nitro-bearing *Astragalus* or their toxins. Am J Vet Res 1980;41:377–382.
20. Pass MA, Muir AD, Majak W, et al. Effect of alcohol and aldehyde dehydrogenase inhibitors on the toxicity of 3-nitropropanol in rats. Toxicol Appl Pharmacol 1985;78:310–315.
21. Majak W, Clark LJ. Metabolism of aliphatic nitro compounds in bovine rumen fluid. Can J Anim Sci 1980;60:319–325.

22. Majak W, Cheng K-J. Identification of rumen bacteria that anaerobically degrade aliphatic nitro toxins. Can J Microbiol 1981;27:646–650.
23. Majak W, Pass MA. Aliphatic-nitro-compounds. In: Cheeke PR, ed. Toxicants of Plant Origin, Vol. 2: Glycosides. CRC, Boca Raton, FL, 1989, pp. 143–159.
24. Anderson RC, Rasmussen MA, Allison MJ. Metabolism of the plant toxins nitropropionic acid and nitropropanol by ruminant microorganisms. Appl Environ Microbiol 1993;59:3056–3061.
25. Gustine DL, Moyer BG, Wangsness PJ, et al. Ruminal metabolism of 3-nitropropanoyl-D-glucopyranoses from crownvetch. J Anim Sci 1977; 44:1107–1111.
26. Pass MA, Majak W, Muir AD, et al. Absorption of 3-nitropropanol and 3-nitropropionic acid from the digestive system of sheep. Toxicol Lett 1984;23:1–7.
27. Gustine DL, Moyer BG. Mechanisms of toxicity of 3-nitropropionic acid in nonruminant animals. In: Smith JA, Hays VW, eds. Proceedings of the 14th International Grasslands Congress. Westview, Boulder, CO, 1983, pp. 736-738.
28. Pass MA, Carlisle CH, Reuhl KR. 3-Nitropropionic acid toxicity in cultured murine embryonal carcinoma cells. Natural Toxins 1994;2:386–394.
29. Matsumoto H, Hylin JW, Miyahara A. Methemoglobinemia in rats injected with 3-nitropropanoic acid, sodium nitrite, and nitroethane. Toxicol Appl Pharmacol 1961;3:493–499.
30. Majak W, Udenberg T, McDiarmid RE, et al. Toxicity and metabolic effects of intravenously administered 3-nitropropanol in cattle. Can J Anim Sci 1981;61:639–648.
31. Muir AD, Majak W, Pass MA, et al. Conversion of 3-nitropropanol (miserotoxin aglycone) to 3-nitropropionic acid in cattle and sheep. Toxicol Lett 1984;20:137–141.
32. Mathews FP. The toxicity of red-stemmed peavine for cattle, sheep, and goats. J Am Vet Med Assoc 1940;97:125–134.
33. MacDonald MA. Timber milkvetch poisoning on British Columbia ranges. J Range Manage 1952;5:16–20.
34. Williams MC, James LF. Toxicity of nitro-containing *Astragalus* to sheep and chicks. J Range Manage 1975;28:260–263.
35. Williams MC, James LF. Livestock poisoning from nitro-bearing *Astragalus*. In: Keeler RF, VanKampen KR, James LF, eds. Effects of Poisonous Plants on Livestock. Academic Press, New York, 1978, pp. 379–389.
36. Williams MC, James LF, Bond BO. Emory milkvetch (*Astragalus emoryanus* var *emoryanus*) poisoning in chicks, sheep, and cattle. Am J Vet Res 1979;40:403–406.
37. James LF, Hartley WJ, Van Kampen KR. Syndromes of *Astragalus* poisoning in livestock. J Am Vet Med Assoc 1981;178:146–150.
38. James LF. Neurotoxins and other toxins from *Astragalus* and related genera. In: Keeler RF, Tu AT, eds. Handbook of Natural Toxins, Vol. 1. Marcel Dekker, New York, 1983, pp. 445–462.
39. Williams MC. Impact of poisonous weeds on livestock and humans in North America. Rev Weed Sci 1994;6:1–27.

40. Williams MC, James LF. Poisoning in sheep from emory milkvetch and nitro compounds. J Range Manage 1976;29:165–167.
41. Williams MC. Toxicological investigations on *Astragalus hamosus* and *Astragalus sesameus*. Aust J Exp Agric Anim Husb 1980;103:162–165.
42. Tarazona JV, Sanz F. Aliphatic nitro compounds in *Astragalus lusitanicus* Lam. Vet Hum Toxicol 1987;29:437–439.
43. Sager RL, Nieto M. Nitrocompuestos organicos alifaticos en dos especies del genero *Astragalus*. An Asoc Quim Argentina 1987;75:5–18.
44. Sager RL, Nieto M. Estudio toxicologico de *Astragalus distinens* Macl. y *Astragalus bergii*. H Rev Arg Prod Anim 1991;11:329–335.
45. Norfeldt S, Henke LA, Morita K, et al. Feeding tests with *Indigofera endecaphylla* Jacq. (creeping indigo) and some observations on its poisonous effects on domestic animals. Univ Hawaii Agric Exp Stat Tech Bull 1951;15:3–23.
46. Britten EJ, Matsumoto H, Palafox AL. Comparative toxic effects of 3-nitropropionic acid, sodium nitrite and *Indigophera endecaphylla* on chicks. Agron J 1959;51:462–464.
47. Shenk JS, Risius ML, Barnes RF. Weanling meadow vole responses to crownvetch forage. Agron J 1974;66:13–15.
48. Gustine DL. Aliphatic nitro compounds in crownvetch: a review. Crop Sci 1979;19:197–203.
49. Bell ME. Toxicology of karaka kernel, karakin, and beta-nitropropionic acid. N Zeal J Sci 1974;17:327–334.
50. Pass MA. Toxicity of plant-derived aliphatic nitrotoxins. In: Colegate SM, Dorling PR, eds. Plant-Associated Toxins: Agricultural, Phytochemical and Ecological Aspects, 4th International Symposium on Poisonous Plants, Fremantle, Western Australia. CAB International, Tucson, AZ, 1994, pp. 541–545.
51. Hamilton BF, Gould DH. Nature and distribution of brain lesions in rats intoxicated with 3-nitropropionic acid: a type of hypoxic (energy deficient) brain damage. Acta Neuopathol 1987;72:286–297.
52. Maricle B, Tobey J, Majak W, et al. Evaluation of clinicopathological parameters in cattle grazing timber milkvetch. Can Vet J 1996;37:153–156.
53. He F, Zhang S, Zhang C, et al. Mycotoxin-induced encephalopathy and dystonia in children. In: Volans GN, Sims J, Sullivan FM, et al., eds. Basic Science in Toxicology. Taylor and Francis, London, 1990, pp. 596–604.
54. Ludolph AC, He F, Spencer PS, et al. 3-Nitropropionic acid—exogenous animal neurotoxin and possible human striatal toxin. Can J Neurol Sci 1991;18:492–498.
55. Ming L. Moldy sugarcane poisoning—a case report with a brief review. Clin Toxicol 1995;33:363–367.
56. He F, Zhang S, Qian F, et al. Delayed dystonia with striatal CT lucencies induced by a mycotoxin (3-nitropropionic acid). Neurology 1995;45:2178–2183.
57. He F, Zhang S, Liu L, et al. Extrapyramidal lesions caused by mildewed sugarcane poisoning (with 3 case reports). Chin J Med 1987;67:395–396.
58. Liu XJ. Investigations on the etiology of mildewed sugarcane poisoning. A review. Chin J Prev Med 1986;20:306–308.

59. Fu Y, He F, Zhang S, et al. Consistent striatal damage in rats induced by 3-nitropropionic acid and cultures of *Arthrinium* fungus. Neurotoxicol Teratol 1995;17:413–418.
60. Lui XJ, Hu WJ, Wang YH, et al. Studies on the mycology and mycotoxins in an outbreak of deteriorated sugar cane poisoning. Chin J Prev Med 1989;23:345–348.
61. Woods AH, Pendleton L. Fourteen simultaneous cases of an acute degenerative striatal disease. Arch Neurol Psychiatry 1925;13:549–568.
62. Verhaart WJC. Symmetrical degeneration of the neostriatum in Chinese infants. Arch Dis Child 1938;13:225–234.

3
The Neurochemistry of 3-Nitropropionic Acid

Norman C. Reynolds, Jr. and Wen Lin

INTRODUCTION

3-Nitropropionic acid (3-NPA) is a widespread, naturally occurring fungal and plant toxin whose administration to animals results in selective morphological brain damage in the striatum *(1–6)*. The basis for regional vulnerability likely reflects some combination of differences in regional blood flow, regional efficiency of mitochondrial energy metabolism, and neuronal response to excitotoxin *(3,7–11)*. The predominate molecular basis for the toxicity of 3-NPA is irreversible inhibition of succinic acid dehydrogenase, an enzyme found in both the Krebs cycle and complex II of the mitochondrial electron transport system *(8,12–14)*. The resultant uncoupling of oxidative phosphorylation severely impairs aerobic neuronal energy metabolism.

Plant Origins

Both 3-NPA and its alcohol cogener, 3-nitropropanol (3-NPOH), are found among plants in several *Astragalus* species (e.g., *A. distortus* or "locoweed") and species of several other genera: *Coronilla*, *Indigofera*, *Lotus* (clover), *Corynecarpus*, *Hiptage*, *Heteropteris*, and *Janusia (8)*. Such plant sources contain miserotoxin, a β-D-glucoside conjugate of 3-NPOH (i.e., 3-nitro-1-propyl-β-D-glucopyranoside), free 3-NPA, and glucose esters of 3-NPA *(15)*. Hydrolysis to free 3-NPA and 3-NPOH in the ruminant gut is an essential step in releasing these toxic compounds *(17)*. Although both 3-NPOH and 3-NPA inhibit succinic acid dehydrogenase in vitro *(16)*, hepatic alcohol dehydrogenase catalysis of 3-NPOH to 3-NPA in rats and ruminants suggests that 3-NPA is the common lethal metabolite in the acute encephalopathy encountered in these species *(16,18)*. Free 3-NPA also occurs in the sugarcane mildew fungus *Arthrinium* and accounts for some animal and human syndromes of acute encephalopathy reported in China

From: *Mitochondrial Inhibitors and Neurodegenerative Disorders*
Edited by: P. R. Sanberg, H. Nishino, and C. V. Borlongan © Humana Press Inc., Totowa, NJ

(19). The specific scientific interest in 3-NPA follows the observation of patients with acute encephalopathy and delayed dystonia observed in China after ingestion of sugarcane infected with *Arthrinium (19)* and the adaptation of the 3-NPA toxin to several models of neurodegenerative diseases *(11,20)*.

In Vitro Synthesis

Commercially available in vitro synthesis of 3-NPA and 3-NPOH has greatly simplified direct access to these reagents for scientific study. 3-NPA is synthesized from β-propiolactone *(21)* and can be recrystallized from chloroform to enhance purity *(22)*. 3-NPOH can be generated by hydrolysis of miserotoxin using β-glucosidase *(23)* or synthesized from 3-bromopropanol *(24)*.

3-NPA AS A MOLECULE

Propionyl Nitro Compounds as a Class

3-NPA (or 3-nitropropanoic acid, a derivative of propane) is a saturated three-carbon carboxylic acid whose nitro substituent on the third carbon provides several unique properties that distinguish 3-NPA from its parent compound propionic (propanoic) acid (*see* Fig. 1). The nitro substituent imparts electroactivity to the compound, increases its acidic nature by an electron withdrawing effect on the carboxylic moiety, and imparts specific nuclear magnetic resonance spectra and reversed phase partition properties unique to this aliphatic nitro compound. Because of the interrelation of reduced forms in biotransformation, these reduced forms, the aldehyde and alcohol derivatives of 3-NPA, should be considered together along with 3-NPA as a family of compounds: the propionyl nitro compounds.

Physical Properties of Propionyl Nitro Compounds

Of the three propionyl nitro compounds, the aldehyde, 3-nitropropionaldehyde (3-NPAL), is unstable and spontaneously decomposes to nitrite and acrolein, CH_2=CH-CHO, at neutral pH *(23)*. In tissue extract studies where the enzyme alcohol dehydrogenase is present, 3-NPAL can be partially oxidized to 3-NPA and therefore can contribute to measured values of 3-NPA *(23)*. As is the case with organic acids in general, liquid chromatographic separation of 3-NPA, 3-NPOH, and 3-NPAL can be accomplished with isocratic reversed phase elution from a silica-based substrate as the stationary phase. Thin-layer chromatography (TLC) has been used with $CHCl_3$:acetone (1:1) containing 1% H_2O as the solvent system and diazotized *p*-nitroaniline spray as the developer. R_f values on silica gel under these conditions yielded 0.17 (3-NPA), 0.22 (3-NPAL semicarbazone), 0.73

	ALIPHATIC COMPOUND	ALIPHATIC NITRO COMPOUND
ALCOHOL	CH_3-CH_2-CH_2-OH Propanol	O_2N-CH_2-CH_2-CH_2-OH 3-nitropropanol
ALDEHYDE	CH_3-CH_2-CHO Propionaldehyde	O_2N-CH_2-CH_2-CHO 3-nitropropionaldehyde
ACID	CH_3-CH_2-COOH Propionic acid	O_2N-CH_2-CH_2-COOH 3-nitropropionic acid

Fig. 1. The addition of a nitro group on carbon 3 of the propionyl aliphatic structure enhances the acidic property of the carboxylic group by an electron-withdrawing effect and provides both electroactivity for electrochemical detection ($E_{1/2}$ = –0.75 to –0.90 V, SCE) and photon absorption for ultraviolet detection (210 nm).

(3-NPOH), and 0.79 (3-NPAL) *(23)*. 3-NPAL semicarbazone is used to trap 3-NPAL by reaction with semicarbazide hydrochloride.

High-performance liquid chromatography (HPLC) has also been used to resolve the propionyl nitro compounds and the semicarbazone using reversed phase isocratic elution with a simple mobile phase of H_2O (up to 5% methanol and pH adjusted with phosphate) and octadecyl silanized silica-based columns with 5 μm pore size (plain and N-CAP'd, 15 cm × 4 mm or 30 cm × 4 mm) protected by small guard columns (10 μm) at flow rates up to 1 mL/min within 35 min *(23,25)*.

Detection by ultraviolet absorption at 210 nm (LCUV) is the detection mode of choice by Majak and colleagues *(23,25)*; however, electrochemical detection (LCEC) is certainly possible in the reductive mode but has not been exploited. LCEC can expand the domain of measurable products to include several other electroactive molecules of experimental interest. All aliphatic nitro compounds display half-wave potentials ($E_{1/2}$) due to the irreversible reduction of the nitro group to hydroxylamine using a four electron transfer in the range $E_{1/2}$ = –0.90 to –0.75 V (SCE) depending upon the nature and concentration of the supporting electrolyte *(26)*. Contributions to electroactivity stem almost exclusively from the nitro group because saturated aliphatic monocarboxylic acids and alcohols are not reducible voltammetrically. On the other hand, most saturated aliphatic aldehydes are easily reducible from $E_{1/2}$ = –1.89 to –1.92 V (SCE) *(27)*. In particular, contributions from the carbonyl function in propionaldehyde show reduction from $E_{1/2}$ = –1.59 to –1.92 with either LiOH or NaOH supporting electrolyte *(28)*. Although the nitro group electron withdrawing influence upon the aldehyde group may well lower the range of voltammetric reduction to some extent, it is unlikely to overlap the range of optimal voltammetry for the nitro substituent *per se*.

Because of its short carbon length, all three nitro compounds are miscible in H_2O, alcohol, or ether as are the unsubstituted parent compounds. The p*K* of propionic acid is 4.87. Although the p*K* of 3-NPA is not reported in Lange's *Handbook of Chemistry*, the p*K* of 2-nitropropionic acid is 3.79, which illustrates the expected relative influence of the nitro substituent in increasing acidity *(29)*.

Nuclear magnetic resonance (NMR) can be used to identify purity of solid propionyl nitro compounds based upon standard spectra *(25)*. In addition, receptor binding interactions can be studied by chemical shift NMR spectra in vitro *(30)* but this approach has not been fully exploited. Another intriguing but essentially untapped technology is the use of water suppressed chemical shift proton magnetic resonance spectroscopy (pMRS), which can identify certain key molecules relating to excitotoxic neurodegenerative processes in vivo, e.g., glutamate, glutamine, lactate, and *N*-acetylaspartate *(2,31,32)*. Scanning at 1/2 Tesla allows simultaneous measurements of several molecules of interest including exogenous molecules (e.g., drugs or 3-NPA) in small nuclear areas such as the striatum (*unpublished data*).

BIOAVAILABILITY

Absorption and Distribution

The toxicity of the propionyl nitro compounds by mouth depends on the form of the compounds ingested in different organisms. Selected species of *Astragalus* with high levels of miserotoxin are especially toxic to ruminants while 3-NPA-containing plants and fungi are typically associated with poisoning of monogastric mammals that include rats and primates *(33)*. Monogastric mammals lack the requisite enteric microorganisms to facilitate the hydrolysis of miserotoxin to 3-NPOH, whereas all mammals are capable of converting 3-NPOH to 3-NPA by hepatic alcohol dehydrogenase *(18)*. The LD_{50} for oral miserotoxin in rats is >2.5 g/kg, whereas the LD_{50} for oral 3-NPOH in rats is 77 mg/kg *(34)*. On the other hand, inhibition of alcohol dehydrogenase in rats prevents 3-NPOH toxicity, suggesting the need for biotransformation of 3-NPOH to 3-NPA to produce the toxicity *(18,35)*.

Observations showing that plants that release 3-NPOH upon hydrolysis are more toxic to ruminants than plants that release 3-NPA can be explained by differential absorption. 3-NPA is more slowly absorbed from the reticulorumen than 3-NPOH and as such is more susceptible to biodegrada-

Table 1
Routes of Administration of 3-NPA

Oral
Systemic injection
Subcutaneous
Intramuscular
Intraperitoneal
Focal injection (striatum)
Direct: microsyringe
Controlled perfusion: microdialysis miniprobes
Continuous perfusion: osmotic minipumps
In vitro cell culture

tion by microorganisms *(36)*. Intraperitoneal injections of 3-NPA and 3-NPOH remove the effect of differential absorption and enteric degradation and show the two propionyl nitro compounds to be equally toxic with an LD_{50} = 61 mg/kg for 3-NPOH and an LD_{50} = 67 mg/kg for 3-NPA *(18,37)*.

The fact that 3-NPA is selective for certain brain regions is not related to a simple process of regional uptake; however, changes in regional blood flow have been postulated based on a distribution of platelet microthrombi in brain regions susceptible to damage by 3-NPA *(13)*. The vulnerability of striatum to 3-NPA induced cytotoxic damage is felt to be due to a high sensitivity of the striatum to mitochondrial dysfunction coupled with a higher level of glutamatergic input *(3)*. Lesions elsewhere in the thalamus and hippocampus may be primarily reactive to vascular hypotension occurring as a systemic response to 3-NPA toxicity *(7,38)*, but all three regions show plasma immunoglobulin G exudate in careful cytologic assessment, suggesting destruction of the blood–brain barrier *(38)*.

Methods of Administration

Different methods of administration of propionyl nitro compounds and miserotoxin can be used to mimic or circumvent enteric absorption, to study acute vs chronic exposure, to maximize or circumvent systemic metabolism and differences in regional vascular perfusion, or to eliminate all aspects of pharmacodynamics to facilitate receptor binding kinetics (*see* Table 1).

Differences in bioavailability in lethal toxin have been studied by comparing responses to ruminant and to monogastric oral ingestion of miserotoxin, 3-NPA, and 3-NPOH *(33,34)*. Oral ingestion of 3-NPOH in rats pretreated with alcohol dehydrogenase inhibitor *(35)* and comparative lethal toxicity of intraperitoneal injections of 3-NPA and 3-NPOH in rats

(35,37) have been used to show that the conversion of 3-NPOH to 3-NPA is essential for lethal toxicity. Further comparisons of 3-NPOH and 3-NPA activity in murine embryonic carcinoma cell culture shows that 3-NPOH does in fact suppress succinic acid dehydrogenase activity but only at much higher concentrations than were required for 3-NPA *(16)*. This is compatible with the lack of toxic bioequivalence of 3-NPOH and 3-NPA in vivo unless 3-NPOH is converted to 3-NPA. Further in vitro analysis suggests that the mechanism of irreversible inhibition is the enzymatic conversion of 3-NPA to 3-nitro acrylic acid, which then covalently reacts with this dehydrogenase enzyme and inactivates it *(38)*.

Subcutaneous injections of 3-NPA into rats have been used to study the contribution of cardiorespiratory failure and breakdown of the blood–brain barrier to morphological changes *(13,38,39)* and decreases in succinic acid dehydrogenase activity (40) in vulnerable brain regions. Alzet® minipumps have been implanted to provide continuous subcutaneous chronic exposure of 3-NPA *(2,41)*.

Single-dose intraperitoneal injections of 3-NPA (30 mg/kg) were used to study age dependence of striatal lesions in rats *(3)*. A combination of intraperitoneal 3-NPA followed by stereotactic infusion of *N*-methyl-D-aspartate (NMDA) was used to study the potentiating effect of 3-NPA-induced metabolic impairment upon NMDA excitotoxin-induced neuronal death *(12)*. Acute single-dose exposure compared with multiple-dose chronic exposure has been studied by both subcutaneous and intraperitoneal routes in rodents to define differences in morphology and in motor performance *(8)*. Chemical preconditioning to minimize successive decreases in energy metabolism was studied by comparing in vivo low-dose (20 mg/kg) intraperitoneal administration with in vitro high-concentration (1 m*M*) hippocampal brain slice responses to electrical stimulation *(14)*. Intramuscular injections were used in baboons to simulate chronic exposure to mitochondrial toxins and to elicit a triad of selective striatal lesions, dyskinesias, and frontostriatal cognitive impairment *(42)*.

Although systemic administration of 3-NPA has the advantage of the neurotoxin crossing the blood–brain barrier *(31)*, stereotactic injections are the most direct method of introducing 3-NPA to the striatum and bypassing several contributions to pharmacodynamics in vivo. Two methods of local injection are possible. Direct injection under positive pressure, either by needle *(2)* or by stereotactically implanted cannulae, applies 3-NPA locally but simultaneously incurs positive pressure mechanical trauma, thereby complicating receptor action with the molecular events of a damage pattern. More rapid injections over shorter periods of time produce less overall damage *(43)*. Gentle application of 3-NPA can be accomplished by introduction through an osmotic miniprobe that perfuses the interstitial space and releases 3-NPA osmotically. Benefits of osmotic miniprobes involve continuous access to the interstitial space for purposes of dialysis of neurotransmitters and metabolic products as well as the instillation of toxin *(44)*. This allows acute and delayed kinetic comparisons of the effects of different toxins such as 3-NPA and the excitotoxin quinolinic acid *(41)*.

In vitro incubation of 3-NPA in brain cell culture or in an enzyme suspension provides direct receptor access and allows simple manipulations of the components in the suspension. Neurotoxicity of 3-NPA in neuronal cell culture showed no differences in lethal responses (LD_{50} = 2.5 m*M*) of hippocampal, striatal, septal, and hypothalamic neurons despite clear differences in cytotoxic responses in vivo. In addition, the effect of 3-NP in these same cultures was quite energy substrate dependent but was clearly attenuated by MK-801, the noncompetitive NMDA antagonist *(45)*. In another in vitro study, the high vulnerability of dopamine-secreting neurons to mild metabolic stress from 3-NPA is clearly relevant to the pathophysiology of Parkinson's disease, which displays selective vulnerability of the dopamine-secreting substantia nigra in vivo *(46)*.

In Vivo Metabolism and Byproducts of Metabolism

3-NPA can be ingested by ruminant and monogastric mammals from plant and fungal sources as free 3-NPA, absorbed and secreted in the urine unchanged *(33)*. Ruminant ingestion of plants also provides conjugated forms of lethal nitro toxins including glucose esters of 3-NPA and a β-D-glucoside conjugate of 3-NPOH called miserotoxin *(15)*. Hydrolysis of conjugates of 3-NPOH and 3-NPA to release the simple propionyl nitro compounds can occur in the presence of ruminant microorganisms that can also produce anaerobic detoxification *(17)*. The absorption of 3-NPOH and 3-NPA from the gut and into the blood precedes a major stoichiometric conversion of 3-NPOH into 3-NPA via hepatic alcohol dehydrogenase *(16,18)* but intraruminal oxidation of 3-NPOH to 3-NPA does not occur to any significant extent *(47)*. Anaerobic detoxification involves the release of inorganic nitrite from 3-NPOH and 3-NPA with subsequent conversion of the nitrite to ammonia *(17)*. Although in vitro studies show the oxidation of 3-NPOH to 3-NPAL by alcohol dehydrogenase and the subsequent formation of both 3-NPA and acrolein, the formation of acrolein from 3-NPAL has not been shown in vivo *(23)*. Nitrite formation in vivo occurs after the conversion of 3-NPOH to 3-NPA *(48)* and 3-NPA to 3-nitroacrylic acid *(49)*.

RECEPTOR SPECIFICITY

Reversible and Irreversible Receptor Binding

In vitro studies have shown that 3-NPA irreversibly inhibits succinic acid dehydrogenase *(49,50)* but reversibly inhibits fumarase and aspartase *(51)*. The molecular basis for the in vitro toxicity of 3-NPA is believed to be limited to the irreversible inhibition of succinic acid dehydrogenase *(8,12–14)*. The resulting deficiency in oxidative phosphorylation in the electron transport chain results in impaired energy metabolism which triggers neuronal death. The nitro group on carbon 3 is the essential structural moiety for the activity of 3-NPA because the parent compound propionic acid, its 3-chloro, 3-mercapto (-SH), and 2-chloro derivatives have no effect on the enzyme activity whatsoever *(16)*. In addition, the dianionic form of the nitro group is believed to be an essential structure for the activity as a succinic acid dehydrogenase inhibitor *(50)*. 3-NPOH does have inhibitory effects on succinic acid dehydrogenase but only at much higher concentrations *(16)*.

The equilibrium constant for the inhibition is $K_i = 2 \times 10^{-4}$ *M*, which suggests that 3-NPA is a weak competitive inhibitor *(49)*. Because the actual inhibition is irreversible and develops slowly and progressively, this suggests a sequence of two steps:

$$(1)\ E + I \rightleftarrows EI\ (\text{fast})$$

$$(2)\ EI + FAD \rightleftarrows EI' + FADH_2 \xrightarrow{O_2} EI'' + FAD + H_2O_2\ (\text{slow})$$

The initial step would involve a fast but reversible enzyme inhibitor adduct formation where E = succinic acid dehydrogenase and I = 3-NPA. Active sites on the enzyme involve a sulfhydryl (-SH) group for covalent reactivity and a flavin adenine dinucleotide (FAD) prosthetic group for oxidation of the 3-NPA to 3-nitroacrylate (I') which covalently binds to the enzyme and becomes reduced with the liberation of H_2O_2 to form a thioether (EI") with carbon 2 of 3-NPA *(49)*. Confirmation of this reaction scheme is the fact that synthetic 3-nitroacrylate inhibits succinic acid dehydrogenase instantly to form an irreversible product. Incubation of 5 μ*M* succinic acid dehydrogenase with either 5 μ*M* of 3-nitroacrylic acid for 2 min at 0°C or 50 μ*M* of 3-NPA for 25 min at 0°C resulted in a 94% inactivation of the enzyme *(49)*.

Comparison of 3-NPA with Other Mitochondrial Inhibitors

Along with 3-NPA, several other mitochondrial inhibitors have been studied to assess independent mechanisms for mitochondrial energy depletion and subsequent secondary excitotoxic effects on selective neuronal deterioration. A partial list of other mitochondrial toxins directly relevant to studies of central nervous system (CNS) degeneration *(32,52)* includes 3-acetyl pyridine (3-AP) *(53,54)*, *N*-methyl-4-phenyl pyridinium (MPP^+) *(55,56)*, malonic acid *(32,57)*, aminooxyacetic acid (AOAA) *(32)*, and azide *(32,52)*. Although all of the inhibitors have been used to study mechanisms of mitochondrial energy depletion in general, some of the inhibitors have been historically associated with specific disease entities; one example is the association of 3-NPA with Huntington's disease because of the production of delayed dystonia *(19)*, age dependence *(31)*, and selective striatal lesions *(1–6,11)*.

1-Methyl-4-phenyl-1,2,3,6-tetrahydropyridine (MPTP) has been used as a model for Parkinson's disease because its conversion to the free radical MPP^+ by monoamine oxidase in the CNS produces a toxic acquired Parkinsonism in primates *(55,58)*. The site of action of MPP^+ is complex I of the mitochondrial electron transport system (ETS), specifically the enzyme NADH-ubiquinone oxidoreductase *(52)*.

3-AP is a nicotinamide adenine dinucleotide antagonist that causes selective degeneration of inferior olives, substantia nigra, and other brainstem nuclei suggestive of human olivoponto–cerebellar degeneration *(53)*. Like 3-NPA, which inhibits succinic acid dehydrogenase and complex II, 3-AP shows age-dependent vulnerability of older rats to neurodeterioration *(54)*.

Malonic acid, like 3-NPA, specifically inhibits succinic acid dehydrogenase and complex II of the ETS although the kinetics of enzyme interaction suggest a reversible inhibition *(57)*. The decrement in ATP production is therefore similar to that produced by 3-NPA when malonic acid is stereotactically introduced to the striatum and severely reduces neuronal oxidative phosphorylation. In addition, like 3-NPA, malonic acid shows age-dependent vulnerability of older rats to neurodeterioration *(31,57)*. Succinic acid, the usual substrate for succinic acid dehydrogenase, is a four-carbon dicarboxylic acid that is reduced to fumaric acid in the Krebs cycle *(59)*. Malonic acid is only a three-carbon dicarboxylic acid whose acidic properties would be similar (*see* Fig. 2) but whose charge distribution and overall dimensions resemble 3-NPA. Stereospecificity of malonic acid for the site of interaction of succinic acid dehydrogenase with 3-NPA is expected but the mechanics of covalent interaction to form a thioether bond are apparently absent. As is the case with the nitro group of 3-NPA, the ability of the distal (ω) carboxylic group of succinic or malonic acid to form a dianionic (partial or formal) charge distribution is likely a key property for enzyme interaction.

Propionic acid	CH_3-CH_2-COOH	$pK_1 = 4.87$
Succinic acid	HOOC-CH_2-CH_2-COOH	$pK_1 = 4.21$, $pK_2 = 5.64$
Malonic acid	HOOC-CH_2-COOH	$pK_1 = 2.86$, $pK_2 = 5.70$

Fig. 2. The presence of carboxylic groups at opposite ends of the aliphatic carbon chain exert electron-withdrawing effects that reduce the p*K* (increase the acidity) for the first hydrogen ion dissociation. The subsequent free carboxylate has the opposite effect, which increases the p*K* for the second hydrogen ion dissociation. Although saturated aliphatic monocarboxylic acids are not electroactive, dicarboxylic acids such as malonic and succinic acids are reducible and electrochemically detectable ($E_{1/2} = -1.80$ to -1.69, SCE).

AOAA inhibits energy metabolism in mitochondria by blocking aspartic acid conversion to malic acid (the malate–aspartate shuttle) *(60)*. The decreased shunting of aspartate into the Krebs cycle reduces malate and therefore reduces energy metabolism by substrate limitation. The mechanism of AOAA substrate limitation to mitochondrial energy metabolism provides an alternative mechanism to oxidative stress. The kinetics of lactate washout in rat models by pMRS suggest that AOAA and malonic acid are equipotent inhibitors of mitochondrial metabolism but much less potent than 3-NPA and MPP^+ *(32)*.

Azide is another metabolic inhibitor that addresses mitochondrial insufficiency by blocking complex IV of the ETS by inhibiting cytochome oxidase. One advantage of azide is its selectivity for striatal lesions *(32),* like that of 3-NPA *(3)* after systemic administration with subsequent delayed dyskinesias followed by a hypokinetic state in rhesus monkeys *(61)*. Unfortunately malonic acid, AOAA, and MPP^+ do not cross the blood–brain barrier *(32)*; therefore their route of administration must be direct injection into cerebrospinal fluid, brain regions of interest, or they must be used in vitro. However, MPTP, the precursor to MPP^+, does cross the blood–brain barrier with selective vulnerability involving the substantia nigra *(62)*. 3-AP also crosses the blood–brain barrier after systemic administration, with selective vulnerability involving the inferior olives and other brainstem nuclei *(54)*.

Postmortem analysis of caudate enzyme activity associated with the ETS of deceased Huntington's disease patients reveals several defects of mitochondrial energy production. Although complex I appeared intact, mild deficiencies in complex IV activity were noted, and major deficiencies in complex II and III were noted (63). Although studies of platelet mitochondria are being pursued for reproducible abnormalities predictable of CNS mitochondrial abnormalities *(63)*, no patterns of abnormalities in the platelets of Huntington's disease patients have yet to be found to be suitable as blueprints in therapeutic interventions designed to improve mitochondrial energy metabolism. Nevertheless, if the reduction of occipital lactate levels can be used as a monitor of successful improvement in energy metabolism in Huntington's disease patients shown by pMRS *(64)*, the use of ubiquinone (coenzyme Q_{10}, an enzyme cofactor of both complex II and III) appears to be promising.

Other Specificity Issues

Successful neuroprotective treatment strategies for several neurodegenerative diseases will rely heavily on our ability to unravel intricate interrelationships between mitochondrial insufficiency and interneuronal excitotoxic mechanisms *(65)*. Specificity issues of pathophysiologic mechanisms must extend beyond specific mitochondrial enzyme inhibition *per se*. Selectivity of striatal responses to subcutaneous administration of 3-NPA can be exaggerated in the striatum and extended to other nuclear areas in the rat such as the thalamus and cerebellar nuclei by the simultaneous administration of amphetamine *(66)*. The combined effects could be due to increased mitochondrial energy depletion, activation of a cortical glutamate response (excitotoxicity), or both.

Specificity of mitochondrial inhibitors also extends to differential effects on neurotransmitter systems, specific ionic current responses, and free radical activity. Studies in rat mesencephalic tissue culture show that 3-NPA reduces both high-affinity dopamine and high-affinity γ-aminobutyric acid (GABA) uptake, whereas malonic acid only reduces high-affinity dopamine uptake and has no appreciable effect on GABA uptake. Blockade of the NMDA subset of glutamate receptors with MK-801 either attenuated or prevented changes in neurotransmitter uptake depending upon relative concentrations *(46)*. Separate or combined uses of metabolic enhancers, glutamate release inhibitors, and NMDA receptor antagonists protect against secondary excitotoxic lesions induced by metabolic toxins *(33,60)*. Impairment of mitochondrial energy metabolism increases potassium conductance and hyperpolarizes the membrane potential. The initial hyperpolarization due to the opening of calcium-activated and ATP-regulated potassium channels gives way to a late depolarization due to ion pump failure *(1,14,38)*. Increased intracellular calcium also results in increased free radicals in mitochondria *(50)*. 7-Nitroindazole, a neuronal nitric oxide synthase inhibitor, attenuates secondary striatal excitotoxic lesions from intrastriatal malonic acid injections and systemic 3-NPA administration. This extends the action of mitochondrial inhibitors to highly reactive NO· and HO· radicals and peroxynitrite in the generation of secondary excitotoxic lesions *(67)*.

REFERENCES

1. Riepe M, Horni N, Ludolph AC, et al. Inhibition of energy metabolism by 3-nitropropionic acid activates ATP-sensitive potassium channels. Brain Res 1992;586:61–66.
2. Beal MF, Brouillet E, Jenkins BG, et al. Neurochemical and histologic characterization of striatal excitotoxic lesions produced by the mitochondrial toxin 3-nitropropionic acid. J Neurosci 1993;13:4181–4192.
3. Bossi SR, Simpson JR, Isacson O. Age dependence of striatal neuronal death caused by mitochondrial dysfunction. NeuroReport 1993;4:73–76.
4. Gould DH, Gustine DL. Basal ganglia degeneration, myelin alterations, and enzyme inhibition in mice by the plant toxin 3-nitropropionic acid. Neuropathol Appl Neurobiol 1982;8:377–393.
5. Hamilton BF, Gould DH. Nature and distribution of brain lesions in rats intoxicated with 3-nitropropionic acid: a type of hypoxic (energy deficient) brain damage. Acta Neuropathol (Berl) 1987;72:286–297.
6. Ludolph AC, Seeling MO, Ludolph AG, et al. 3-Nitropropionic acid decreases cellular energy levels and causes neuronal degeneration in cortical explants. Neurodegeneration 1992;1:21–28.
7. Hong E, Castillo C, Rivero I, et al. Vasodilator and antihypertensive actions of 3-nitropropionic acid. Pro West Pharmacol Soc 1990;33:209–211.
8. Ludolph AC, He F, Spencer PS, et al. 3-Nitropropionic acid—exogenous animal neurotoxin and possible human striatal toxin. Can J Neurol Sci 1991; 18:492–498.
9. Ludolph AC, Seeling M, Ludolph AG, et al. ATP deficits and neuronal degeneration induced by 3-nitropropionic acid. Ann NY Acad Sci 1992;648: 300–302.
10. Brouillet E, Hantraye P, Ferrante RJ, et al. Chronic mitochondrial energy impairment produces selective striatal degeneration and abnormal choreiform movements in primates. Proc Natl Acad Sci USA 1995;92:7105–7109.
11. Beal MF. Neurochemistry and toxin models in Huntington's disease. Curr Opin Neurol 1994;7:542–547.
12. Simpson JR, Isacson O. Mitochondrial impairment reduces the threshold for in vivo NMDA-mediated neuronal death in the striatum. Exp Neurol 1993;121:57–64.
13. Hamilton BF, Gould DH. Nature and distribution of brain lesions in rats intoxicated with 3-nitropropionic acid: a type of hypoxic (energy deficient) brain damage. Acta Neuropathol (Berl) 1987;72:286–297.
14. Riepe MW, Niemi WN, Megow D, et al. Mitochondrial oxidation in rat hippocampus can be preconditioned by selective chemical inhibition of succinic dehydrogenase. Exp Neurol 1996;138:15–21.

15. Stermitz FR, Yost GS. Analysis and characterization of nitro compounds from Astragalus species. In: Keeler RF, VanKampen KR, James LF, eds. Effects of Poisonous Plants on Livestock. Academic Press, New York, 1978, pp. 371–378.
16. Pass MA, Carlisle CH, Reuhl KR. 3-nitropropionic acid toxicity in cultured murine embryonal carcinoma cells. Natural Toxins 1994;2:386–394.
17. Majak W, Cheng K-J, Hall JW. Enhanced degradation of 3-nitropropanol by ruminal micoorganisms. J Anim Sci 1986;62:1072–1080.
18. Pass MA, Muir AD, Majak W, et al. Effect of alcohol and aldehyde dehydrogenase inhibitors on the toxicity of 3-nitropropanol in rats. Toxicol Appl Pharmacol 1985;78:310–315.
19. He F, Zhang S, Qian F, et al. Delayed dystonia with striatal CT lucencies induced by a mycotoxin (3-nitropropionic acid). Neurology 1995;45:2178–2185.
20. Beal MF, Hyman BT, Koroshetz WJ. Do defects in mitochondrial energy metabolism underlie the pathology of neurodegenerative diseases? Trends Neurosci 1993;16:125–131.
21. Gresham TL, Jansen JE, Shaver FW, et al. Beta-propiolactone XIII. Reactions with sodium nitrite, sodium dithionite, sodium cyanide, sodium thiocyanate, sodium succinimide, and aryl sulfinic acids and their salts. J Am Chem Soc 1952;74:1323–1325.
22. Hansen TJ. Ames mutagenicity tests on purified 3-nitropropionic acid. Food Chem Toxicol 1984;22:399–401.
23. Benn MH, McDiarmid RE, Majak W. In-vitro biotransformation of 3-nitropropanol (miserotoxin aglycone) by horse liver alcohol dehydrogenase. Toxicol Lett 1989;47:165–172.
24. Yost GS. Toxic secondary metabolites of plants. Thesis, Colorado State University, Fort Collins, CO, 1977.
25. Muir AD, Majak W. Quantitative determination of 3-nitropropionic acid and 3-nitropropanol in plasma by HPLC. Toxicol Lett 1984;20:133–136.
26. Kolhoff IM, Lingane JJ. Nitrocompounds. Polarography. Interscience, New York, 1952, II:746–776.
27. Plambeck JA. Practice of classical voltammetry and polarography. Electroanalytical Chemistry. John Wiley Sons, New York, 1982, pp. 282–306.
28. Kolhoff IM, Lingane JJ. Carbonyl compounds. Ibid. 652–697.
29. Dean JA. Lange's Handbook of Chemistry, 14th edit. McGraw-Hill, New York, 1992.
30. James TL. Nuclear magnetic resonance in biochemistry. Academic Press, New York, 1975, pp. 65–94.
31. Brouillet E, Jenkins BG, Hyman BT, et al. Age dependent vulnerability of the striatum to the mitochondrial toxin 3-nitropropionic acid. J Neurochem 1993;60:356–359.
32. Jenkins BG, Brouillet E, Chen Y-C I, et al. Non-invasive neurochemical analysis of focal excitotoxic lesions in models of neurodegenerative illness using spectroscopic imaging. J Cereb Blood Flow Metab 1996;16:450–461.
33. Majak W, McDiarmid RE. Detection and quantitative determination of 3-nitropropionic acid in bovine urine. Toxicol Lett 1990;50:213–220.

34. Majak W, Pass MA, Madryga FJ. Toxicity of miserotoxin and its aglycone (3-nitropropanol) to rats. Toxicol Lett 1983;19:171–178.
35. Pass MA, Muir AD, Majak W, et al. Effect of alcohol and aldehyde dehydrogenase inhibitors on the toxicity of 3-nitropropanol in rats. Toxicol Appl Pharmacol 1985;78:310–315.
36. Pass Ma, Majak W, Muir AD, et al. Absorption of 3-nitropropanol and 3-nitropropionic acid from the digestive system of sheep. Toxicol Lett 1984;23:1–7.
37. Pass MA, Majak W, Yost GS. Lack of a protective effect of thiamine on the toxicity of 3-nitropropanol and 3-nitropropionic acid in rats. Can J Anim Sci 1988;68:315–320.
38. Nishino H, Shimano Y, Kumazaki M, et al. Hypothalamic neurons are resistant to the intoxication with 3-nitropropionic acid that induces lesions in the striatum and hippocampus via damage in the blood–brain barrier. Neurobiology 1995;3:257–267.
39. Hamilton BF, Gould DH. Correlation of morphologic brain lesions with physiologic alterations and blood-brain barrier impairment in 3-nitropropionic toxicity in rats. Acta neuropathol (Berl) 1987;74:67–74.
40. Gould DH, Wilson MP, Hamar DW. Brain enzyme and clinical alterations induced in rats and mice by nitroaliphatic toxicants. Toxicol Lett 1985;27: 83–89.
41. Reynolds NC, Lin W, Cameron CM, et al. Differential responses of extracellular GABA to intrastriatal perfusions of 3-nitropropionic acid and quinolinic acid in a freely moving laboratory rat. Brain Res 1997;778:140–149.
42. Palfi S, Ferrante RJ, Brouillet E, et al. Chronic 3-nitropropionic acid treatment in baboons replicates the cognitive and motor deficits of Huntington's disease. J Neurosci 1996;16(90):3019–3025.
43. Roberts RC, Ahn A, Swartz KJ, et al. Intrastriatal injections of quinolinic acid or kainic acid: differential patterns of cell survival and the effects of data analysis on outcome. 1993;124:274–282.
44. Bazzett TJ, Becker JB, Kaatz KW, et al. Chronic intrastriatal dialytic administration of quinolinic acid produces selective neural degeneration. Exp Neurol 1993;120:177–185.
45. Fink SI, Ho DY, Sapolsky RM. Energy and glutamate dependency of 3-nitropropionic acid neurotoxicity in culture. Exp Neurol 1996;138:298–304.
46. Zeevalk GD, Derr-Yellin E, Nicklas WJ. Relative vulnerability of dopamine and GABA neurons in mesencephalic culture to inhibition of succinate dehydrogenase by malonate and 3-nitropropionic acid and protection by NMDA receptor blockade. J Pharmacol Exp Ther 1995;275:1124–1130.
47. Majak W, Cheng K-J, Hall JW. The effect of cattle diet on the metabolism of 3-nitropropanol by ruminal microorganisms. Can J Anim Sci 1982;62:855–860.
48. Muir AD, Majak W, Pass MA, et al. Conversion of 3-nitropropanol (miserotoxin aglycone) to 3-nitropropionic acid in cattle and sheep. Toxicol Lett 1984;20:137–141.
49. Coles CJ, Edmondson DE, Singer TP. Inactivation of succinate dehydrogenase by 3-nitropropionate. J Biol Chem 1979;254:5161–5167.
50. Alston TA, Mela L, Bright HJ. 3-Nitropropionate, the toxic substance of *Indigofera*, is a suicide inactivator of succinate dehydrogenase. Proc Natl Acad Sci USA 1977;74:3767–3771.

51. Porter DJT, Bright HJ. 3-Carbanionic substrate analogues bind very tightly to fumarase and aspartase. J Biol Chem 1980;255:4772–4780.
52. Beal MF. Aging, energy, and oxidative stress in neurodegenerative diseases. Ann Neurol 1995;38:357–366.
53. Deutsch AY, Rosin DL, Goldstein M, et al. 3-Acetyl pyridine-induced degeneration of the nigrostriatal dopamine system: an animal model of olivo- ponto-cerebellar atrophy-associated parkinsonism. Exp Neurol 1989;105:1–9.
54. Schulz JB, Henshaw DR, Jenkins BG, et al. 3-Acetyl pyridine produces age dependent excitotoxic lesions in rat striatum. J Cereb Blood Flow Metab 1994;14:1024–1029.
55. Zuddas A, Oberto G, Vaglini F, et al. MK-801 prevents 1-methyl-4-phenyl-1,2,3,6-tetrahydropyridine-induced parkinsonism in primates. J Neurochem 1992;59:733–739.
56. Lange KW, Loschmann P-A, Sofic E, et al. The competitive NMDA antagonist CPP protects substantia nigra neurons from MPTP-induced degeneration in primates. Nauyn Schmiedebergs Arch Pharmacol 1993;348:586–592.
57. Beal MF, Brouillet E, Jenkins B, et al. Age dependent striatal excitotoxic lesions produced by the endogenous mitochondrial inhibitor malonate. J Neurochem 1993;61:1147–1150.
58. Langston JW, Ballard P, Tetrud JW, et al. Chronic Parkinsonism in humans due to a product of meperidine-analog synthesis. Science 1983;219:979–980.
59. Skulachev VP. Energy transformations in the respiratory chain. Curr Top Bioenerg 1971;4:127–190.
60. Beal MF, Swartz KJ, Hyman BT, et al. Amino oxyacetic acid results in excitotoxic lesions by a novel indirect mechanism. J Neurochem 1991; 57:1068–1073.
61. Mettler FA. Choreoathetosis and striopallidonigral necrosis due to sodium azide. Exp Neurol 1972;32:291–308.
62. Burns RS, Chiueh CC, Markey SP, et al. A primate model of Parkinsonism: selective destruction of dopaminergic neurons in the pars compacta of the substantia nigra by *N*-methyl-4-phenyl-1,2,3,6-tetrahydropyridine. Proc Natl Acad Sci USA 1983;80:4546–4550.
63. Gu M, Gash MT, Mann VM, et al. Mitochondrial defect in Huntington's disease caudate nucleus. Ann Neurol 1996;39:385–389.
64. Koroshetz WJ, Jenkins B, Rosen B, et al. Evidence for a metabolic disorder in Huntington's disease. Neurology 1994;44:A338.
65. Schulz JB, Matthews RT, Henshaw DR, et al. Neuroprotective strategies for treatment of lesions produced by mitochondrial toxins: implications for neurodegenerative diseases. J Neurosci 1996;71:1043–1048.
66. Bowyer JF, Clausing P, Schmned L, et al. Parenterally administered 3-nitropropionic acid and amphetamine can combine to produce damage to terminals and cell bodies in the striatum. Brain Res 1996;712:221–229.
67. Schulz JB, Matthews RT, Jenkins BG, et al. Blockade of neuronal nitric oxide synthase protects against excitotoxicity in vivo. J Neurosci 1995;15: 8419–8429.

II

Mitochondrial Dysfunctions

Models of Neurodegeneration and Mechanisms of Action

4
In Vitro Studies of 3-Nitropropionic Acid

Gail D. Zeevalk

CELLULAR SUBSTRATES FOR 3-NITROPRIOPIONIC ACID

3-Nitropropionic Acid as a Suicide Inhibitor of Succinate Dehydrogenase

In the 1950s a trailing species of indigo, *Indigofera endecaphylla*, introduced into Hawaii as a forage or cover crop, was found to produce sickness in dairy cattle. Chemical analysis of the plant revealed a simple three-carbon, nitrogen-containing acid, 3-nitropropionic acid (3-NPA, Fig. 1) *(1)*, identical to hiptagenic acid, as the responsible agent. 3-NPA is widely distributed in nature. It has been isolated from plant species of *Indigofera*, *Hiptage*, *Viola*, *Corynecarpus*, and *Astragalus*. The fungi *Aspergillus flavus*, *A. orysae*, *Penicillium astrovenetum*, and *Arthrinium* synthesize the nitroalkane. Many examples in the literature and reviewed in other chapters of this book demonstrate the neurotoxic consequences of consumption of legumes containing 3-NPA or plants such as sugarcane mildewed by contaminating *Arthrinium*. Biochemical studies of the cellular substrates for 3-NPA reveal multiple cellular targets, but its action as an inhibitor of succinate dehydrogenase (SDH) appears to clearly be its most deleterious attribute. In a brief abstract by Hollocher (1973) *(2)*, 3-NPA was first put forth as an irreversible inhibitor of SDH. Detailed studies by Alston et al. *(3)* expanded on this proposal to demonstrate the irreversible nature of the inhibition of SDH by 3-NPA. When 3-NPA was added to respiring rat liver mitochondria, the rate of O_2 consumption decreased exponentially to zero. Succinate addition did not restore respiratory activity. Oxidation of NAD-linked substrates was not affected by 3-NPA, demonstrating a relatively selective action on succinate oxidation.

Inactivation of SDH by 3-NPA requires that the nitroalkane be in the carbanion form. The pK_a for nitropropionate carbanion is 9.1. The K_i for inactivation by the dianion is approx 10 μM as compared with a K_i of 200 μM

From: *Mitochondrial Inhibitors and Neurodegenerative Disorders*
Edited by: P. R. Sanberg, H. Nishino, and C. V. Borlongan © Humana Press Inc., Totowa, NJ

$O_2N\text{-}CH_2\text{-}CH_2\text{-}COO^-$	$^-OOC\text{-}CH_2\text{-}CH_2\text{-}COO^-$	$^-OOC\text{-}CH_2\text{-}COO^-$
3-nitropropionic acid	succinate	malonate

Fig. 1. Structures of the endogenous substrate for SDH, succinate, and the irreversible and reversible SDH inhibitors, 3-nitropropionic acid and malonate, respectively.

for the monoanion form *(4)*. Substantial reprotonation occurs at physiological pH. The mitochondrial milieu, which is the site of interaction for 3-NPA and SDH, is more alkaline than the cytosol and would likely promote carbanion formation. Alston and colleagues *(3)* proposed that 3-NPA was first oxidized by SDH. This would form 3-nitroacrylate. The carbanion of 3-NPA would then form an N-5 adduct with the flavin of SDH. The studies of Coles et al. *(4)* supported the dehydration of 3-NPA by SDH to 3-nitroacrylate to form the true inhibitor species. Beef heart mitochondrial electron transport particles (ETPs) incubated with 3-NPA dianion developed a slow irreversible inactivation. Rate of oxidation of 3-NPA by SDH was 0.1% of the rate of succinate oxidation. Direct addition of 3-nitroacrylate to the mitochondrial ETPs produced a very rapid and irreversible inhibition, as would be predicted if this were the inhibiting species. Coles *(4)* studies argued, however, against a nucleophilic addition to N-5 of the covalently bound flavin component of the enzyme. Absorption and fluorometric changes produced by the interaction of 3-NPA with SDH more closely resembled changes occurring at the active substrate site rather than alkylation of N-5 of flavin. As shown in Fig. 2, Coles *(4)* proposed a two-step inactivation of SDH by 3-NPA. In step 1, the dianion of 3-NPA is oxidized to 3-nitoacrylate by a two-electron transfer to the flavin component. In step 2, the thiol group of SDH interacts with 3-nitroacrylate to form a thioether and the flavin group is reoxidized by the respiratory chain. Such a mechanism for inhibition would classify 3-NPA as a true suicide inhibitor, i.e., a compound that is relatively inactive *per se*, but reacts with the enzyme to form a product that in turn irreversibly inactivates it.

Other Cellular Targets of 3-NPA

3-NPA has been reported to inhibit a number of different cellular enzymes. When added to partially purified acetylcholinesterase (AChE) prepared from rat brain, 3-NPA was a fairly potent inhibitor *(5)*. Kinetic analysis of the ACHE inhibition by 3-NPA revealed that inhibition was reversible and competitive. The enzyme-inhibitor dissociation constant (K_i) for brain ACHE in the presence of 3-NPA was 18 μM. Mohammed et al. *(6)* reported

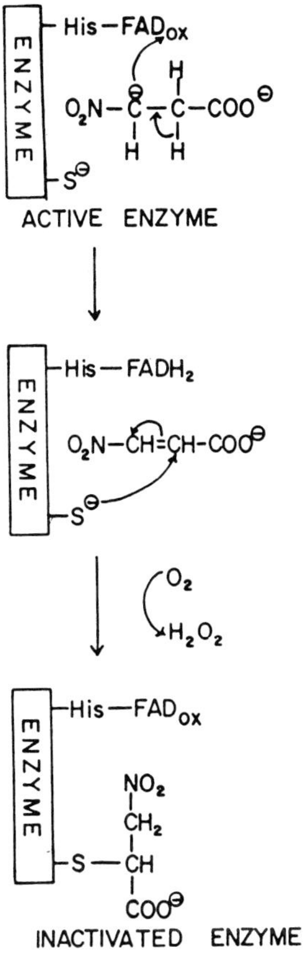

Fig. 2. Postulated mechanism for the inactivation of succinate dehydrogenase by 3-NPA. (Reproduced with permission of the American Society for Biochemical and Molecular Biology from Coles et al., J Biol Chem 1979;254:5166.)

the inhibition of rat brain monoamine oxidase by 3-NPA. Kinetic analysis suggested a noncompetitive type of inhibition with a K_i of approx 8 μ*M*. Recovery of enzyme activity upon dialysis indicated that the inhibition was reversible. Similar to the inhibition of SDH, the carbanion form of 3-NPA was found to be a potent competitive inhibitor of both fumerase and aspartase *(7)*.

The presence of 3-NPA in neural tissue is thus likely to exert a number of differing biochemical effects in addition to the effects of 3-NPA on energy metabolism. For example, inhibition of AChE or monoamine oxidase may result in elevated levels of the neurotransmitters acetylcholine, dopamine, serotonin, and norepinephrine during the initial stages of 3-NPA intoxication. The degree of the rise in neurotransmitter levels would ultimately be determined by the extent of inhibition of the enzyme as well as the effects on energy metabolism (ATP/ADP ratios and secondary effects on neurotransmitter synthesis), and the temporal profile of toxicity. Although 3-NPA may serve to dampen the activity of several cellular enzymes, the loss of neurons due to exposure most likely rests with its irreversible inhibition of SDH. Inhibition of SDH would have the expected result of lowering high-energy phosphate levels. Erecińska and Nelson *(8)* reported a rapid decrease in creatine phosphate/creatine ratios and a less pronounced decrease in ATP/ADP ratios. Lactate/pyruvate ratios were elevated, indicating that oxidation of NADH produced by glycolysis was impaired. Perturbation of amino acid metabolism by 3-NPA was also observed, most notably a decrease in tissue levels of aspartate. Reducing equivalents from NADH produced during glycolysis need to enter the mitochondria via the malate/aspartate shuttle. The decrease in tissue aspartate by 3-NPA could impede the reoxidation of NADH and further compromise metabolism and ATP production.

OTHER INHIBITORS OF SDH

Malonate

Malonate has been recognized as an inhibitor of respiration since the early 1900s, when Lund first observed the inhibition of frog muscle respiration by malonate (*see* ref. *[9]* for review). Its action as a competitor with succinate for succinate oxidation was reported by Quastel and Whethan in 1928. The use of this compound to inhibit SDH was instrumental in unraveling the sequence of the tricarboxylic acid cycle. As with 3-NPA, the active form for inhibition of SDH is the dianion. The pK_{a2} for malonate is 5.17 and, therefore, the completely ionized species exists at physiological pH. However, below pH 7.4, the amount of the monoanion or carbonic acid form can increase appreciably and this can impact on the rate and degree of penetration into the cell. The K_is for inhibition of SDH by malonate in various homogenates or mitochondrial preparations vary between 5 and 50 μ*M* (*see* ref. *[9]* for details). In contrast, inhibition of succinate oxidation by malonate in whole cells or tissue slices is very weak. Webb *(9)* attributes this discrepancy to one or more of several possible factors including permeability, the enzyme environment in the cell vs artificial media, the concentration of succinate in the cell, or the rate of succinate oxidation. Although often viewed as a selective inhibitor of SDH, malonate can have inhibitory effects on other metabolizing enzymes, i.e., fumerase, malate dehydrogenase, and oxaloacetate decarboxylase. The potency for inhibition of these enzymes is not as great as with SDH. Malonate may form stable complexes with metal cations, most notably Mg^{2+} and Ca^{2+}, and deplete cellular concentrations of free ions. Some reports of inhibition of enzymes, other than SDH, by malonate attribute inhibition to depletion of metal ions. Malonate has also been shown to be a competitive inhibitor of malate transport *(10)*, which could disturb oxidation of glycolytically derived NADH as described previously for 3-NPA. In neuronal cells, the toxic consequences of malonate appear related to competition with succinate for SDH, as studies have demonstrated that toxicity due to malonate can be overcome by addition of excess succinate *(11)*.

Rat brain contains a substantial concentration of free malonate, 192 nmol/g wet wt *(12)*. Malonate is thought to be derived from fatty acid oxidation, which provides the precursor acetyl-CoA *(13)*. Acetyl-CoA is converted to malonyl-CoA, which in turn can form malonate. A rare condition of malonyl-CoA deficiency *(14)* has been described. One severely affected child had high urinary levels of malonate and succinate *(15)*. Central nervous system (CNS) manifestations were mental retardation and seizures, although it is not clear whether CNS disturbances were due to metabolic acidosis, seizure activity, or metabolic impairment.

Methylmalonate

Methylmalonate exists in brain and is formed from methylmalonyl-CoA mutase *(16)*. SDH from rat brain mitochondria was inhibited by methylmalonate with a K_i value of 4.5 m*M* *(16,17)*. Inhibition was competitive and reversible. Cerebral brain slices incubated with methylmalonate showed increased lactate formation and glucose utilization consistent with an increase in anaerobic metabolism due to inhibition of aerobic respiration *(16)*. Methylmalonate was toxic to striatal and cortical neurons in vitro *(19)* and when injected into the striatum in vivo *(16)*. It is unclear at present whether the toxic effects of methylmalonate are directly due to methylmalonate or to secondary formation of malonate from hydrolysis. Deficiency in methylmalonyl-CoA mutase, an inherited metabolic disorder, results in methylmalonic acidemia and hypoglycemia. The outcome may be fatal if not treated promptly. Mental retardation presents in survivors *(20)*, but the underlying cause of CNS involvement is unclear.

Neuronal Vulnerability to SDH Inhibitors In Vitro

Prior to consideration of the neuronal consequences of in vitro administration of inhibitors of SDH, some discussion of the in vivo vulnerability of neurons to SDH inhibition is warranted. In vivo studies address two issues with regard to neuronal vulnerability: that of interregional susceptibility (rank order of sensitivity among different brain regions, i.e., striatum vs cortex, vs cerebellum, etc.) and that of subregional or cellular susceptibility (cell populations within the same region). Whole animal studies that examine the neurotoxic consequences of SDH inhibition fall into two categories: those that administer the inhibitory agent systemically, i.e., intraperitoneally, subcutaneously, intramuscularly, or orally, and those that administer it centrally. Systemic application of 3-NPA produces neuropathology that is manifested in a heirarchy of regional vulnerability, with the striatum showing the greatest vulnerability as discussed in detail in other chapters in this book. The question of interregional vulnerability appears not to be related to differential inhibition of SDH. Histochemical staining indicates that SDH activity is uniformly depressed throughout the brain following systemic 3-NPA administration *(21)*. Systemic administration of 3-NPA, however, can produce a number of secondary effects, such as decreased arterial pH and bicarbonate and loss of blood–brain barrier integrity *(22)*. The loss of integrity of the blood–brain barrier may be important to the issue of selective regional vulnerability in vivo. In a study by Hamilton and Gould *(22)*, albumin extravasation was noted in striatum following subcutaneous administration of 3-NPA and the amount of extravasation correlated with the extent of striatal damage. No leakage of albumin or cell damage was found in cortex. Nishino et al. *(23)* also provide data to suggest that breakdown of the blood–brain barrier may be a contributing factor for the specific vulnerability of the striatum. A separate but related issue regards the differential susceptibility on a subregional or cellular level. This has been demonstrated in numerous studies using 3-NPA or malonate exposure where there is selective loss of subpopulations of neuropeptide containing γ-aminobutyric acid-ergic (GABAergic) projection neurons in the striatum that resembles the loss of striatal neurons seen in Huntington's disease *(24,25)*. It is also evident when 3-NPA is injected into other brain regions *(26)*. Intrahippocampal injection of 3-NPA between the CA1 and CA3 regions produces loss of neurons with a selective vulnerability similar to what is found in ischemia: CA1> CA3 > dentate gyrus. Thus, in vivo, susceptibility to 3-NPA occurs on both an interregional and a subregional level. Similar issues may be addressed in vitro. Scrutiny of in vivo and in vitro findings may shed light on the reasons for the regional and cellular susceptibilities of neurons to SDH inhibition.

Comparisons of the findings from in vitro studies of neuronal vulnerability to SDH inhibition by 3-NPA or malonate are complicated by the different types of cultures employed (explants, mixed neuronal/glial cultures, neuronal enriched cultures), culture conditions such as media supplementation, age of cultures at the time of treatment, and concentration and exposure times used for the toxins. Despite these differences, it is clear that the majority of neuronal populations studied thus far succumb to in vitro treatment by 3-NPA. *See* Table 1 for a summary of the different culture types tested in vitro. The first report of toxicity to neurons in vitro was by Ludolph et al. (1992) *(27)*. Cortical explants were exposed to 3-NPA for periods of time up to 4 h. ATP, adenylate energy charge (AEC), SDH inhibition, and histological damage were monitored during the time of exposure. Histological damage first appeared by 180 min of treatment. Partial inhibition of SDH was found by 15 min (approx 35% of control), but full inhibition required 2–4 h of exposure. This slow evolution of inhibition would be consistent for a suicide inhibitor. As mentioned previously, the rate of oxidation of 3-NPA to 3-nitroacrylate by SDH, the true inhibitor species, was only 0.1% of the rate of succinate oxidation. ATP and AEC levels were down by 120 min (although not statistically different), but were clearly depressed by 240 min. These findings support the concept that SDH inhibition and high-energy phosphate levels are important in the evolution of 3-NPA-induced histological damage. However, acute morphological changes were used as the end point and it is unclear how this relates to irreversible damage.

When cultures of cortical or striatal neurons were exposed to 1–2 m*M* 3-NPA for 48 h, irreversible damage as determined by counts of trypan blue labeled cells was observed *(28)*. Because both cultures were treated in a similar fashion with regard to culture conditions, time of exposure, and concentration of toxin, it is possible to make some statements regarding relative vulnerability. One caveat to this is that a uniform set of culture conditions may not be optimal for all neuronal types and may influence results. A concentration of 1–2 m*M* 3-NPA for 48 h produced maximal cell loss in both striatal and cortical cultures (approx 55–60%). Extrapolating from the graphs, the EC_{50}s for cell loss by 3-NPA were approx 0.4 and 0.9 m*M* in striatal and cortical cultures, respectively. If such comparisons are valid, this would suggest that striatal neurons are relatively more vulnerable than their cortical counterparts. An interesting observation from the dose–response study by Behrens et al. *(28)* was that the percentage of cell death plateaued above 1 m*M* and represented approx 60% of the cultured neurons in both systems. Conversely, approx 40% of the cultured striatal and cortical neurons were refractile to 3-NPA. This would argue that there are both inter- and subregional differences in response to 3-NPA in vitro.

The order of vulnerability of different interregional neuronal populations in vitro to 3-NPA treatment may be viewed in a study by Fink et al. *(29)*. Cultures from five different brain regions–striatum, septum, hippocampus,

Table 1
Summary of In Vitro Studies and Exposure Conditions for 3-NPA Toxicity in Neurons

Culture	Treatment age in vitro (div)	Exposure time (h)	3-NPA concentration (m*M*)	End point measurement	Sensitivity to 3-NPA	Reference
Cortical explant	15	0–4	0.25–1	Acute swelling	s	*(27)*
Cortical	10–12	14–48	1–2	Cell counts	s	*(28)*
Striatal	10–12	14–48	1–2	Cell counts	s	*(28)*
Hippocampal	19–21	18–22	0.1–10	Cell counts	s	*(29)*
Hypothalamic	19–21	18–22	0.1–10	Cell counts	s	*(29)*
Cerebellar	19–21	18–22	0.1–10	Cell counts	s	*(29)*
Cerebellar Granule cells	5–21	24–216	0.01–1	Cell counts	s	*(32)*
Mesencephalic	12	24	0.1–0.5	High-affinity uptake	s	*(33,34)*
Hippocampal	7	48	0–15	Cell counts	s	*(46)*

div, days in vitro; s, sensitive; ns, not sensitive, approx $EC_{50} > 5$ m*M*.

hypothalmus, and cerebellum–were exposed to various concentrations of 3-NPA for 18–22 h and toxicity determined by cell counts. As shown in Fig. 3 and Table 2, striatal neurons, found to be particularly vulnerable in vivo following systemic administration of 3-NPA, fell somewhere in between hippocampal and hypothalamic neurons (most vulnerable) and septal or cerebellar neurons (least vulnerable) with regard to sensitivity to 3-NPA toxicity. Similar to what was reported by Behrens et al. *(28)*, a substantial portion of striatal neurons were relatively resistant to 3-NPA and this varied among the other cultures tested. Again, differences in inter-regional and subregional vulnerability to 3-NPA appear to exist in vitro, as well as in vivo: however, this study would suggest that in vitro, striatal neurons are not uniquely sensitive to direct challenge with 3-NPA. An interesting component of this study was the finding that 3-NPA toxicity was substrate dependent. Greater toxicity to 3-NPA was observed at 3.5 m*M* glucose as compared with 20 m*M* glucose containing medium. The authors interpret this as the need for glucose to maintain glycolytic ATP production during the inhibition of aerobic metabolism by 3-NPA. Aside from the general necessity for maintaining ATP processes, high glucose levels may alter the electrophysiological response of neurons to 3-NPA. When cultured hippocampal neurons were exposed to 3-NPA in high-glucose medium (10 m*M*), they underwent an initial prolonged hyperpolarization mediated by activity of ATP-sensitive K^+ channels. This was subsequently followed by depolarization *(30)*. At 4 m*M* glucose, no hyperpolarizations were observed and the onset to depolarization in the presence of 3-NPA was much more rapid. Activation of ATP-sensitive K^+ channels and hyperpolarization may be a protective mechanism in ischemia *(31)*. During 3-NPA exposure, activation of ATP-sensitive K^+ channels and hyperpolarization may tip the balance between a neuron's resisting or succumbing to a metabolic stress.

In contrast to the relative resistance of cerebellar cultures to 3-NPA exposure reported by Fink et al. *(29)* (EC_{50} >10 m*M*), Weller and Paul *(32)* found that cerebellar granule cell cultures were sensitive to 3-NPA. Exposure of cultures to 3-NPA for 24 h on d 8 in vitro resulted in cell loss with an EC_{50} for 3-NPA of 250 μ*M*. Even greater sensitivity was seen in 21-d-old cultures (EC_{50} = 50 μ*M*). One notable difference between culture conditions in the study by Fink et al. *(29)* as compared with that by Weller and Paul *(32)* was the glucose concentration in the medium: 20 m*M* vs 5 m*M*, respectively. Given the glucose dependency of 3-NPA toxicity observed by Fink and colleagues, this may explain the difference in vulnerability of cerebellar neurons in the two systems.

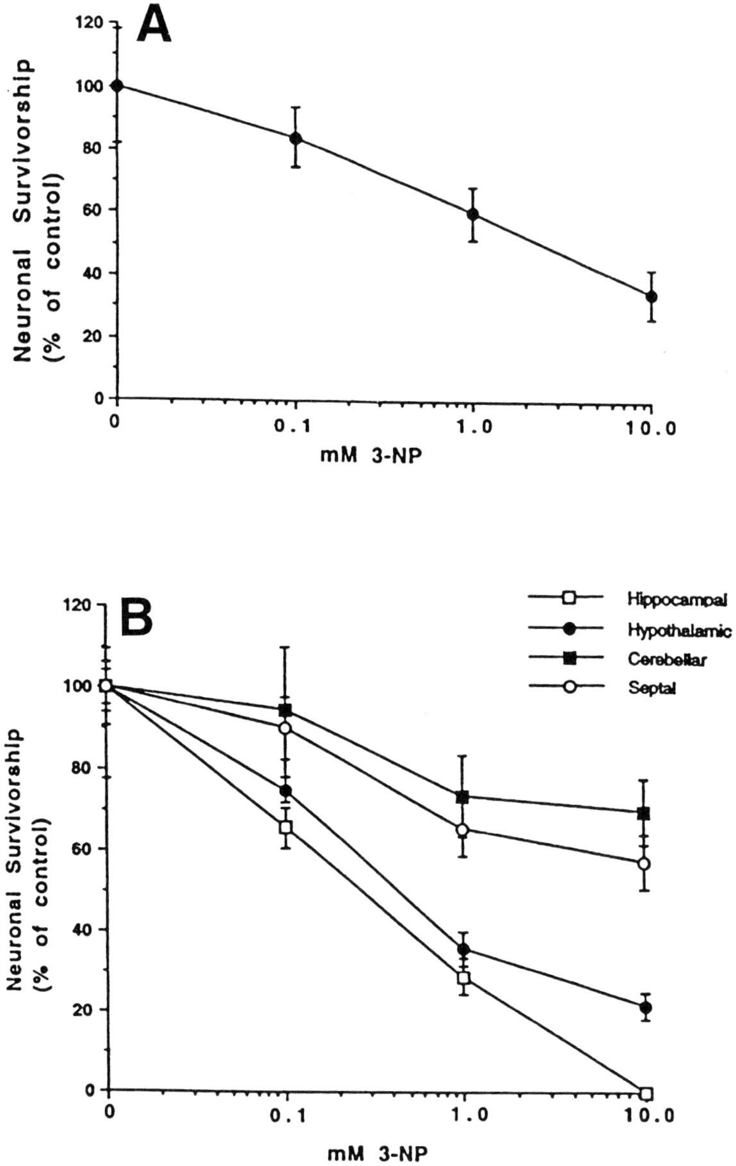

Fig. 3. Effects of 3-NPA on cultured neurons. (Data represented as percentage of control; control, 0.0 m*M* 3-NPA/20.0 glucose.) **(A)** Neuronal survival in mixed striatal cultures exposed to 3-NPA in 20.0 m*M* glucose medium; n = 5–6/point. **(B)** Survival in mixed cultures from various brain regions exposed to 3-NPA in 20.0 m*M* glucose medium; n = 6–14/point except septum n = 23–38/point. (Reproduced with permission of Academic Press from Fink et al., Exp Neurol 1996;138:300.)

Table 2
Median Lethal Dose of 3-NPA after Neuronal Incubation at Various Glucose Concentrations

Brain region	Glucose concentration 20.0 m*M*	3.0–4.0 m*M*	0.2 m*M*
Striatal	2.5[a]	0.11	<0.1
Septal	>10.0	0.1	<0.1
Hippocampal	0.3		<0.1
Hypothalamic	0.6		<0.1
Cerebellar	>10.0		<0.1

Note: Mixed cultures from various brain regions were exposed to 3-NPA and indicated concentrations of glucose. Median lethal doses are calculated by logarithmic regression analysis of data presented in Figs. 1a and 1b (20 m*M* glucose) or data from experiments performed simultaneously on sister cultures (3.0 and 0.2 m*M* glucose).

[a]Numbers indicate median lethal dose of 3-NPA (m*M*).

Published by permission of Academic Press from Fink et al., Exp Neurol 1996;138:300.

Mesencephalic neurons representing presumptive midbrain structures such as the substantia nigra were also sensitive to 3-NPA toxicity *(33)*. Mesencephalic cultures are an important source of dopamine neurons, the major catecholaminergic population lost in Parkinson's disease. Cultures were exposed to 3-NPA for 24 h, allowed to recover for 2 d, and toxicity to the dopamine population was assessed by measurement of the high affinity uptake for [^{3}H]dopamine. 3-NPA produced a dose-dependent loss of uptake with an EC_{50} of 210 µ*M*. ATP levels and SDH activity were down approx 60% following 3 h of exposure to 0.5 m*M* 3-NPA, demonstrating the effects of the nitroalkane on energy metabolism. Selective vulnerability of dopamine and GABA neurons within the mesencephalic culture system was studied by exposing cultures to either 3-NPA or malonate for 24 h followed by simultaneous measurement of the high-affinity uptake for [^{3}H]dopamine and [^{14}C]GABA after 2 d of recovery *(34)*. As shown in Fig. 4, both the mesencephalic dopamine and GABA populations were equally sensitive to 3-NPA intoxication. In contrast, use of the competitive SDH inhibitor malonate (Fig. 5) revealed a greater vulnerability within the dopamine population; dopamine neurons showed a dose-dependent loss in response to malonate, whereas GABAergic neurons were relatively resistant. Anaerobic metabolism of glucose to lactate should increase when oxidative metabolism is inhibited. After 3 h of treatment, 3-NPA increased lactate levels by 78%, whereas the increase in lactate by malonate was only 19%. This suggests that malonate

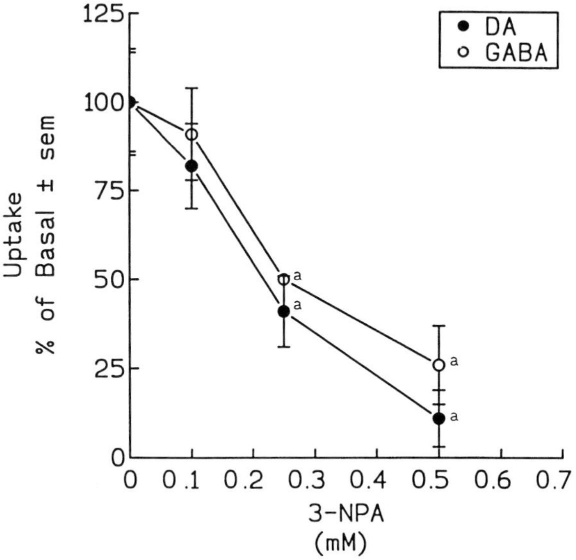

Fig. 4. Mesencephalic cultures were treated with various concentrations of 3-NPA on d 12 in vitro. After 24 h, the medium was removed, and the cultures washed and fed with conditioned medium and allowed to recover for 48 h before simultaneous measurement of [^{3}H]dopamine and [^{14}C]GABA uptake. Data are the means ± SEM (bars) from three to four separate experiments done in duplicate. $^a p$, 0.05 or better, different from control. (Reproduced with permission of Williams & Wilkins from Zeevalk et al., J Pharmacol Exp Ther 1995;275:1126.)

produced less severe effects on oxidative metabolism. Thus, dopamine neurons as compared with GABA neurons in vitro were more susceptible to toxicity when metabolic impairment was modest. The high degree of sensitivity of dopamine neurons to malonate toxicity also held true in vivo. Intrastriatal infusion of malonate in rats produced greater loss of striatal dopamine as compared with GABA when examined 1 wk following infusion *(35)*. Loss of dopamine from terminals in the striatum was permanent and translated into retrograde loss of dopamine neurons in the ipsilateral substantia nigra after 1 mo *(36)*. The resistance of a subpopulation of striatal *(28,29)* or cortical *(28)* neurons in vitro to SDH inhibition alluded to a subregional vulnerability, but these studies did not investigate a relationship between classes of neurons and sensitivity. Studies in mesencephalic cultures show a difference in sensitivity to metabolic stress in two different neurotransmitter populations. In vivo studies refine this further by demonstrating the selective loss of certain neuropeptide containing GABAergic neurons in striatum *(24)*.

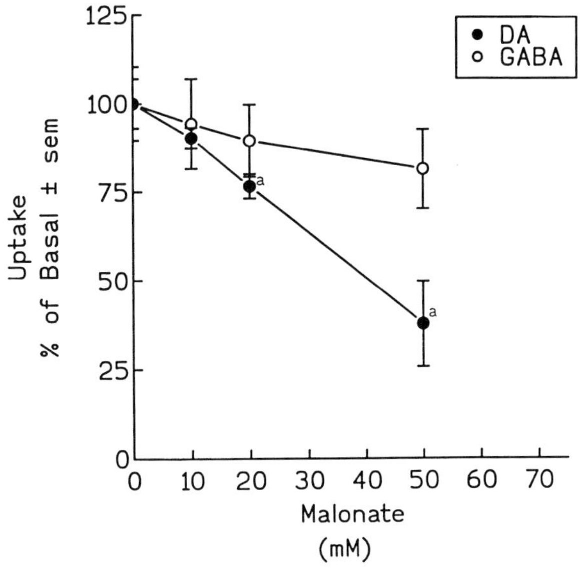

Fig. 5. Twelve-day in vitro mesencephalic cultures were treated with various concentrations of malonate as described in the legend to Fig. 4. Data are from four to five separate experiments done in duplicate. [a]*p*, 0.05 or better, different from control. (Reproduced with permission of Williams & Wilkins from Zeevalk et al., J Pharmacol Exp Ther 1995;275:1126.)

In Vitro Studies of the Mechanisms of 3-NPA or Malonate Toxicity

Our present understanding of the mechanisms involved in neuronal loss due to 3-NPA or malonate is limited by the scarcity of available data and the current focus on a few specific, albeit potentially important, mechanisms that have emerged from the ischemia literature. These include the involvement of excitotoxic mechanisms, oxidative stress, and programmed cell death. Numerous studies point to the involvement of glutamate receptors, particularly the *N*-methyl-D-aspartate (NMDA) subtype, in ischemic damage (*see* ref. *[37]* for review). Because 3-NPA or malonate exposure imposes a metabolic stress in neurons, examination of the involvement of excitotoxicity in these insults is reasonable. Although some discrepancies exist among and within studies, a general consensus emerges. Glutamate receptors and particularly the NMDA subtype contribute to 3-NPA or malonate toxicity in vitro with the qualifier that the contribution of excitotoxic mechanisms appears inversely related to the severity of the metabolic stress.

Table 3
Mechanisms Associated with 3-NPA Toxicity from In Vitro Studies

Reference	Excitotoxic	Oxidative stress	Apoptotic
(27)	Yes	ND	ND
(28)	No	ND	Yes
(29)	Yes[a]	ND	ND
(32)	Yes	ND	ND
(33,34)	Yes	Yes	ND
(46)	Yes for neurons undergoing necrosis	ND	Yes for some neurons

ND, not determined.
[a]MK-801 protected; however, competitive antagonists did not.

In other words, taking into account dose and exposure, when lower concentrations and/or shorter exposure times are used, glutamate antagonists protect, whereas with higher concentrations and/or longer exposure times, glutamate antagonists become less or ineffective (*see* Table 3). Acute tissue damage in cortical explants caused by 1 m*M* 3-NPA for 4 h was modestly attenuated by kynurenate, a nonselective glutamate antagonist, or MK-801, a noncompetitive NMDA antagonist and more effectively attenuated by the combination of MK-801 plus 6-cyano-7-nitroquinoxaline-2,3-dione (CNQX), a non-NMDA antagonist *(27)*. Fink et al. *(29)* reported protection of striatal cultures with MK-801 vs 1 m*M* 3-NPA for 18–22 h. In contrast, Behrens et al. *(28)* did not observe protection with MK-801 plus CNQX in either striatal or cortical cultures treated with 1–2 m*M* 3-NPA for 48 h. The different exposure times for 3-NPA employed in these studies may have dictated the outcome. The influence of concentration and exposure for 3-NPA or malonate (and presumably the severity of metabolic impairment) on revealing an excitotoxic component to toxicity is best exemplified in the studies by Weller and Paul *(32)* and Zeevalk et al. *(34,35)*. Cerebellar granule cultures pretreated with MK-801 or 2-amino-5-phosphonopentanoic acid (APV) (competitive NMDA antagonist) prior to 3-NPA exposure were completely protected when exposure times were limited to 24 h and concentrations were ≤300 μ*M*. Glutamate receptor antagonists were less effective vs higher concentrations of 3-NPA and ineffective if exposure times were increased to 72 h *(32)*. These findings are consistent with those reported for mesencephalic cultures *(33,34)*. MK-801 protected against a 24 h exposure to low concentrations of 3-NPA, but were less effective as the concentration increased *(33)*. Blocking NMDA receptors during a 24-h treatment with malonate, which affects metabolism less severely, was fully protective *(34)*. Thus, in vitro evidence suggests that glutamate receptors participate in the early events following metabolic impairment and can substantially contribute to damage if energy impairment is modest. On the other hand, more profound or prolonged inhibition of energy metabolism likely involves additional deleterious and parallel occurring mechanisms that would render NMDA receptor blockade an ineffective protective strategy.

There are numerous reports from the ischemia literature to support a role for free radicals or oxidative stress in damage to neurons due to metabolic inhibition. Because of the energy stress imposed by 3-NPA or malonate, it is likely that free radicals play a role in this toxicity as well. In vivo studies, addressed in more detail in other chapters of this book, advance a role for oxidative stress in 3-NPA or malonate toxicity. Copper/zinc superoxide dismutase overexpressing transgenic mice were less susceptible to 3-NPA-induced toxicity *(38)*, as were human glutathione peroxidase overexpressing transgenics *(39)* and nitric oxide synthase knockout mice *(40)* in response to malonate. In vitro studies in mesencephalic culture demonstrate the importance of the endogenous antioxidant glutathione in modulating susceptibility of dopamine neurons to malonate. When cellular levels of total glutathione were reduced by pretreatment with buthionine sulfoxamine, an inhibitor of glutathione synthesis, dopamine neurons were more sensitive to malonate treatment *(39)*. Reduction of cellular glutathione levels also rendered mesencephalic GABA neurons, previously shown to be more resistant to malonate *(34)*, equally as sensitive as the dopamine population to malonate induced toxicity *(41)*. In addition, agents that trap free radicals protect against malonate in vivo *(42)* or in vitro *(41)*. Excitotoxicity and oxidative stress may be interactive events that participate in the destruction of neurons during metabolic inhibition. Overstimulation of glutamate receptors can generate free radicals *(43)*, and conversely, free radicals can enhance release of excitatory amino acids *(44,45)*.

Evidence for a "programmed cell death" initiated by 3-NPA was demonstrated in cortical and striatal cultures. Twenty-four h of exposure to 2 m*M* 3-NPA resulted in DNA fragmentation into nucleasomal lengths (laddering). Cell body shrinkage was noted beginning at approx 4–5 h of incubation with 3-NPA. The presence of inhibitors of protein or RNA synthesis during 48 h of treatment with 3-NPA were protective, whereas glutamate receptor antagonists were ineffective neuroprotectants. These findings contrast with those reported for cortical explants exposed to 1 m*M* 3-NPA. In the latter case, somal swelling and vacuolization, characteristics of necrosis, were evident between 120 and 180 min and were partially attenuated by glutamate receptor antagonists when examined at 4 h. As mentioned previously, dose and exposure times can influence the contribution of different mechanisms to irreversible damage caused by metabolic poisons. In addition, a recent study suggests that factors within an individual neuron may determine whether it responds to metabolic difficulties by descent down an apoptotic or necrotic pathway *(46)*. When cultured hippocampal neurons were viewed at various times (8–48 h) during exposure to 5 m*M* 3-NPA, some neurons underwent a rapid necrotic cell death that was NMDA receptor mediated, inasmuch as it was prevented by MK-801 and potentiated by low micromolar levels of extracellular glutamate. Another population of neurons, however, underwent a slow, apoptotic death that was prevented by inhibition of protein synthesis, but not by NMDA receptor blockade. Interestingly, blockade of apoptotic cell death with cycloheximide increased the number of necrotic neurons, whereas protection from necrotic cell death with MK-801 resulted in an increase in the number of cells with apoptotic profiles. This suggests that a single neuron has the potential to respond to a metabolic insult by following an apoptotic or necrotic pathway to cell death. The factors that influence which pathway is taken remain to be fully elucidated.

CONCLUDING REMARKS

Although biochemical assessment of 3-NPA or malonate defines the existence of several cellular targets, the primary action of these compounds as cell toxicants is most likely due to their ability to disturb energy metabolism via inhibition of SDH. Enhancement of glucose uptake capability by introduction of defective Herpes simplex virus vectors overexpressing the rat brain glucose transporter protected cultured hippocampal neurons from 3-NPA treatment *(47)*. This would be consistent with cell loss by 3-NPA due to an energy insult. In vivo studies of systemically applied 3-NPA show the greater vulnerability of striatum as compared with other brain regions to 3-NPA intoxication. The underlying reason(s) for the heightened vulnerability of striatum in vivo is at present unclear, but appears to be related to more than just cellular vulnerability to 3-NPA *per se*. In vitro studies do not support a greater susceptibility of striatal neurons to SDH inhibition. Cultured neurons from many different brain regions are sensitive to either 3-NPA or malonate. In one study that employed similar times of exposure and concentrations of toxins to neurons from several different brain regions *(29)*, striatal neurons did not exhibit a unique sensitivity. Selective blood–brain barrier breakdown in the region of the striatum may contribute to the vulnerability of this structure to 3-NPA in vivo *(22,23)*. On the other hand, in vitro studies utilize immature neurons that may respond differently than their adult counterparts to SDH inhibition. Neuronal circuitry may also be a factor in vulnerability in vivo that is not mimicked in vitro. When specifically looked for, in vitro studies did show differences in the cellular susceptibility among neuronal populations. In the case of mesencephalic dopamine neurons, this held true in vivo *(35)* as well as in vitro *(34)*. In vitro studies of neuronal vulnerability suggest that a number of different factors may contribute to a neurons susceptibility to SDH inhibition. Some factors that may be important are glucose availability or a neurons ability to utilize glucose to maintain ATP levels; the level and activity of antioxidant systems that may allow some neurons to better withstand an oxidative stress; and the presence, density, and subcomposition of glutamate receptors that may modulate an excitotoxic component to toxicity. Prior experience may also contribute to neuronal vulnerability. Hippocampal slices from rats exposed to hypoxia showed increased tolerance and greater functional recovery when first exposed in vivo to a single intraperitoncal injection of 3-NPA (20 mg/kg, 1 h–3 d prior) *(48)*. Lastly, in vitro studies suggest that exposure conditions or the state of a neuron or its environment at a particular point in time may impact on the type of mechanism or relative contribution of a particular mechanism to toxicity subsequent to 3-NPA or malonate exposure. If the process of cell death involves multiple, parallel signals that ultimately funnel down to a few cell death pathways, targeting convergence points may most effectively counter neurodegeneration due to metabolic poisons.

REFERENCES

1. Morris MP, Pagan C, Warmke HE. Hiptagenic acid, a toxic component of *Indigofera endecaphylla*. Science 1954;119:322–323.
2. Hollocher TC. Ninth International Congress Biochemistry 1973; Abstr.2r9, p. 109.
3. Alston TA, Mela L, Bright HJ. 3-Nitropropionate, the toxic substance of *Indigofera*, is a suicide inactivator of succinate dehydrogenase. Proc Natl Acad Sci USA 1977;74:3767–3771.
4. Coles CJ, Edmondson DE, Singer TP. Inactivation of succinate dehydrogenase by 3-nitropropionate. J Biol Chem 1979;254:5161–5167.
5. Osman MY. Effect of β-nitropropionic acid on rat brain acetylcholinesterase. Biochem Pharmacol 1982;31:4067–4068.
6. Mohammed YS, Mahfouz MM, Osman MY. Effect of β-propiolactone and β-nitropropionic acid on rat brain monoamine oxidase. Biochem Pharmacol 1977;26:62–63.
7. Porter DJT, Bright HJ. 3-Carbanionic substrate analogues bind very tightly to fumerase and aspartase. J Biol Chem 1980, 255:4772–4780.
8. Erecińska M, Nelson D. Effects of 3-nitropropionic acid on synaptosomal energy and transmitter metabolism: relevance to neurodegenerative brain diseases. J Neurochem 1994;63:1033–1041.

9. Webb JL. Malonate. In: Enzyme and Metabolic Inhibitors. Academic Press, New York, 1966, pp. 1–245.
10. Koeppen, AH, Riley KM. Effect of free malonate on the utilization of glutamate by rat brain mitochondria. J Neurochem1987;48:1509–1515.
11. Greene JG, Porter RHP, Eller RV, et al. Inhibition of succinate dehydrogenase by malonic acid produces an "excitotoxic" lesion in rat striatum. J Neurochem1993;61:1151–1154.
12. Mitzen EJ, Koeppen AH. Malonate, malonyl-coenzyme A, and acetyl-coenzyme A in developing rat brain. J Neurochem 1984;43:499–506.
13. Riley KM, Dickson AC, Koeppen AH. The origin of free brain malonate. Neurochem Res 1991;16:117–122.
14. Brown GK, Scholem RD, Bankier A, et al. Malonyl coenzyme A decarboxylase deficiency. J Inher Metab Dis 1984;7:21–26.
15. Haan E, Scholem R, Croll H, et al. Malonyl coenzyme A decarboxylase deficiency. Clinical and biochemical findings in a second child with a more severe enzyme defect. Eur J Pediatr 1986;144:567–570.
16. Narasimhan P, Sklar R, Murrell M, et al. Methylmalonyl-CoA mutase induction by cerebral ischemia and neurotoxicity of the mitochondrial toxin methylmalonic acid. J Neurosci 1996;16:7336–7346.
17. Dutra JC, Dutra-Filho CS, Cardozo SEC, et al. Inhibition of succinate dehydrogenase and β-hydroxybutyrate dehydrogenase activities by methylmalonate in brain and liver of developing rats. J Inher Metab Dis 1993;16:147–153.
18. Wajner M, Dutra JC, Cardoso SE, et al. Effect of methylmalonate on in vitro lactate release and carbon dioxide production by brain of suckling rats. J Inher Metab Dis 1992;15:92–96.
19. McLaughlin BA, Nelson D, Silver IA, et al. Methylmalonate toxicity in primary neuronal cultures. Neuroscience 1998;86:279–290.
20. Rosenberg LE, Fenton WA. Disorders of propionate and methylmalonate metabolism. In: Scriver CR, Beaudet AL, Sly WS, Valle D, eds. The Metabolic Basis of Inherited Disease, 6th edit. McGraw-Hill, New York, 1997, pp. 845–854.
21. Gould DH, Wilson MP, Hamar DW. Brain enzyme and clinical alterations induced in rats and mice by nitroaliphatic toxicants. Toxicol Lett 1985;27:83–89.
22. Hamilton BF, Gould DH. Correlation of morphologic brain lesions with physiologic alterations and blood–brain impairment in 3-nitropropionic acid toxicity in rats. Acta Neuropathol 1987;74:67–74.
23. Nishino H, Shimano Y, Kumazaki M, et al. Chronically administered 3-nitropropionic acid induces striatal lesions attributed to dysfunction of the blood-brain barrier. Neurosci Lett 1995;186:161–164.
24. Beal MF, Brouillet E, Jenkins BG, et al. Neurochemical and histologic characterization of striatal excitotoxic lesions produced by the mitochondrial toxin 3-nitropropionic acid. J Neurosci 1993;13:4181–4192.
25. Palfi S, Ferrante RJ, Brouillet E, et al. Chronic 3-nitropropionic acid treatment in baboons replicates the cognitive and motor deficits of Huntington's disease. J Neurosci 1996;16:3019–3025.

26. Pang Z, Umberger GH, Geddes JW. Neuronal loss and cytoskeletal disruption following intrahippocampal administration of the metabolic inhibitor malonate, lack of protection by MK-801. J Neurochem 1997;66:474–484.
27. Ludolph AC, Seelig M, Ludolph A, et al. 3-Nitropropionic acid decreases cellular energy levels and causes neuronal degeneration in cortical explants. Neurodegeneration 1992;1:155–161.
28. Behrens MI, Koh J, Canzoniero LMT, et al. 3-Nitropropionic acid induces apoptosis in cultured striatal and cortical neurons. NeuroReport 1995;6:545–548.
29. Fink SL, Ho DY, Sapolsky RM. Energy and glutamate dependency of 3-nitropropionic acid neurotoxicity in culture. Exp Neurol 1996;138:298–304.
30. Riepe M, Hori N, Ludolph AC, et al. Inhibition of energy metabolism by 3-nitropropionic acid activates ATP-sensitive potassium channels. Brain Res 1992;586:61–66.
31. Ben-Ari Y, Krnjevic K, Crepel V. Activators of ATP-sensitive K+ channels reduce anoxic depolarization in CA3 hippocampal neurons. Neuroscience 1990;37:55–60.
32. Weller M, Paul SM. 3-Nitropropionic acid is an indirect excitotoxin to cultured cerebellar granule neurons. Eur J Pharmacol 1993;248:223–228.
33. Zeevalk GD, Derr-Yellin E, Nicklas WJ. NMDA receptor involvement in toxicity to dopamine neurons in vitro caused by the succinate dehydrogenase inhibitor 3-nitropropionic acid. J Neurochem 1995;64:455–458.
34. Zeevalk GD, Derr-Yellin E, Nicklas WJ. Relative vulnerability of dopamine and GABA neurons in mesencephalic culture to inhibition of succinate dehydrogenase by malonate and 3-nitropropionic acid and protection by NMDA receptor blockade. J Pharmacol Exp Ther 1995;275:1124–1130.
35. Zeevalk GD, Manzino L, Hoppe J, et al. In vivo vulnerability of dopamine neurons to inhibition of energy metabolism. Eur J Pharmacol 1997; 320:111–119.
36. Sonsalla PK, Manzino L, Sinton CM, et al. Inhibition of striatal energy metabolism produces cell loss in the ipsilateral substantia nigra. Brain Res 1997; 773:223–226.
37. Choi DW, Rothman SM. The role of glutamate neurotoxicity in hypoxic-ischemic neuronal death. Annu Rev Neurosci 1990;13:171–182.
38. Beal MF, Ferrante RJ, Henshaw R, et al. 3-Nitropropionic acid neurotoxicity is attenuated in copper/zinc superoxide dismutase transgenic mice. J Neurochem 1995;65:919–922.
39. Zeevalk GD, Bernard LP, Albers DS, et al. Energy stress-induced dopamine loss in glutathione peroxidase-overexpressing transgenic mice and in glutathione-depleted mesencephalic cultures. J Neurochem 1997;68:426–429
40. Schulz JB, Huang PL, Matthews RT, et al. Striatal malonate lesions are attenuated in neuronal nitric oxide synthase knockout mice. J Neurochem 1996; 67:430–433.
41. Zeevalk GD, Bernard LP, Nicklas WJ. Role of oxidative stress and the glutathione system in loss of dopamine neurons due to impairment of energy metabolism. J Neurochem 1998;70:1421–1430.

42. Schulz JB, Matthews RT, Jenkins BG, et al. Improved therapeutic window for treatment of histotoxic hypoxia with a free radical spin trap. J Cereb Blood Flow Metab 1995;15:948–952.
43. Lafon-Cazal M, Pietri S, Culcasi M, et al. NMDA-dependent superoxide production and neurotoxicity. Nature 1993;364:534–537.
44. Pellegrini-Giampietro DE, Cherici G, Alesiani M, et al. Excitatory amino acid release from rat hippocampal slices as a consequence of free-radical formation. J Neurochem 1988;51:1960–1963.
45. Pellegrini-Giampietro DE, Cherici G, Alesiani M, et al. Excitatory amino acid release and free radical formation may cooperate in the genesis of ischemia-induced neuronal damage. J Neurosci 1990;10:1035–1041.
46. Pang Z, Geddes JW. Mechanisms of cell death induced by the mitochondrial toxin 3-nitropropionic acid: acute excitotoxic necrosis and delayed apoptosis. J Neurosci 1997;17:3064–3073.
47. Ho DY, Saydam TC, Fink SL, et al. Defective Herpes simplex virus vectors expressing the rat brain glucose transporter protect cultured neurons from necrotic insult. J Neurochem 1995;65:842–850.
48. Riepe MW, Kasischke K, Gericke CA, et al. Increase of hypoxic tolerance in rat hippocampal slices following 3-nitropropionic acid is not mediated by endogenous nerve growth factor. Neurosci Lett 1996;211:9–12.

5

Cognitive and Motor Deficits Produced by Acute and Chronic Administration of 3-Nitropropionic Acid in Rats

Gary L. Dunbar, Deborah A. Shear, Jie Dong, and Kristi L. Haik-Creguer

INTRODUCTION

The mitochondrial toxin 3-nitropropionic acid (3-NPA) has received substantial attention recently as a potential tool for studying the types of neuropathological and behavioral deficits observed in Huntington's disease (HD). The use of 3-NPA in animal models of HD was prompted by findings that accidental ingestion of this toxin produces striatal degeneration that resembles what is observed in the brains of HD patients. In the mid-1970s, 3-NPA was isolated as the toxic substance in locoweed and milkvetches responsible for the widespread poisoning of domestic livestock *(1,2)*. However, its significance as a potential tool for gaining a better understanding of HD was catapulted by reports of its effects on children in China who were exposed to it after ingesting moldy sugarcane *(3,4)*. Some of these children displayed dystonia, or severe plastic rigidity, including a tonic flexion of the arms and legs. The computed tomography (CT) scans of these children revealed bilateral basal ganglia hypodensity *(3,5)*. These findings suggested that 3-NPA could be a useful tool for studying mechanisms of neuronal damage, particularly for neurodegenerative diseases, such as HD. Although several laboratories have demonstrated that acute intrastriatal injections and chronic systemic administration of 3-NPA accurately reproduce many of the neuropathological characteristics of HD *(6–13)*, relatively little is known about the behavioral effects of 3-NPA, particularly its effect on cognitive processing. Interest in our laboratory has focused on the use of 3-NPA in an animal model of neurodegeneration that mimics some of the behavioral as well as the neuropathological symptoms of HD. This chapter focuses on what is known about the behavioral effects of acute intrastriatal and chronic

From: *Mitochondrial Inhibitors and Neurodegenerative Disorders*
Edited by: P. R. Sanberg, H. Nishino, and C. V. Borlongan © Humana Press Inc., Totowa, NJ

systemic injections of 3-NPA in animal models of HD. We present some of the recent findings in our laboratory within the context of what is already known about the effects of 3-NPA on cognitive and motor ability.

BEHAVIORAL EFFECTS OF 3-NPA

As can be ascertained from Table 1, most of the behavioral studies using 3-NPA have focused on motor abnormalities. This is not surprising considering that the choreiform motor disturbances observed in HD patients are the single most obvious symptom of this disease. Some of the earliest descriptions of 3-NPA-induced motor disturbances in laboratory animals came from the work of Gould and his colleagues *(14–16)*. (*See also* Chapter 2.) Their initial observation of the behavioral effects of 3-NPA involved single or repeated injections of a relatively high dose (120 mg/kg, ip) in mice *(14)*. This produced initial hypoactivity, followed by episodes of abnormal movements and tremors, and then by a period of severe hypoactivity. This pattern of depressed motor activity, bouts of abnormal movements, and recumbency (inability to maintain normal posture) was also observed in rats that were given an acute, subcutaneous injection of a relatively high dose (30 mg/kg) of 3-NPA *(15)*. A more detailed description of the motor disturbances that are produced by 3-NPA was subsequently given by Hamilton and Gould *(16)*. They gave rats subcutaneous injections of 3-NPA using either a single injection at 30 mg/kg or four daily injections at 10 mg/kg. They described the clinical signs produced by both of these injection paradigms as consisting of three stages. In Stage I, intoxicated rats were somnolent, but showed normal gait and postural reactions when aroused. Stage II was characterized by rhythmic "paddling" movements of the forelimbs and barrel rotations (360° rotations of the long axis of the body). These episodes of transient motor abnormalities were common in rats injected acutely with the high dose, but were reduced or absent in rats chronically administered the lower dose. Stage III was characterized by recumbency (loss of normal posture) and was common to both groups of treated rats. This last stage corresponded most strongly to bilateral lesions in the caudate–putamen, hippocampus, and thalamus.

The most comprehensive analyses of the effects of 3-NPA on locomotor activity has come from the work of Sanberg and his colleagues *(10,11,17,18)*. They found that intraperitoneal injections of 10 mg/kg 3-NPA given every 4 d for 4 wk cause hypoactivity in both 6- and 10-wk-old rats, with the older rats displaying uncoordinated movement, rigidity, and recumbency *(10)*. Similarly, bilateral injections of 3-NPA into the striatum (at either 500 or 750 nmol concentrations) produced an even more profound hypoactivity in

Table 1
Studies of the Effects of 3-NPA on Cognitive or Motor Functions in Animals

Study	Species	Method	Major Findings
Gould and Gustine 1982 *(14)*	Mouse	120/mg/kg/d, i.p., single or repeated injections	Hypoactivity, followed by episodes of hyperactivity and abnormal movements, followed by severe hypoactivity
Gould et al., 1985 *(15)*	Mouse Rat	140–150 mg/kg, s.c. 30 mg/kg, s.c., single injections	Hypoactivity, followed by abnormal movements (paddling, rolling) followed by recumbency
Hamilton and Gould, 1987 *(16)*	Rat	30 mg/kg, s.c., single dose or 10 mg/kg/d for 1–4 d	Somnolent stage followed by hyperactivity, with paddling movements and barrel rolls, followed by hypoactivity and recumbency
Koutouzis et al., 1994a *(10)*	Rat	10 mg/kg/d, i.p., every 4 d	Bradykinesia in 6-wk-old rats; rigidity and recumbency in 10-wk-old rats; hypoactivity in both groups
Koutouzis et al., 1994b *(11)*	Rat	Bilateral intrastriatal injections	Hypoactivity and passive-avoidance retention deficits
Borlongan et al., 1995a *(17)*	Rat	10 mg/kg, i.p., every 4 d for 28 d	Hypoactivity and passive-avoidance retention deficits
Borlongan et al., 1995b *(18)*	Rat	10 mg/kg, i.p., every 4 d for either 18 d or 28 d	Sustained hyperactivity (for rats given 3-NPA every 4 d for 8 d) or change to hypoactivity (for rats given 3-NPA every 4 d for 28 d)
Brouillet et al., 1995 *(9)*	Macaque Baboon	8 mg/kg/d, i.m., for either 3–6 wk or 10 mg/kg/d increased to 28 mg/kg over 4 mo	Apomorphine-induced choreiform movements and dystonia; prolonged treatment (4 mo) caused *spontaneous* dystonia and dyskinesia
Palfi et al., 1996 *(22)*	Baboon	14 mg/kg/d, i.m., increased to 33 mg/kg/d over a 20-wk period	Apomorphine-induced dykinesia, dystonia, and choreiform movements from wk 3 to wk 11; spontaneous abnormal movements thereafter; congnitive deficits on an ovject detour retrieval task (ORDT) during wk 3–6
Nishino et al., 1996 *(12)*	Rat	20 mg/kg/d, s.c., for 2–3 d	Forepaw "paddling," barrel rolls, recumbency, and severe dyspnea

14-wk-old rats *(11)*. In addition, both doses produced significant deficits for retention of a passive avoidance task. Hypoactivity and passive avoidance retention deficits were also observed in both 14- and 28-wk-old rats given 10 mg/kg of 3-NPA subcutaneously every 4 d for 4 wk *(17)*. Interestingly, these retention deficits were observed during nocturnal testing, but not when the tests were conducted during the daytime. A comprehensive profile of the effects of 3-NPA on locomotor activity was obtained in a fourth study *(18)*. In this study, it was found that 14-wk-old rats that were given IP injections of 3-NPA at 10 mg/kg every fourth day for 4 wk showed an initial level of hyperactivity during peak activity periods on injection d 4 and 8, but showed significant hypoactivity on injection d 16, 20, 24, and 28. Rats that were given 3-NPA injections on d 4 and 8, and then saline injections on d 12, 16, 20, 24, and 28, maintained a higher than normal level of locomotor activity throughout the entire 4-wk period. Collectively, the findings from Sanberg's laboratory demonstrate that both systemic and intrastriatal administration of 3-NPA can produce locomotor abnormalities and deficits for retention of a passive avoidance task. In addition, the activity levels of rats that are chronically treated with relatively low levels of systemically administered 3-NPA can change from hyperactive to hypoactive over a short period of time.

Taken together, the findings from the laboratories of both Gould and Sanberg suggest that relatively high doses of 3-NPA (e.g., an acute subcutaneous injection at 30 mg/kg or chronic, systemic injections at 10 mg/kg daily for 1–4 d) produce acute abnormal motor movements, including forepaw "paddling" and barrel rotations, while lower doses (e.g., 10 mg/kg every fourth day) can produce a transition of hyperactivity to hypoactivity, in the absence of acute abnormal movements. Recent findings *(12)* indicate that daily injections of an intermediate dose of 3-NPA (20 mg/kg, s.c.) produce these acute motor abnormalities after 2 or 3 d in about half of the 3-NPA-treated rats. To what extent these acute motor movements are analogous to motor disturbances observed in HD is questionable. Similar acute motor disturbances are observed after acute, intrastriatal injections of quinolinic acid (QA) in rats *(19,20)*. However, because the early QA-induced forepaw "paddling" and barrel rotations dissipate readily within 8 h of the QA injection, they are not generally considered to be analogous to the choreiform movements of HD patients *(21)*. It is particularly relevant to note that chronic, intrastriatal administration of QA at doses that produce striatal lesions that are comparable to those produced by acute QA injections, cause transient hyperactivity, but do not cause this forepaw "paddling" and barrel rotations *(21)*. In this respect, it seems unlikely that the acute, abnormal motor movements produced by high doses of 3-NPA are analogous to the choreiform movements of HD, and it is more likely that they represent the seizure-like activity that is induced by acute intrastriatal injections of excitotoxins, such as QA.

Recent behavioral assessments of the effects of chronic, systemic administration of QA have been conducted with primates *(8,22)*. In one study *(8)* an aged macaque and two aged baboons that were treated with daily injections of 3-NPA (8 mg/kg, i.m.) showed increased motor activity, choreiform movements, and foot and limb dystonia when challenged with injections of apomorphine. In addition, two preadolescent baboons that were given 3-NPA injections, starting with a dose of 10 mg/kg that was progressively increased to 28 mg/kg over a 4-mo period, showed spontaneous dystonia and dyskinesia after prolonged 3-NPA treatment. In a second study *(22)*, four baboons were given progressively larger doses of 3-NPA (10–30 mg/kg, i.m.) at levels that overcame the effects of tolerance, but avoided acute intoxication over a 20-wk period. Although only apomorphine-induced dyskinesia, dystonia, and choreiform movements were observed during the first 11 wk of 3-NPA treatments, spontaneous abnormal movements (e.g., foot and leg dystonia) were observed starting at wk 11 and persisting through wk 20. In addition to the motor assessments, an object retrieval detour task (ORDT) was used to assess the cognitive ability of these 3-NPA-treated primates. The baboons that received 3-NPA had difficulty redirecting their arm movements to retrieve a food reward in the ORDT at 3–6 wk after the injections had begun. The results from the ORDT task suggest that 3-NPA causes deficits in a motor learning task prior to the onset of spontaneous motor abnormalities. This is analogous to some of the early cognitive deficits reported in HD that occur prior to the onset of the choreiform movements *(23–26)*.

Collectively, studies of the behavioral effects of 3-NPA in laboratory animals suggest that this toxin mimics many of the behavioral as well as the neuropathological symptoms observed in HD patients. Although some of the acute motor abnormalities resulting from 3-NPA intoxication may not be analogous to the motor dysfunction of HD, the progression of hyperactivity to hypoactivity observed in the chronic, systemically treated animals does reflect a process similar, albeit greatly accelerated, to that which occurs in HD. In addition, the observed deficits in retention of a passive avoidance task in rats and the ORDT in primates mimic some of the cognitive deficits observed in HD patients.

RECENT EVIDENCE OF COGNITIVE AND MOTOR DEFICITS IN 3-NPA-TREATED RATS

Recent work in our laboratory has supported findings of HD-like cognitive and motor deficits, using both systemic and intrastriatal administration of 3-NPA in rats. Of particular interest to us was whether 3-NPA treatments produced cognitive impairments on a spatial learning task specifically designed to test for deficits in both working and reference memory. Recent findings by Levy and colleagues *(27)* have indicated that the primate striatum is activated during tasks for spatial working memory. Similarly, Robbins and colleagues *(25,26)* have found that spatial working memory is significantly compromised in HD patients, and that the onset of this impairment occurs in the early stages of the disease, prior to the onset of choreiform movements. Given these findings, we designed two experiments to test the effects of either intrastriatally or systemically administered 3-NPA on working memory deficits in a spatial learning task, in addition to tests of various motor functions.

In the first study, we compared the effects of bilateral intrastriatal injections of a 750 nmol solution of 3-NPA with a 200 nmol solution of QA. Our choice of these doses was based on previous work that indicated that they produced significant behavioral deficits *(11,28)*, as well as our pilot work which indicated that these doses caused lesions of similar size, at least when measured by the areas within the striatum that were metabolically depressed (as revealed by reductions in cytochrome oxidase staining).

Prior to surgery, the rats were trained to traverse a narrow, meter-long balance beam. The rats were then given bilateral intrastriatal injections of either the neurotoxin or vehicle. Following a 2-wk postoperative recovery period, the rats were tested on the following behavioral tasks: (1) radial-arm-water-maze (RAWM), (2) balance-beam, (3) grip-strength, (4) paw-placement, and (5) open-field.

The apparatus for the RAWM task was a modification of the one developed by Burešova et al. *(29)*. It consisted of a water tank (140 cm diameter) with eight Plexiglas-enclosed channels that radiate from an open central area (40 cm in diameter). Platforms that could be raised and lowered remotely by the experimenter were located in the central area and at the ends of each channel (*see* Fig. 1). In the "up" positions, the platforms were just 1 cm below the water surface. In the "down" position the platforms were approx 30 cm below the water surface.

Two versions of the RAWM task were used *(30)*. The spatial version consisted of raising the platforms at the ends of the same four channels during each trial (channels 2, 4, 7, and 8, clockwise) while the remaining chan-

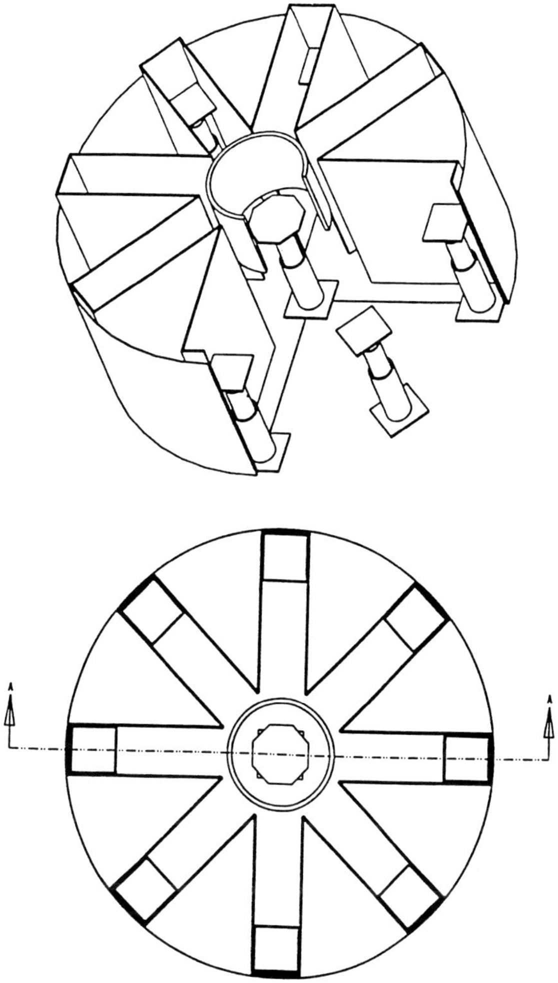

Fig. 1. Schematic drawing of the radial-arm-water-maze (RAWM) apparatus.

nels never contained raised platforms. Several extramaze cues (e.g., window, table, posters) were visible and in the same locations relative to the "correct" channels during all trials of the spatial version of the RAWM. In the nonspatial version, the four channels that contained the platforms were determined randomly for each trial, so that the spatial location could not be used as a cue. Instead, the cues were different objects that were visibly suspended over the channels. The same four cues (i.e., cross, cup, glove, ball) signified which channels had raised platforms, while the four other cues (triangle, circle, washcloth, square) always marked channels that did not contain raised platforms. A black curtain encircled the maze to block out all other extramaze cues, in an effort to facilitate the use of nonspatial cues for this version of the RAWM task.

For both versions of the task, a trial consisted of placing a rat on the platform in the central area of the maze and then releasing the platform to its "down" position after 10 s. The rat then was given 3 min to find the four raised platforms. Entries into channels that never contained the raised platform (on the spatial version) or channels marked by cues that were always associated with channels that never contained a raised platform (on the nonspatial version) were considered "reference" memory errors. Entries into channels that the rat had previously visited during a trial were considered "working" memory errors. The trial was complete after the rat had visited all four platforms, or until 3 min had elapsed. Rats that did not find all the platforms within the 3 min were guided by hand to the unvisited platforms and allowed to stay on each of these for a 10-s "orientation" period. All rats were given four trials each day (two trials on each version of the RAWM), alternating between the spatial and nonspatial versions with a 2-h intertrial interval between each *(30)*.

As can be seen in Fig. 2, both the QA- and the 3-NPA-treated rats had significantly fewer completed trials on the nonspatial version of the RAWM, whereas only the 3-NPA-treated rats had fewer completed trials on the spatial version. However, both treatments caused a significant increase in reference memory errors on the spatial version of the task, but only the 3-NPA-treated group had more spatial working memory errors (Fig. 3). Moreover, the 3-NPA treatments caused significantly more reference memory and working memory errors than did the QA injections. These results indicate that the injections of 750 nmol of 3-NPA cause more severe cognitive deficits than did administration of 200 nmol of QA.

Results of the motor task revealed significant between-group differences for the balance-beam, grip-strength, and open-field tasks, but not on measures of swim speeds or performance on the paw-placement task. The negative results on the latter tasks indicate that the impairments on the RAWM task were mnemonic in nature and were not due to difficulties in swimming (as indicated by the similar swim speeds) or vision (as tested by the paw-placement task). On the balance-beam task, both QA- and 3-NPA-treated rats took longer to initiate movements (Fig. 4A), but only the 3-NPA-treated rats had fewer completed trials (i.e., were unable to traverse the beam within the allotted 3-min interval) (Fig. 4B). On the grip-strength task, rats used their forepaws to hang onto a 2 mm diameter steel rod that was suspended

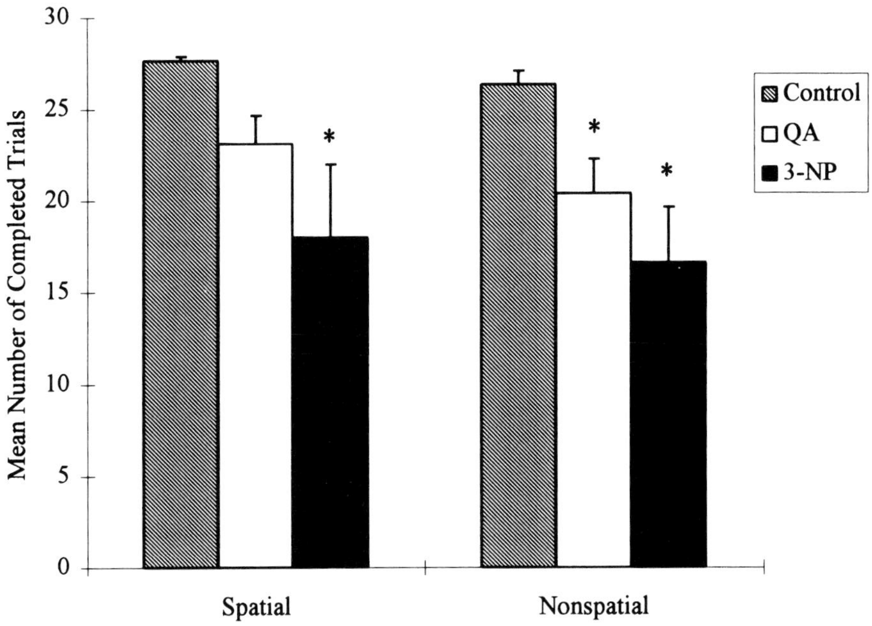

Fig. 2. Both the 3-NPA- and QA-treated rats completed significantly fewer trials in the nonspatial version of the RAWM, but only the 3-NPA-treated rats completed fewer trials in the spatial version of the RAWM ($*p \leq 0.05$).

one-half meter above a soft mattress. The dependent measure for this task was the length of time that the animal was able to hang onto the rod before dropping to the mattress. Surprisingly, 3-NPA-treated rats had much longer hang times than either the QA-treated or the control rats, at both 2 and 4 wk after surgery (Fig. 5). Finally, on the open-field task, both the QA- and the 3-NPA-treated rats were hyperactive (traveled more distance) during testing at 2 wk postsurgery, but only the QA rats were hyperactive during testing at 4 wk postsurgery (Fig. 6). The activity levels of the 3-NPA-treated rats were no different than the controls at the later testing period. The results from these motor tasks corresponded to those from the cognitive test, suggesting that the 3-NPA treatments caused more severe motor abnormalities than did treatments with QA.

Histological evaluation of the tissue indicated that the QA treatments produced a slightly larger lesion in the striatum, but the 3-NPA lesion was more destructive. Although both treatments caused similar amounts of striatal atrophy and enlargement of the lateral ventricles, only the 3-NPA treatments produced necrotic cores within the striatum. These necrotic cores have been

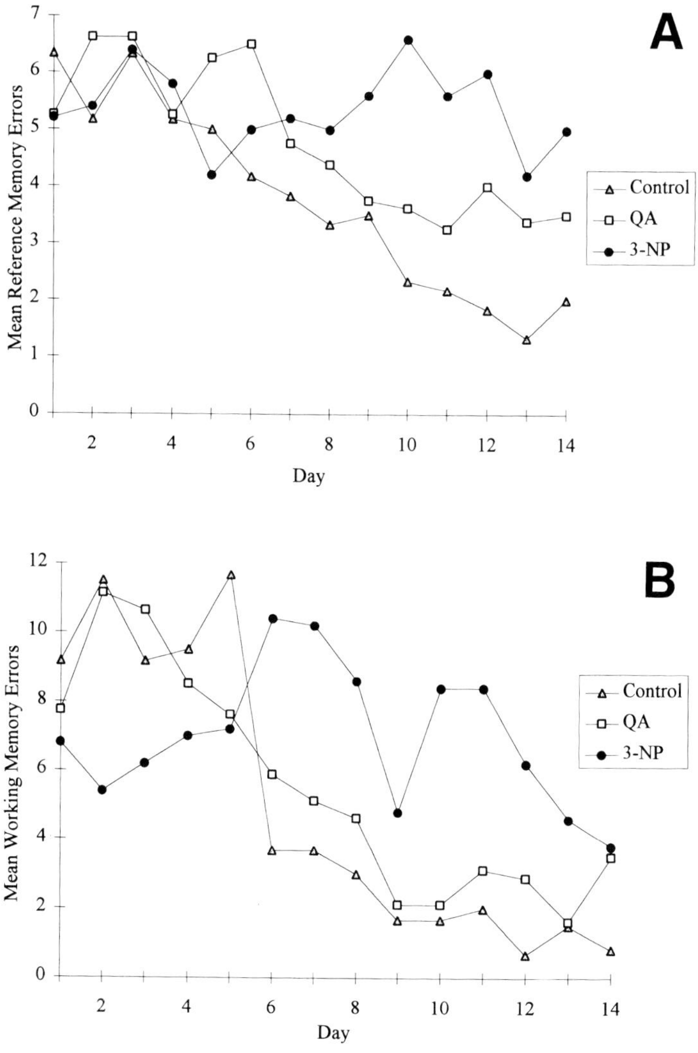

Fig. 3. Both 3-NPA- and QA-treated rats made more reference memory errors **(A)** than control rats in the spatial version of the RAWM, but only the 3-NPA-treated rats made significantly more working memory errors **(B)** on this task.

described in detail by Beal et al. *(6)*, and are probably the reason for the greater cognitive and motor deficits produced by the 3-NPA lesions.

Collectively, the results of the cognitive and motor measures in our study support the conclusion of Sanberg and colleagues *(11,17,18)* that QA treatments may mimic some of the earlier stages of HD, while 3-NPA treatments

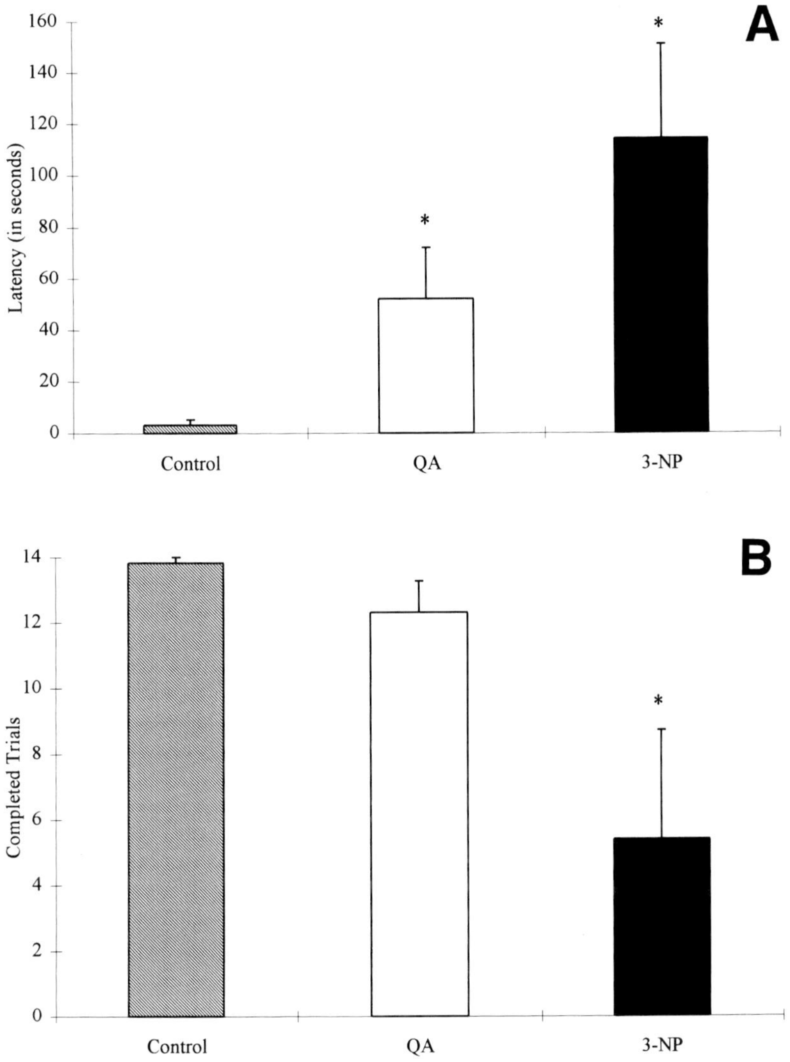

Fig. 4. Both the 3-NPA- and QA-treated rats took significantly longer to initiate movement **(A)** on the balance-beam task, but only the 3-NPA-treated rats completed significantly fewer trials **(B)** ($p < 0.05$).

better simulate the later stages. This was particularly apparent when comparing the performances of the QA- and 3-NPA-treated rats on the balance-beam, grip-strength, and open-field measures. Whereas most of the QA-treated rats were able to traverse the balance beam, several of the 3-NPA-

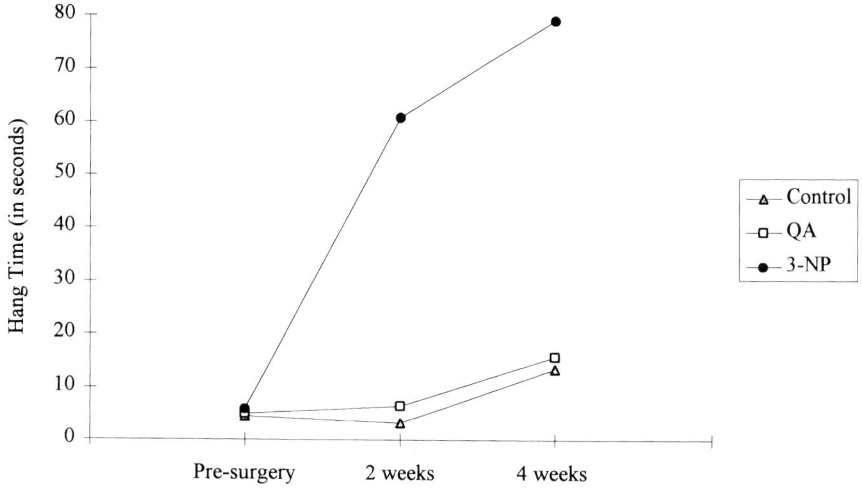

Fig. 5. Acute intrastriatal injections of 3-NPA produced a progressive increase in hang times on the grip-strength task as compared to control and QA-treated rats.

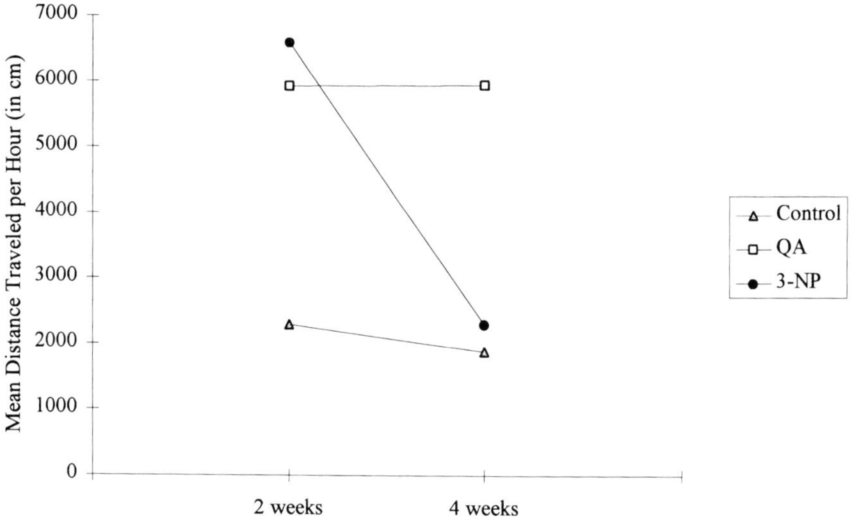

Fig. 6. Intrastriatal injections of both QA and 3-NPA caused an increase in activity levels at 2 wk postlesion but only QA rats showed hyperactivity at 4 wk postlesion.

treated rats froze in a rigid posture at the suspended end of the beam and all these rats had a difficult time progressing along the narrow beam. This rigidity, which is characteristic of the later stages of HD *(31)*, was also evident in the grip-strength test. The 3-NPA-treated rats displayed a tonic rigidity that tended to increase over time (i.e., from wk 2 to wk 4, postsurgery). Finally, the open-field results were similar to those observed by Borlongan and colleagues *(18)*, with 3-NPA-treated rats demonstrating some initial hyperactivity earlier after the treatment, but returning to control levels by 4 wk postsurgery. Although the 3-NPA-induced hypoactivity reported by Koutouzis et al. *(11)* was not observed in our study, the 3-NPA-treated rats did become progressively less active over time.

In a second study, we investigated the effects of chronic, systemic injections of 3-NPA on cognitive and motor functions in young and aged rats. In this study we compared the effects of 28 d of subcutaneous infusions of either 3-NPA (15 mg/kg/d) or vehicle from osmotic minipumps in a group of 2-mo-old and a group of 14-mo-old rats. Two weeks following pump implantation, the rats were tested on the spatial version of the RAWM task and on the balance-beam and open-field tasks. The results indicated that 3-NPA caused significant reference and working memory errors for both age groups on the RAWM task (Fig. 7), although these deficits did not appear to be as severe as those created by the acute, intrastriatal injections. Similarly, the 3-NPA treatments caused significant impairment for both age groups on the balance-beam task, with 3-NPA treated rats making more footslips than controls (Fig. 8). However, it was also apparent that the chronic 3-NPA treatment did not produce the same degree of rigidity as did the intrastriatal injections, as most of the chronically treated rats were able to traverse the narrow beam. Finally, the chronic administration of 3-NPA produced only minor, transient levels of hypoactivity at 1 and 2 wk after implantation (Fig. 9A). However, the 3-NPA-treated rats did show a consistent reduction in the number of rearings (the number of times the rat stood up on his hind legs) during the open-field tests, especially the aged rats on wk 2 and 4 following implantation (Fig. 9B).

The neuroanatomical damage produced by chronic infusion of 3-NPA was similar in both age groups. Both groups had significant depletion of metabolic activity in the dorsolateral striatum, using optical densitometry measures of cytochrome oxidase-stained brain sections. However, the extent of these lesions were smaller and more variable than those seen with the acute intrastriatal injections of 3-NPA. Furthermore, none of the 3-NPA lesions in either group of rats contained a necrotic core, devoid of labeled cells. Neuron counts near the lesion sites revealed a significant loss of Nissl-stained neurons, but sparing of NADPH-diaphorase-labeled neurons. Although there was a trend toward a greater 3-NPA-induced loss of neurons in the aged group, no significant differences were observed between the two age groups.

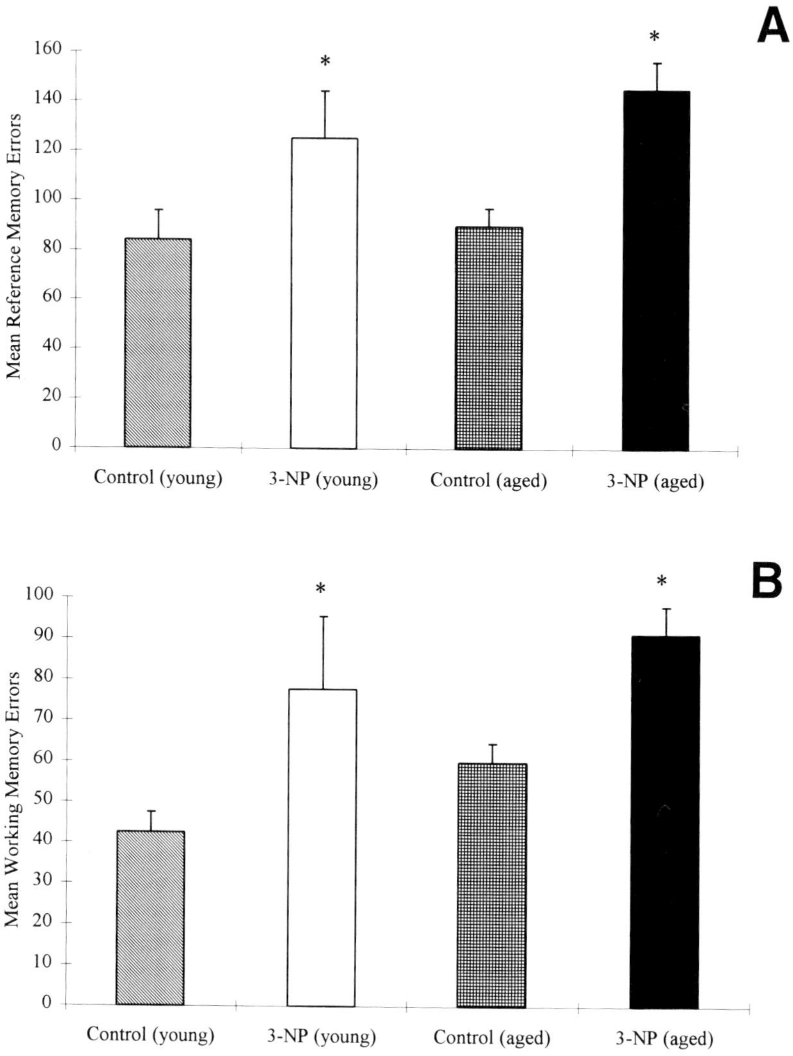

Fig. 7. Chronic administration of 3-NPA caused significant spatial reference **(A)** and working **(B)** memory errors for both age groups on the RAWM task (*$p < 0.05$).

These results suggest that chronic infusions of 3-NPA can produce significant cognitive and motor deficits. Furthermore, the effects of 3-NPA appear to be age dependent *(8,32)*, disrupting spatial learning performance and reducing the number of rearings in the open-field more in aged than in young rats. The results also indicate that chronic subcutaneous infusions of 3-NPA induce neuronal loss and metabolic activity within the striatum,

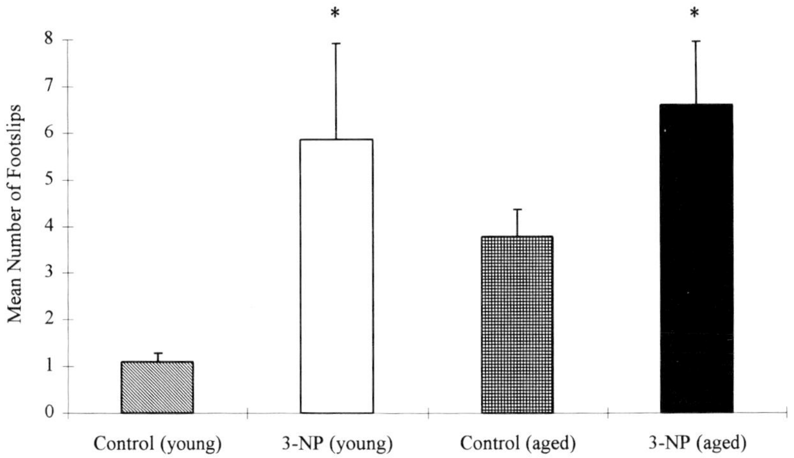

Fig. 8. Chronic administration of 3-NPA caused significantly more footslips for both age groups on the balance-beam task (*$p < 0.05$)

although we did not observe any significant age-dependent effects for these measures. In comparison with the consequences of intrastriatal injections of 3-NPA, the chronic administration of 3-NPA at the doses used in this study produced much less severe behavioral and neuroanatomical deficits.

USE OF 3-NPA IN ANIMAL MODELS OF HD

There are four important questions that need to be addressed when considering the use of 3-NPA to assess behavior in an animal model of HD: (1) What advantage does the use of 3-NPA offer over the use of QA? (2) Should the 3-NPA be administered intrastriatally or systemically? (3) Should the 3-NPA be administered acutely or chronically? and (4) What is the optimal dose to use? The answers to these questions are not easy and depend a great deal on what stage of HD you are trying to mimic and what types of behaviors you intend to assess. Nevertheless, the information obtained from the behavioral studies reviewed here can provide some guidance.

Whether to use 3-NPA or QA to mimic behavioral symptoms of HD would depend primarily on which stage of HD you are interested in modeling. Obviously, because 3-NPA readily crosses the blood–brain barrier it can be administered systemically, whereas QA must be delivered intrastriatally. However, if ease of administration is not a critical factor, the available evidence would suggest that QA would be the better choice for mimicking the earlier cognitive and motor deficits of HD, while 3-NPA pro-

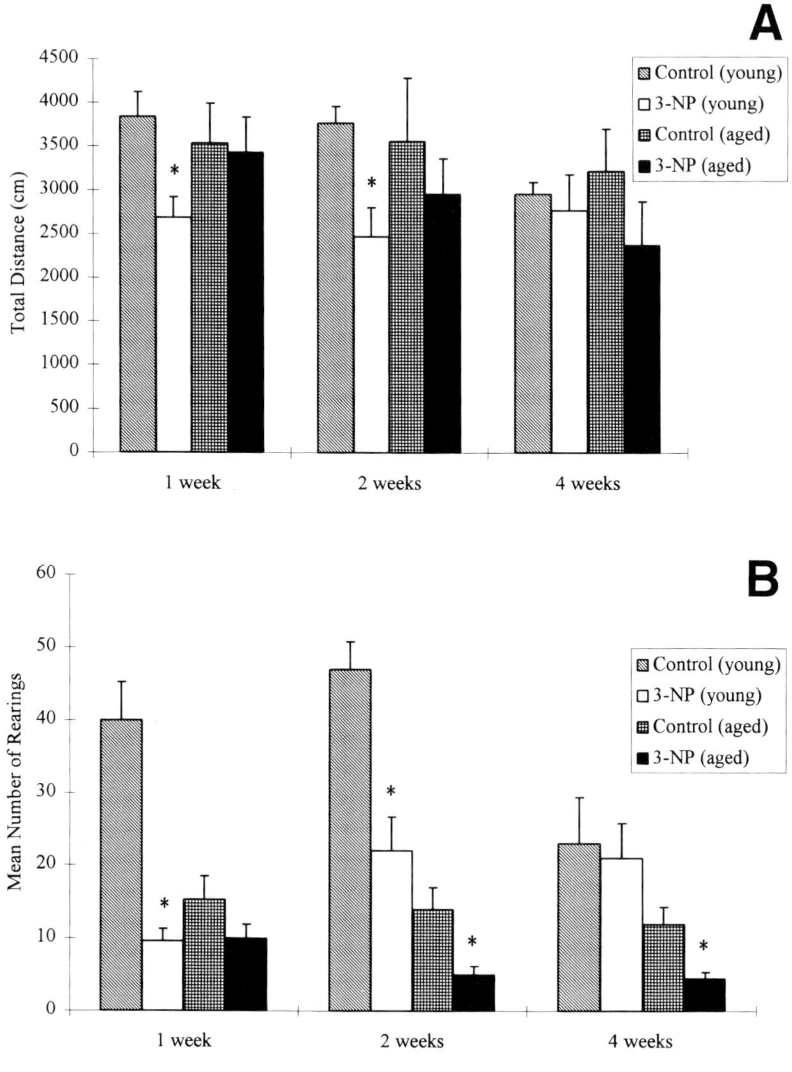

Fig. 9. Chronic administration of 3-NPA produced a transient decrease in total distance traveled **(A)** at wk 1 and 2 following pump implantation, along with a consistent reduction in rearing activity **(B)** (*$p < 0.05$).

vides a better behavioral model for the later stages (or juvenile onset) of HD. This is because 3-NPA produces profound hypoactivity, dystonia, and rigidity, which are characteristic of the later stages (and the juvenile onset) of HD. The hyperactivity that characterizes earlier stages of HD may be simulated by reducing the dose or limiting the number of injections of a low dose of 3-NPA *(18)*, but the striatal damage in such rats is significantly smaller or nonexistent. Conversely, QA produces the hyperactivity that is characteristic of earlier stages of HD, but it is difficult to mimic the hypoactivity, dystonia, and rigidity of the later stages of HD in rats administered QA either acutely or chronically. Thus, the use of QA seems to mimic the earlier stages of HD, whereas 3-NPA provides a better model for the later stages of the disease.

Whether to use intrastriatal or systemic administration of 3-NPA depends primarily on how important it is to ensure that the toxin does not affect peripheral organs or central structures other than the striatum. It is unlikely that systemically administered 3-NPA selectively affects the striatum without having any (however small) effect on other central nervous system (CNS) structures, the peripheral nervous system (PNS), or other peripheral organs. Subcutaneous injections of high doses of 3-NPA have been shown to cause some damage to the thalamus and hippocampus *(16)*. However, lower doses of 3-NPA seem to affect the striatum more selectively, and because the effects of HD are not confined to the striatum *(33,34)*, restricting the administration of 3-NPA to the striatum may not be a critical variable. Given this, it seems that systemic administration provides a more appropriate model of HD. Work in our laboratory corresponds to the findings by Koutouzis and colleagues *(11)* that intrastriatal injections of 3-NPA produce a more severe lesion with a necrotic core and more profound behavioral deficits. Behaviorally, these deficits may mimic the very last stages of HD, but the neuroanatomical correlates do not resemble what occurs in HD. Thus, systemic administration of 3-NPA appears to provide the better model of HD.

Whether or not to use acute or chronic administration of 3-NPA depends on how important it is to mimic the progressive nature of HD. The obvious advantage of using acute injections of 3-NPA is that it can produce behavioral abnormalities almost instantly. However, as discussed previously, the degree to which some of these early motor abnormalities are analogous to motor dysfunction in HD is highly questionable. The rhythmic "paddling" of the forepaws and the barrel rotations observed with acute injections at behavior-altering doses are more akin to the tonic-clonic seizure-like activity observed after acute injections of other neurotoxins, such as QA. Given this, the progressive nature of HD is more accurately replicated by chronic administration of 3-NPA.

Selecting the optimal dose of 3-NPA for assessing behavior in an animal model of HD is still a largely unresolved issue. Assuming the use of a model using chronic, systemically administered 3-NPA, a dose in the range of 10 mg/kg has proven to be effective in producing HD-like behavioral deficits *(10,17,18)*. However, the use of progressively higher doses of 3-NPA has also proven to be effective in producing HD-like behavioral deficits, and this procedure allows the experimenter to guard adequately against acute intoxication as well as tolerance *(22)*. Unfortunately, delivery of progressively higher doses of 3-NPA via osmotic pumps or polymer implants has yet to be developed. In lieu of these developments, we and others are using more frequent systemic injections (e.g., twice daily at smaller doses) to produce HD-like deficits without the behavioral aberrations produced by acute intoxication and that allows for flexible adjustment of doses over time. Clearly, more research is needed in this area before an optimal range of doses is established for this animal model of HD.

CONCLUSIONS

As a tool for mimicking important behavioral symptoms of HD, 3-NPA has proven to be very effective. It produces motor abnormalities and cognitive deficits, including working memory deficits, that resemble those produced by HD. The behavioral deficits produced by 3-NPA resemble those found in the later stages of HD, whereas use of QA produces behavioral changes that mimic more of the earlier symptoms of HD. The most accurate simulation of HD is produced when the 3-NPA is administered chronically and systemically. Although the refinement of the transgenic models of HD may soon make neurotoxic models obsolete, the use of 3-NPA presently offers the best simulation of the later stages of HD. Further work is needed to determine optimal dose ranges and to develop delivery mechanisms that can administer a progressive increase in dose over longer time periods. Also, research involving combined treatments of low doses of 3-NPA with other toxins, such as QA, may provide an even better model of HD.

ACKNOWLEDGMENT

This work was supported by NIH Grant 1-R15-NS30694-01A3.

REFERENCES

1. Williams MC, James LF. Toxicity of nitro-containing *Astralagus* to sheep and chicks. J Range Manage 1975;29(4):260–263.
2. James LF, Hartley J, Williams MC, et al. Field and experimental studies in cattle and sheep poisoned by nitro-bearing *Astragalus* or their toxins. Am J Vet Res 1980;41:377–382.
3. Ludolph AC, He F, Spencer PS, et al. 3-Nitropropionic acid-exogenous animal neurotoxin and possible human striatal toxin. Can J Neurol Sci 1991;18:492–498.

4. Lui X, Luo X, Hu W. Studies on the epidemiology and etiology of moldy sugar cane poisoning in China. Biomed Environ Sci 1992;5:161–177.
5. Ming L. Moldy sugar cane poisoning - a case report with a brief review. J Toxicol Clin Toxicol 1995;33:363–367.
6. Beal MF, Brouillet E, Jenkins B, et al. Neurochemical and histologic characterization of striatal excitotoxic lesions produced by the mitochondrial toxin 3-nitropropionic acid. J Neurosci 1993a;13:1481–1492.
7. Beal MF, Brouillet E, Jenkins B, et al. Age-dependent striatal excitotoxic lesion produced by endogenous mitochondrial inhibitor malonate. J Neurochem 1993b;81:1147–1150.
8. Brouillet E, Jenkins BG, Hyman BT, et al. Age-dependent vulnerability of the striatum to the mitochondrial toxin 3-nitropropionic acid. J Neurochem 1993;60:356–369.
9. Brouillet E, Hantraye P, Ferrante R J, et al. Chronic mitochondrial energy impairment produces selective striatal degeneration and abnormal choreiform movement in primates. Proc Natl Acad Sci USA 1995;92:7105–7109.
10. Koutouzis TK, Borlongan CV, Scorcia T, et al. Systemic 3-nitropropionic acid: longterm effects on locomotor behavior. Brain Res 1994a; 646:242–246.
11. Koutouzis TK, Borlongan CV, Freeman TB, et al. Intrastriatal 3-nitropropionic acid: a behavioral assessment. NeuroReport 1994b; 5:2241–2245.
12. Nishino H, Fujimoto I, Shimano Y, et al. 3-Nitropropionic acid produces striatum selective lesions accompanied by iNOS expression. Journal of Chemical Neuroanatomy 1996;10: 209–212.
13. Wüllner U, Young AB, Penney JB, et al. 3-Nitropropionic acid toxicity in the striatum. J Neurochem 1994;63:1772–1781.
14. Gould DH, Gustine DL. Basal ganglia degeneration, myelin alterations, and enzyme inhibition induced in mice by the plant toxin 3-nitropropionic acid. Neuropathol Appl Neurobiol 1982;8:377–393.
15. Gould DH, Wilson MP, Hamar DW. Brain enzyme and clinical alterations induced in rats and mice by nitroaliphatic toxicants. Toxicol Lett 1985;27:83–90.
16. Hamilton BF, Gould DH. Nature and distribution of brain lesions in rats intoxicated with 3-nitropropionic acid: a type of hypoxic (energy deficient) brain damage. Acta Neuropathol 1987;72:286–297.
17. Borlongan CV, Koutouzis TK, Randall TS, et al. Systemic 3-nitropropionic acid: behavioral deficits and striatal damage in adult rats. Brain Res 1995a; 36:549–556.
18. Borlongan CV, Koutouzis TK, Freeman TB, et al. Behavioral pathology induced by repeated systemic injections of 3-nitropropionic acid mimics the motoric symptoms of Huntington's disease. Brain Res 1995b;697:254–257.
19. Marrannes R Wauquier A. Episodic barrel rotations induced by intrastriatal injection of quinolinic acid in rats. Inhibition by anti-convulsants. Pharmacol Biochem Behav 1988;31:153–162.
20. Sanberg PR, Calderon SF, Giordano M, Tew JM, Norman AB. The quinolinic acid model of Huntington's disease: locomotor abnormalities. Exp Neurol 1989;105:45–53.

21. Bazzett TJ, Falik RC, Becker JB, et al. Chronic intrastriatal administration of quinolinic acid produces transient hypermotility in the rat. Brain Res 1994;39:69–73.
22. Palfi S, Ferrante RL, Brouillet E, et al. Chronic 3-nitropropionic acid treatment in baboons replicates the cognitive and motor deficits of Huntington's disease. J Neurosci 1996;16:3019–3025.
23. Diamond R, White RF, Myers RH, et al. Evidence of presymptomatic cognitive decline in Huntington's disease. J Clin Exp Neuropsychol 1992;14:961–975.
24. Foroud T, Siemers E, Kleindorfer D, et al. Cognitive scores in carriers of Huntington's disease gene compared to noncarriers. Ann. of Neurol 1995;37:657–664.
25. Lange KW, Sahakian BJ, Quinn NP, et al. Comparison of executive and visuospatial memory function in Huntington's disease and dementia of Alzheimer type matched for degree of dementia. J Neurol Neurosurg Psychiatry 1995;58:598–606.
26. Lawrence AD, Sahakian BJ, Hodges JR, et al. Executive and mnemonic functions in early Huntington's disease. Brain 1996;119:1633–1645.
27. Levy R, Friedman HR, Davachi L, et al. Differential activation of the caudate nucleus in primates performing spatial and nonspatial working memory tasks. J Neurosci 1997;17:3870–3882.
28. Block F, Kunkel M, Schwarz M. Quinolinic acid lesion of the striatum induces impairment in spatial learning and motor performance in rats. Neurosci Lett 1993;149:126–128.
29. Burešová O, Bureš J, Oitzl MS, et al. Radial maze in the water tank: an aversively motivated spatial working memory task. Physiol Behav 1985; 34:1003–1005.
30. Pitsikas A, Algeri S. Deterioration of spatial and nonspatial reference and working memory in aged rats: protective effects of life-long calorie restriction. Neurobiol Aging 1992;13:369–373.
31. The Huntington's Disease Collaborative Research Group. A novel gene containing a trinucleotide repeat that is expanded and unstable on Huntington's disease chromosomes. Cell 1993;72:971–983.
32. Bossi SR, Simpson JR, Isacson O. Age dependence of striatal neuronal death caused by mitochondrial dysfunction. NeuroReport 1993;4:73–76.
33. Mann DMA, Oliver R, Snowden JS. The topographical distributions of brain atrophy in Huntington's disease and progressive nuclear palsy. Acta Neuropathol 1993;85:553–559.
34. Vonsattel JP, Myers RH, Stevens TJ, et al. Neuropathological classification of Huntington's disease. J Neuropathol Exp Neurol 1985;44:559–577.

6

Comparative Study on 3-Nitropropionic Acid Neurotoxicity

Cesario V. Borlongan, Toru Shimizu, and Paul R. Sanberg

INTRODUCTION

The complex etiologies and mechanisms of cell death associated with, as well as protective/reparative techniques for several neurodegenerative disorders, have been examined using animal models. We have witnessed during the last 5 yr advances in gene knockout animal models and how these animal systems can pave the way for better understanding of human diseases. For example, the discovery of the gene underlying Huntington's disease (HD) opens the possibility that genetic therapy may be the next logical step toward finding a cure for this disease. HD is a progressive debilitating disorder associated with severe degeneration of basal ganglia neurons, especially the intrinsic neurons of the striatum, and characterized by involuntary abnormal choreiform movements and progressive dementia. In this chapter, cross-species models of HD are discussed, which recently received critical attention from researchers interested in the involvement of impaired energy metabolism in the evolution of the disease. The preceding chapters dealt largely with the utility of rodent model of 3-nitropropionic acid (3-NPA), which has been demonstrated as closely resembling the neurobiological and clinical symptoms of the disease. In spite of the many valuable data garnered from the 3-NPA rodent model, there is growing concern that such a model may not fully encompass the human symptoms of HD, and some researchers even have welcomed the idea of concentrating laboratory experiments on nonhuman primate models, i.e., monkeys (Fig. 1). The view of validating laboratory findings in monkeys is strengthened by the fact that many experimental treatment modalities have to be tested in monkeys prior to FDA approval for human clinical use. Although such a hierarchical phylogenetic routine for bringing laboratory science to the clinic appears to dictate our thinking of what is "applied science" or "applicable to humans,"

From: *Mitochondrial Inhibitors and Neurodegenerative Disorders*
Edited by: P. R. Sanberg, H. Nishino, and C. V. Borlongan © Humana Press Inc., Totowa, NJ

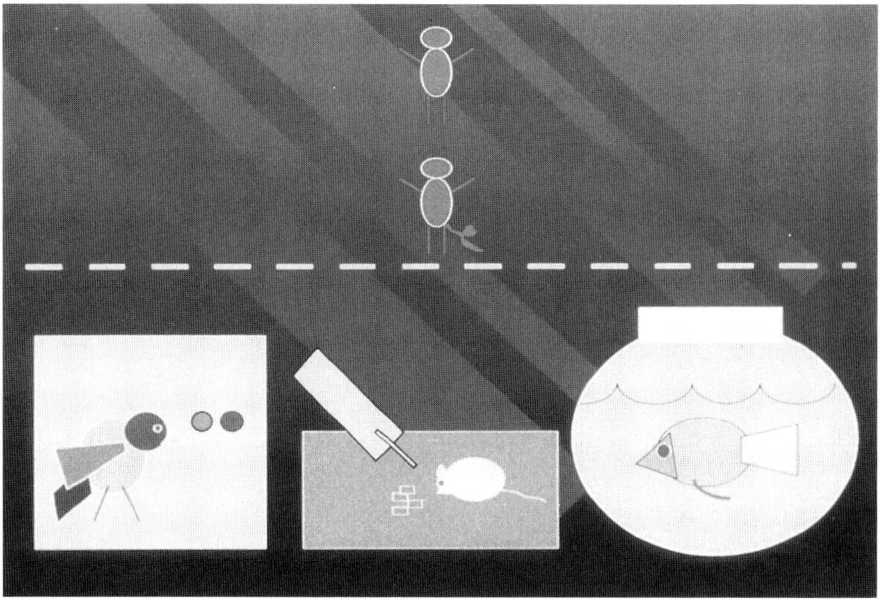

Fig. 1. Hierarchical study of human disorders. The monkey is viewed as the animal that is closest to humans. The *dotted line* demarcates the monkey studies as applicable to humans, whereas other animal studies (pigeon, rat, or goldfish) need to be validated first in monkeys before being accepted as having relevance to clinical applications.

any disregard of the scientific knowledge that we can obtain from experiments with simple, nonprimate animals will obscure our quest for making sense of the "beast" in us. If we limit ourselves to studying closely related species, we will never truly appreciate the beauty of humankind. The convenience of having to relate to a species similar to ours cannot replace the search for the minute details of pathophysiological mechanisms underlying neurodegeneration and regeneration. For example, the faulty firing of a neuron, which can trigger maladaptive behavior, can best be observed in a single-celled organism. Also, for the sake of argument, we need to remember that many Nobel Prize winners devoted years of scientific investigation to nonprimate organisms. The following are included in this group, to name a few: Sir Emil Adolf Behring (the first nobel laureate) studied guinea pigs and cattles to show that diphtheria and other bacteria-mediated diseases can be treated by conferring immunity to the afflicted individual through serum therapy. Sir Henrik Dam (1943 Nobel laureate) conducted pioneering investigations in chicks that led to the discovery of vitamin K. The studies of Sirs

Karl Von Frisch, Konrad Lorenz, and Nikolaas Tinbergen (1973 Nobel co-laureates) on bees, sticklebacks, fighting fish, seagulls, and wild geese formed the basis of their theories on imprinting and animal social behavior that later formed the basis of using methods of ethology in the study of human behavior. Thus, a human disorder may be examined equally well in organisms that otherwise occupy a lower hierarchical phylogenetic status. When we encounter a problem we often ask ourselves: What is the bottom line? The species at the bottom of the phylogenetic hierarchy may well be the best organism through which to find solutions to human diseases. The 3-NPA models of rodent, pigeon, and goldfish are as important as that of the monkey, and can provide us with important scientific data that form the tenets of our appreciation of human order and disorder. We cannot abandon our basic research on simple, nonprimate animals; they provide an elegant scientific resource that can reveal the beauty of human coexistence with other animals.

> Fighting behaviour in fish and bonding behaviour in wild geese soon became the main objects of my research. Looking again at these things with a fresh eye, I realized how much more detailed a knowledge was necessary, just as my great co-laureate Karl von Frisch found new and interesting phenomena in his bees after knowing them for several decades, so, I felt, the observation of my animals should reveal new and interesting facts.
>
> - from The Autobiography of Sir Konrad Lorenz

THE RODENT MODEL OF 3-NPA

We first reported that systemic administration of 3-NPA, an inhibitor of the mitochondrial citric acid cycle, results in a progressive locomotor deterioration in rodents resembling that of HD *(1)* (Fig. 2). In addition, by manipulating the time course of 3-NPA injections, sustained hyperactivity (early HD) or hypoactivity (advanced HD) can be replicated. These data suggest that this animal model can be used to test experimental treatments for HD across different stages of the disease (e.g., neural transplantation as described in Chapter 19). The 3-NPA model offers many technical advances over conventional excitotoxin (ibotenic acid, quinolinic acid, or kainic acid) models of HD. The main advantage of 3-NPA over the excitotoxins is that selective striatal lesions can be accomplished via systemic administration, and therefore the animal is not exposed to surgical trauma associated with intraparenchymal injections of the neurotoxin (Table 1). The systemic 3-NPA model is regarded also as a more improved model than intrastriatal 3-NPA model because, in addition to the former's technical advantages as mentioned previously, systemic 3-NPA administration has higher selectiv-

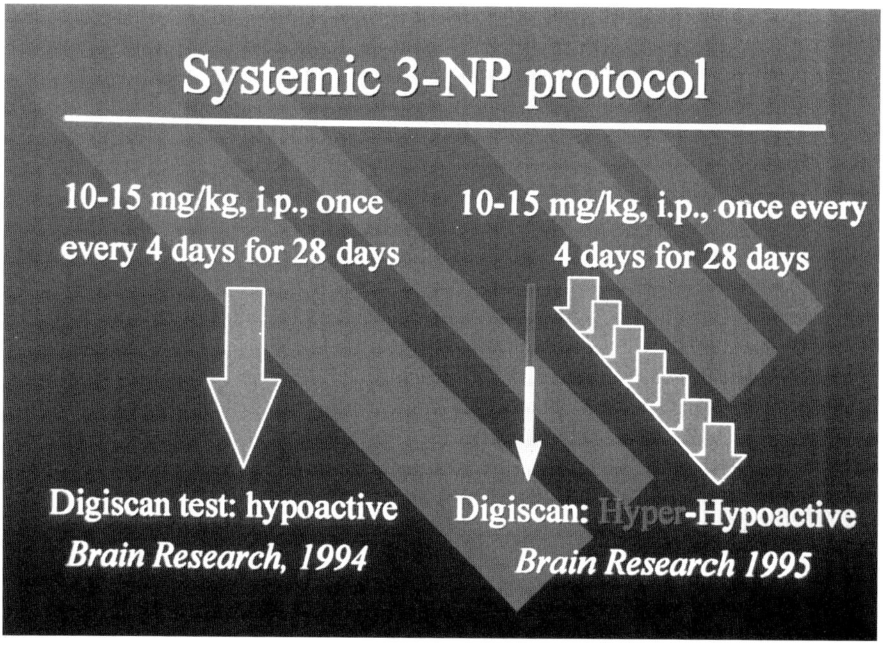

Fig. 2. Progressive behavioral changes induced by chronic systemic 3-NPA. We reported that following chronic injections with 3-NPA (10–15 mg/kkg, i.p., for 28 d), treated animals display significant hypoactivity (*left panel*, adapted from ref. *[3]*), and the same dosing regimen can produce initially hyperactive animals that later would show hypoactivity when tested over time (*right panel*, adapted from ref. *[1]*).

ity in producing striatal lesions compared to intrastriatal 3-NPA injection which produces widespread extrastriatal damage *(1–4)*. In the argument that we recently presented *(5)*, systemic administration of 3-NPA in rodents can parallel the evolution of HD whereas the intraparenchymal injections of excitotoxins (and also noted in intrastriatal 3-NPA infusion) *(3,4)* can only mimic specific stages of the disease. In the case of the excitotoxins, the general observation in treated rats is a significant hyperactivity *(6)*, whereas in the intrastriatally 3-NPA injected rats, there is a significant hypoactivity *(2–4)*. These 3-NPA induced behavioral abnormalities in rats have been noted also in nonhuman primates (monkeys), and in addition the hallmark choreic movements can be demonstrated in monkeys *(7)*. These behavioral differences may be rooted in the inherent species-specific differences and may be due, partly, to the rat being quadruped while the monkeys, like humans, are biped *(1,5)* (Fig. 3). It is possible that the motor coordination

Table 1
Advantages of Systemic 3-NPA Over Intraparenchymal 3-NPA

3-NPA injections: Systemic	vs Intraparenchymal
• Hyper-hypoactive animal: "HD"	• Hypoactive animal: early HD
• Small, localized leason in the dorsolateras aspect of striatum	• Necrotic striatal area, extrastriatal damage: cortex, hippocampus
• Less trauma to the animal	• Tramatic and may interfere with treatment strategies

and body balance in quadrupeds are resistant to moderate levels of 3-NPA neurotoxicity, but they may be easily altered in lesioned biped animals (or in humans). Accordingly, the hyperactivity in rats may well be the correlate of chorea.

Another reason for the incomplete manifestation of chorea in rats may derive from the anatomical differences between rats and monkeys or humans. The homologous structure of caudate–putamen (two distinct structures in monkeys and humans) is the single "striatum" in rats. Because HD primarily comprises putaminal damage and systemic 3-NPA has been suggested to produce similar putaminal damage in monkeys *(7–11)*, the demarcation of the caudate from the putamen seems to inhibit compensatory mechanisms of the caudate to the putamen, and thus chorea is observed. In contrast, because the caudate and the putamen are intertwined in the rat, the putaminal damage induced by 3-NPA may be partially offset by the spared caudate, thus blocking the choreic manifestation. However, even when the striatum is severely damaged in rats that were exposed to intrastriatal 3-NPA, no choreic movements were noted *(2–4)*, suggesting that the elusive initiation of chorea in rats cannot be reproduced by varying the degree of striatal damage.

The observation of progressive motor abnormality in the rodent model of 3-NPA has recently been replicated in the monkey *(10,11)*. Although the absence of chorea in the rat may seem to limit its utility in investigating experimental treatment strategies for amelioration of choreic movements, the availability of computerized analyses (such as the Digiscan system) (Fig. 4) for general locomotor activity in the rat offers a highly objective evaluation of behavioral changes compared to subjective motor examination of monkeys. In addition, because the financial cost for purchasing and housing rats is significantly less than that of the monkeys, validation of results through replications, as well as through large sample size, can be easily conducted in

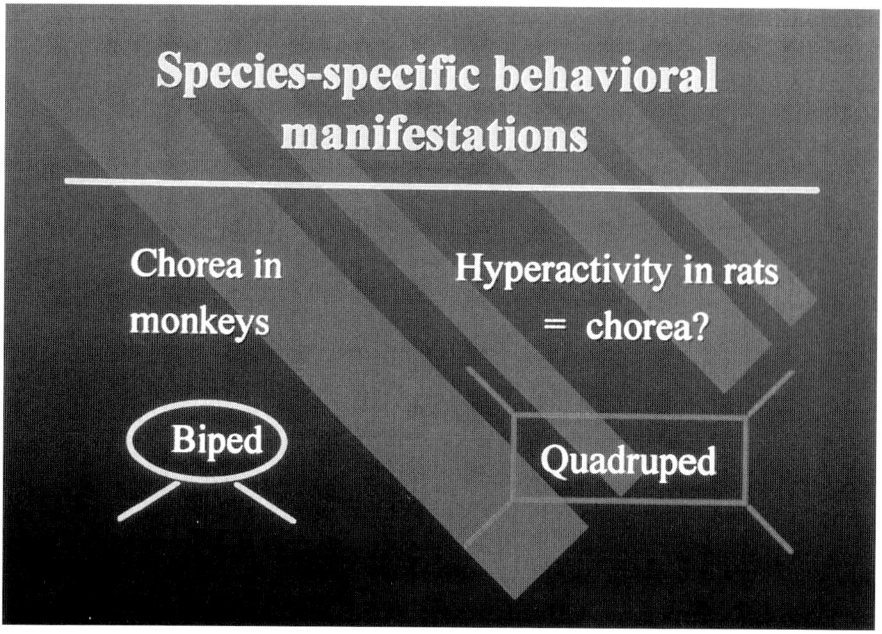

Fig. 3. Chorea as a function of species-specific body balance. Biped animals such as humans and monkeys show chorea, but quadrupeds such as rats have not been documented to display such abnormal movement. Compared to bipeds, the more distributed body balance in quadrupeds may exert resistance against neurotoxicity that can induce chorea.

rats, whereas experiments with monkeys are generally limited to a few animals. Both rodent and monkey models have also been used to examine 3-NPA-induced impairments in cognitive tasks *(10–13)*. Likewise, because many already validated cognitive (learning and memory) tasks are available for the rodent, objective analyses and replication of experiments can be conducted in a reliable manner. Thus, the rodent model of 3-NPA, like the monkey model, has its limitations, but the advantages it offers more than compensate for valid extrapolations of laboratory data to the clinic.

THE PIGEON MODEL

Because of the biped mode of the pigeon and our desire to observe the choreic movements hypothesized to occur exclusively in biped animals, we initiated investigations on the effects of 3-NPA on this animal. Our interest in neural transplantation also prompted us to investigate the pigeon as a better model of neurodegeneration and as a subsequent platform for trans-

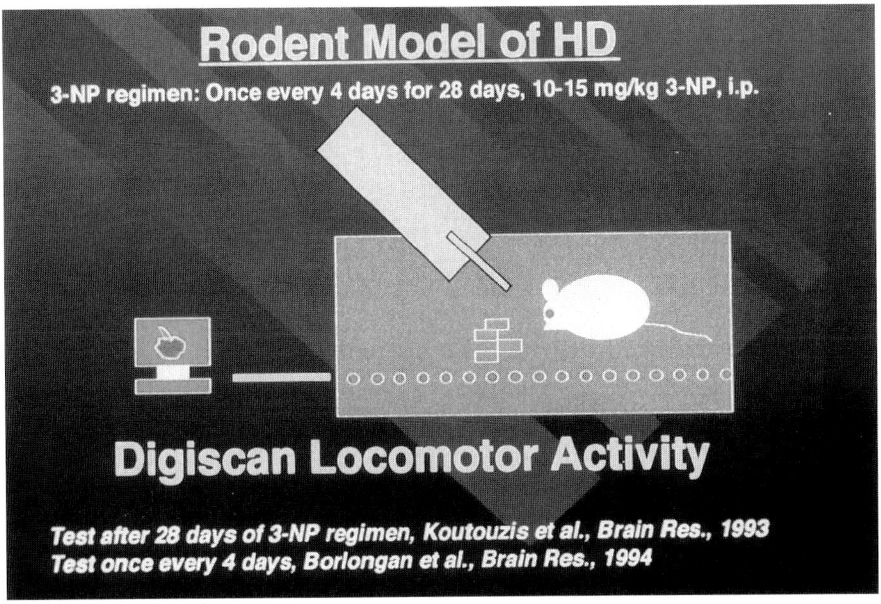

Fig. 4. Digiscan apparatus setup for testing rat locomotor behavior.

plantation therapy. Of note, one report used the chicken as a model of Parkinsonism and the same model has been used to examine the effects of neural transplantation *(14)*. Finally, the availability of more sophisticated learning and memory tasks in pigeons should provide more interesting data on the dementia or cognitive impairment that is a major index of HD, as well as other neurodegenerative disorders.

Our pigeon model consisted of acutely injecting 3-NPA (30 mg/kg, i.p.) and placing the animal in a Skinner box (Fig. 5). A digital camera was used to videotape the behavioral patterns of the pigeon for a period of 1 h following the toxin injection. We noted tremor and hyperkinetic movements in the upper body (characterized by stereotypic head tilting and shaking), accompanied by a wobbly gait and rigidity in the lower body (legs and feet) *(13)* (Fig. 6). However, these behavioral abnormalities were observed only during the first hour following 3-NPA injection and the animal appeared normal thereafter. Pigeons that were injected twice (30 mg/kg, i.p., once daily for 2 consecutive days) also displayed the same behavioral alterations, but also recovered after the second injection. The results indicate that the pattern of shaking and head tilting in pigeons treated with 3-NPA seems to resemble the choreic movements of HD.

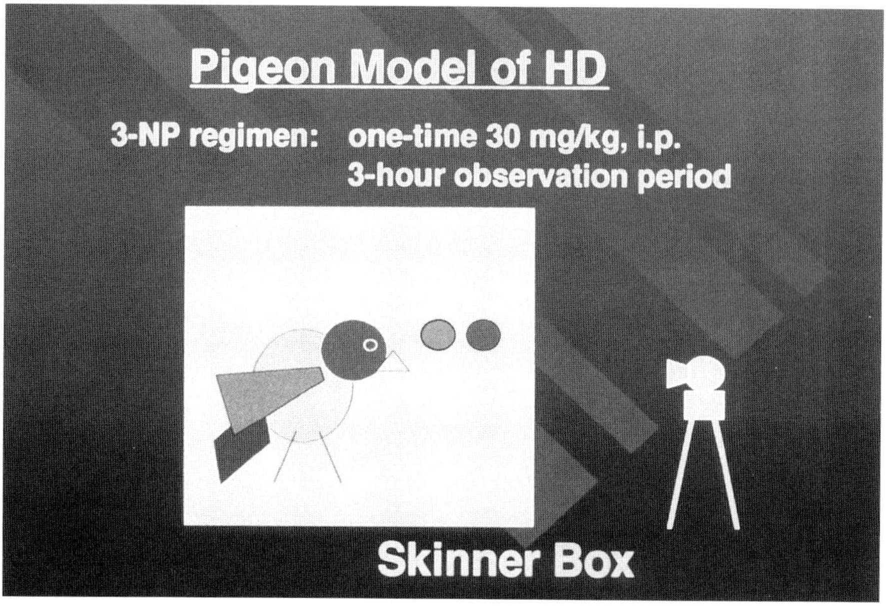

Fig. 5. Apparatus for testing pigeon behavior.

Anatomical investigation of the brains from pigeons treated with 3-NPA revealed that the area parahippocampalis (APH) and paleostriatum showed significant cell loss. Of note, the avian hippocampus, including the APH and paleostriatum, has been suggested to be the homologue of the mammalian basal ganglia *(15)*. Interestingly, we also noted similar basal ganglia damage in rats exposed to middle cerebral artery occlusion (an experimental model of stroke) *(16)*. Because of overlapping mechanisms (free radical formation, oxidative stress) underlying many neurodegenerative disorders, it is not surprising to see similar anatomical substrates affected by these diseases.

Long-term studies in pigeons using chronic 3-NPA injection are now underway. Similar to rat experiments, the cost of setting up pigeon studies is less than that of monkey studies. Compared to rats and monkeys, the pigeon has been the primary animal choice for investigating cognitive behavior.

THE GOLDFISH MODEL

The goldfish has been used as a model of Parkinson's disease *(17,18)*. It was demonstrated that goldfish show parkinsonian syndrome following 1-methyl-4-phenyl-1,2,3,6-tetrahydropyridine (MPTP) injection, and that

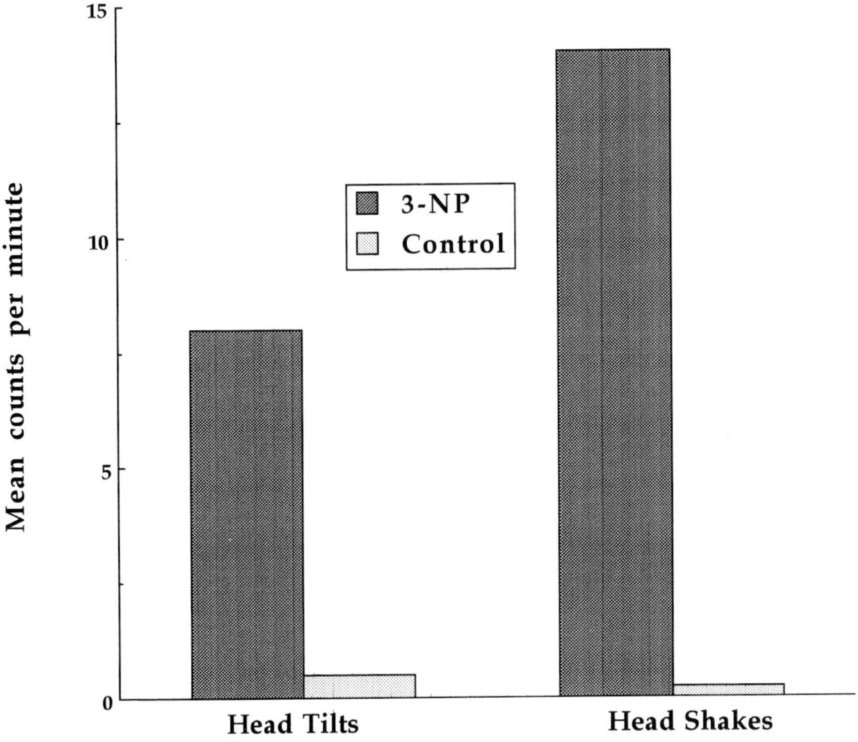

Fig. 6. Choreic-like movements in pigeons. Tremor and hyperkinetic movements (head shaking and tilting) in the upper body and rigidity in the lower body were noted.

the anti-Parkinsonian drugs L-dopa or L-deprenyl block the MPTP toxicity *(19)*. Recently, it was shown that a specific brain area, the nucleus pars medialis, is selectively destroyed in MPTP-treated goldfish, which may be the homologue of the mammalian substantia nigra *(20)*. We became aware of these interesting studies after our pilot studies with our own goldfish model of HD. We initially based our experiments on a 3-NPA model of goldfish model in an attempt to demonstrate choreic movements. As stated earlier, our working hypothesis is that bipeds may show chorea more than quadrupeds. We previously tested 3-NPA treated rats in a swimming pool to negate the effect of its being a quadruped. Following chronic injections of 3-NPA (10–15 mg/kg once every 4 d or 28 d, i.p.), the rats were placed in the swimming pool and observed for about 10 min. The 3-NPA treated rats sank many more times than normal control rats, but choreic movements were

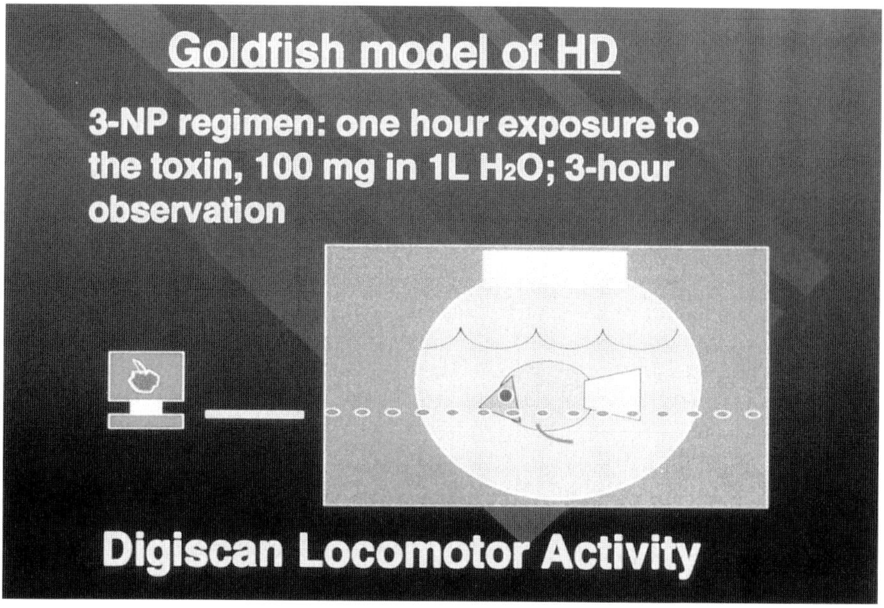

Fig. 7. Apparatus for testing goldfish behavior. Following the 3-NPA dosing regimen, the aquarium was placed inside the Digiscan activity chamber overnight.

not observed. Because swimming is not a natural behavior for the rats, we then proceeded to observe a more appropriate species, the goldfish.

Goldfish were exposed to 3-NPA (1 h in an aquarium filled with 1 L of water and 100 mg of 3-NPA; Fig. 7). After this treatment regimen, the animal was transferred to another aquarium (3-NPA free) that was placed in the Digiscan apparatus. A 3-h test period was allowed and general locomotor activity data were recorded. The goldfish exposed to 3-NPA displayed a transient behavioral pathology similar to that seen in 3-NPA treated rats, in that the goldfish showed an initial hyperactivity (early HD) from 1 h to 3 h post-3-NPA exposure, a normalization of motor behavior at about 4 h post-exposure, and significant hypoactivity from 5 h to 12 h post-exposure (Fig. 8). Interestingly, the 3-NPA treated goldfish reverted to near normal locomotor activity by 24 h post-exposure.

The reversal of behavioral abnormalities in 3-NPA treated goldfish may be explained by the highly regenerative capability of goldfish neurons. In the goldfish, the development of neurons continues well into adulthood *(21)*, and when injured, the central nervous system (CNS) has the capacity to regenerate new neurons spontaneously *(22)*. Indeed, it has recently been docu-

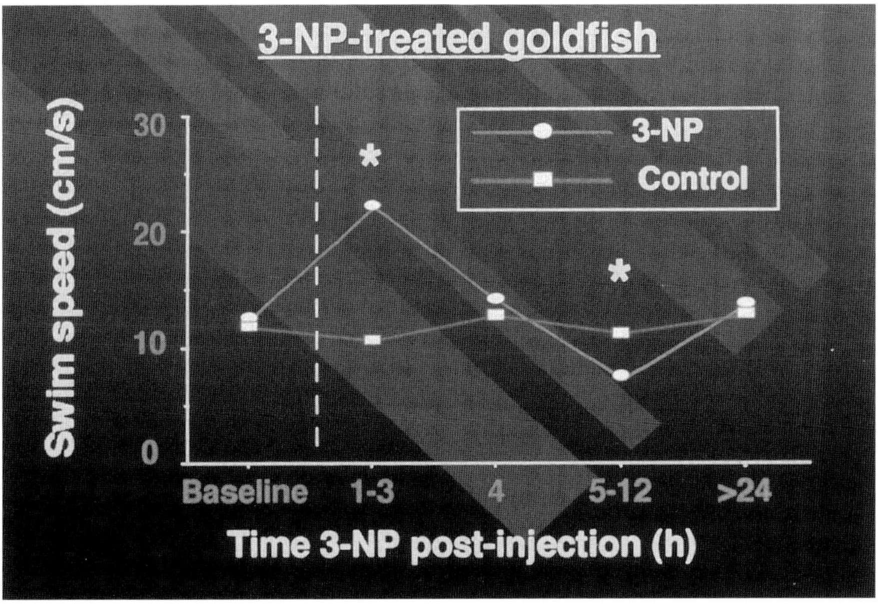

Fig. 8. Progressive locomotor activity changes in the goldfish. A similar pattern of hyperactivity, transient normalization, and hypoactivity was observed over a period of 24 h following exposure of the goldfish to 3-NPA. The recovery of behavior after 24 h was probably due to the capability of the goldfish CNS to regenerate neurons.

mented that optic tectal cells in the adult goldfish spontaneously regenerate after injury, and that myosin light-chain kinase may mediate such neuronal recovery *(23)*, and the administration of protein kinase C inhibitors prevents such regeneration *(24)*. Many mechanisms thought to underlie the regenerative capacity of goldfish neurons have been postulated, including the expression of the goldfish homologue of the *Notch* gene, which is a developmental signaling molecule in neuroepithelial cells *(21)* and the expression of E587 antigen, which is a cell recognition molecule normally present during development for regulation of axon–glial interactions *(25)*.

It would be interesting to examine whether chronic injections of 3-NPA in the goldfish can cause an overload of the toxin in the CNS which in turn can trigger a breakdown in the capacity of the neurons to regenerate. This is an interesting rationale of why the goldfish model may offer insights to processes of neuronal degeneration and regeneration, in that we can learn the threshold of neurotoxicity required to induce degeneration, as well as the factors involved in protection against neurotoxicity.

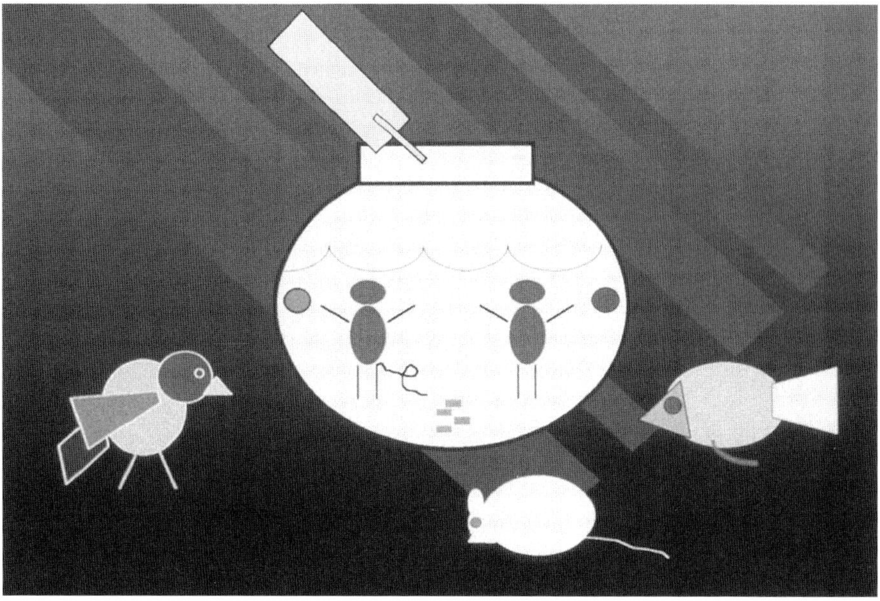

Fig. 9. A different view of human coexistence with animals.

SUMMARY

Many of the studies presented here are preliminary data, but they are proof of cross-species similarities in behavior and in cellular vulnerability to CNS injury. The existence of differences across species, such as the absence of chorea in the rat or the regenerative capacity of the goldfish neurons, should not be viewed as pure limitations of the animal models, but rather they should be exploited to reveal important factors and mechanisms that led to their being resistant to neurotoxicity. Neurodegeneration may be best understood if man tries to see the beast within him (Fig. 9).

Man imitates man. This is an orderly conduct of a disorder.

- Anonymous

REFERENCES

1. Borlongan CV, Koutouzis TK, Freeman TB, Cahill DW, Sanberg PR. Behavioral pathology induced by repeated systemic injections of 3-nitropropionic acid mimics the motoric symptoms of Huntington's disease. Brain Res 1996a;697:254–257.

2. Borlongan CV, Koutouzis TK, Freeman TB, Cahill DW, Sanberg PR. Systemic 3-nitropropionic aicd: behavioral deficits and striatal damage in rats. Brain Res Bull 1995;36:549–556.
3. Koutouzis TK, Borlongan CV, Scorcia TA, Sanberg PR. Systemic 3-nitropropionic acid: long-term effects on locomotor behavior. Brain Res 1994a; 646:242–246.
4. Koutouzis TK, Borlongan CV, Freeman TB, Sanberg PR. Intrastriatal 3-nitropropionic acid: a behavioral assessment. NeuroReport 1994b;5:2241–2245.
5. Borlongan CV, Nishino H, Sanberg PR. Systemic but not intraparenchymal administration of 3-nitropropionic acid mimics the neuropathology of Huntington's disease: a speculative explanation. Neurosci Res 28:185–189, 1997.
6. Sanberg PR, Calderon SF, Girodano M, Tew JM, Norman AB. The quinolinic acid model of Huntington's disease: Locomotor abnormalities. Exp Neurol 1988;105:45–53.
7. Brouillet E, Hantraye P, Ferrante RJ, Dolan R, Leroy-Willig A, Kowall NW, Beal MF. Chronic mitochondrial enegy impairment produces selective striatal degeneration and abnormal choreiform movements in primates. Proc Natl Acad Sci USA 1995;92:7105–7109.
8. Brouillet E, Jenkins BG, Hyman BT, Ferrante RJ, Kowall NW, Srivastava R, Roy DS, Rosen BR, Beal MF. Age-dependent vulnerability of the striatum to the mitochondrial toxin 3-nitropropionic acid. J Neurochem 1993;60:356–359.
9. Ferrante RJ, Hantraye P, Brouillet E, Kowall NW, Beal MF. Striatal pathology of impaired mitochondrial metabolism in primate profiles of Huntington's disease. Soc Neurosci Abstr 1993;19:408.
10. Palfi S, Ferrante RJ, Brouillet E, Beal MF, Dolan R, Guyot MC, Peschanski M, Hantraye P. Chronic 3-nitropropionic acid treatment in baboons replicates the cognitive and motor deficits of Huntington's disease. J Neurosci 1996;16: 3019–3025.
11. Palfi S, Riche D, Brouillet E, Guyot MC, Mary V, Wahl F, Peschanski M, Stutzmann JM, Hantraye P. Riluzole reduces incidence of abnormal movements but not striatal cell death in a primate model of progressive striatal degeneration. Exp Neurol 1997;46:135–141.
12. Dong J, Shear DA, Gundy CD, Creguer-Haik K, Dunbar GL. Behavioral and histological comparisons of two animal models of Huntington's disease. Soc Neurosci Abstr 1996;22:227.
13. Borlongan CV, Koutouzis TK, Poulos SG, Saporta S, Hauser RA, Cahill DW, Freeman TB, Sanberg PR. Bilateral intrastriatal transplantation of rat fetal striatal tissue into the 3-nitropropionic acid rat model of Huntington's disease. Soc Neurosci 1996;22:580.
14. Yanai J, Silverman WF, Shamir D. An avian model for the reversal of 6-hydroxydopamine induced rotating behaviour by neural grafting. Neurosci Lett 1995;187:153–156.
15. Butler AB, Hodos W. Comparative Vertebrate Neuroanatomy Evolution and Adaptation. Wiley-Liss, New York, 1996.
16. Borlongan CV, Koutouzis TK, Jorden JR, Martinez R, Rodriguez AI, Poulos SG, Freeman TB, McKeown P, Cahill DW, Nishino H, Sanberg PR. Neural

transplantation as an experimental treatment modality for cerebral ischemia. Neurosci Biobehav Rev 1997;21:79–90.
17. Youdim MB, Dhariwal K, Levine M, Markey CJ, Markey S, Caohuy H, Adeyemo OM, Pollard HB. MPTP-induced "Parkinsonism" in the goldfish. Neurochem Int 1992;20(Suppl):275–278.
18. Pollard HB, Dhariwal K, Adeyemo OM, Markey CJ, Caohuy H, Levine M, Markey S, Youdim MB. A parkinsonian syndrome induced in the goldfish by the neurotoxin MPTP. FASEB J 1992;6:3108–3016.
19. Adeyemo OM, Youdim MB, Markey SP, Markey CJ, Pollard HB. L-deprenyl confers specific protection against MPTP-induced Parkinson's disease-like movement disorder in the goldfish. Eur J Pharmacol 1993;240:185–193.
20. Pollard HB, Kuijpers GA, Adeyemo OM, Youdim MB, Goping G. The MPTP-induced Parkinsonian syndrome in the goldfish is associated with major cell destruction in the forebrain and subtle changes in the optic tectum. Exp Neurol 1996;142:170–178.
21. Sullivan SA, Barthel LK, Largent BL, Raymond PA. A goldfish Notch-3 homologue is expressed in neurogenic regions of embryonic adult and regenerating brain and retina. Dev Genet 1997;20:208–223.
22. Hanna GF, Nawar NN, Sharma SC. Regeneration of ascending spinal axons in goldfish. Brain Res 1998;791:235–245.
23. Jian X, Szaro BG, Schmidt JT. Myosin light chain kinase: expression in neurons and upregulation during axon regeneration. J Neurobiol 1996;31:379–391.
24. Heacock AM, Agranoff BW. Protein kinase inhibitors block neurite outgrowth from explants of goldfish retina. Neurochem Res 1997;22:1179–1185.
25. Giordano S, Laessing U, Ankerhold R, Lottspeich F, Stuermer CA. Molecular characterization of E587 antigen: an axonal recognition molecule expressed in the goldfish central nervous system. J Comp Neurol 1997;377:286–297.

7

Mechanisms of 3-Nitropropionic Acid Neurotoxicity

James W. Geddes, Vimala Bondada, and Zhen Pang

INTRODUCTION

3-Nitropropionic acid (3-NPA) was originally identified as a component of moldy sugarcane, produced by a fungus, *Arthrinium* sp. Ingestion of the mildewed sugarcane results in motor impairment and muscle weakness that is neither progressive nor reversible, and is accompanied by bilateral damage to the striatum *(1,2)* (*see* also Chapters 1, 2, 5, 10, and 12). In rats, repeated systemic administration of 3-NPA results in motor impairment and consistent morphologic injury to striatum (Fig. 1). Hippocampal damage is less consistent, neocortical damage is infrequent, and the cerebellum is spared *(3–5)*. The striatal damage induced by repeated systemic 3-NPA resembles many of the neuropathologic aspects of Huntington's disease *(6,7)*, (*see also* Chapters 2, 5, 10, and 12), particularly when low doses of 3-NPA are used *(8)*. In addition, the regional damage caused by chronic 3-NPA resembles that produced by severe global ischemia *(9)* and hypoglycemia *(10)*. Thus, 3-NPA is a useful tool for understanding neuronal damage resulting from late-onset neurodegenerative disorders such as Huntington's disease, as well as acute insults such as ischemia and hypoglycemia.

The major neurochemical action of 3-NPA is irreversible inhibition of the mitochondrial enzyme succinate deydrogenase *(11)*, present in the tricarboxylic acid cycle and a component of complex II of the electron transport chain (*see also* Chapter 3). This well-defined action of 3-NPA contrasts with many other mechanisms used to induce degeneration such as anoxia, hypoxglycemia, and β-amyloid, in which the primary mechanisms are complex or less well understood. Adding to the utility of 3-NPA as a tool for understanding neurodegeneration is that this toxin can be administered systemically, intracerebrally, or utilized in vitro. In contrast, ischemia is diffi-

From: *Mitochondrial Inhibitors and Neurodegenerative Disorders*
Edited by: P. R. Sanberg, H. Nishino, and C. V. Borlongan © Humana Press Inc., Totowa, NJ

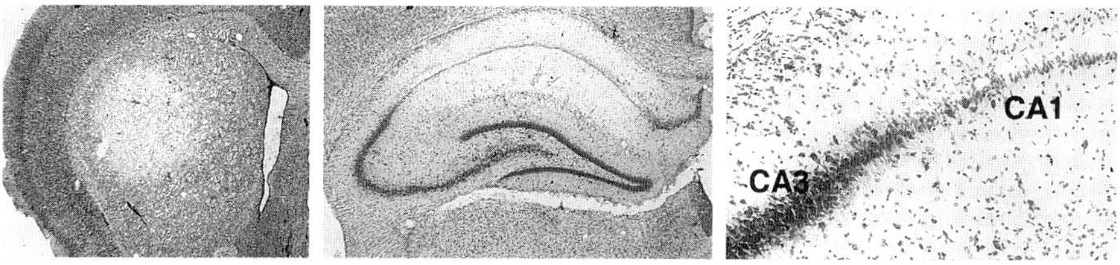

Fig. 1. Cresyl violet staining of hippocampus and striatum following systemic 3-NPA. In a rat treated with 3-NPA (20 mg/kg daily for 5 d), cresyl violet staining indicates neuron loss in the dorsolateral striatum and CA1 of hippocampus.

cult to produce in vitro, and β-amyloid is toxic in vitro *(12)*, but often does not cause neurodegeneration in vivo *(13)*.

Despite knowing the primary action of 3-NPA, it is unclear why repeated systemic administration results in preferential damage to striatum. The striatal vulnerability is not the result of an enrichment of succinate dehydrogenase in this location *(14)*, and systemic administration of 3-NPA does not result in a greater impairment of succinate dehydrogenase activity in striatum vs other brain regions *(11,15)*. One other simple explanation could be a greater sensitivity of striatal neurons to the effects of this mitochondrial toxin. However, cultured striatal, hippocampal, septal, and hypothalamic neurons are similarly sensitive to 3-NPA *(16)*. In addition, hippocampus is as sensitive as striatum to direct injection of a related toxin, malonate, a reversible inhibitor of succinate dehydrogenase *(14)*. Another possible mechanism for the striatal vulnerability to 3-NPA is selective damage to the striatal blood–brain barrier *(17–19)*. This is discussed in Chapters 5, 8, and 11.

INDIRECT EXCITOTOXIC MECHANISMS

An important clue to the mechanism underlying 3-NPA toxicity was the observation by Hamilton and Gould *(17)* that the neuronal damage produced by systemic 3-NPA resembled that produced by kainic acid, sustained stimulation of the perforant path, hypoglycemia, and ischemia. They proposed an excitotoxic mechanism. Excitotoxicity refers to the overstimulation of neurons with glutamate, or toxins such as kainate, resulting in neuronal cell death *(20)*. Excitotoxicity is largely mediated by ionotropic glutamate receptors, with activation of *N*-methyl-D-asparate (NMDA) receptors resulting in more rapid neuron death than non-NMDA receptors *(21,22)*. In cultures, the neuron death involves a rapid swelling, mediated by the influx of

Na^+ and Cl^-, and a delayed component resulting from Ca^{2+} influx *(21)*. Although excitotoxin-induced neuron death is typically necrotic (or oncotic *[23]*), recent studies also indicate an apoptotic component *(24,25)*. The contribution of apoptotic mechanisms to excitotoxic neuron death is a subject of considerable interest and debate *(26–29)*.

In vivo, excitotoxic insult results in the death of neurons whose cell bodies and dendrites are located near the excitotoxin, but a sparing of axons passing through the region *(30)*. Neurons containing nicotinamide adenine dinucleotide phosphate diaphorase (NADPH-diaphorase, recently identified as nitric oxide synthase) are also relatively resistant to excitotoxic insult, although these cells may also be resistant to apoptosis *(31)*. Excitotoxic mechanisms are thought to underlie neuronal damage following acute insults, such as seizures and hypoxia–ischemia, in which neurons are exposed to high levels of extracellular glutamate *(32,33)*.

The idea that excitotoxic neuron death could also be triggered indirectly, in the absence of a large elevation in extracellular glutamate, was first proposed by Henneberry and colleagues *(34)*. Their "indirect" excitotoxic hypothesis proposes that impairment of energy metabolism results in the loss of ATP, inability to maintain ion pumps (Na^+, K^+-ATPase), membrane depolarization, and removal of the voltage-dependent Mg^{2+} block of the NMDA receptor. This allows glutamate to chronically activate the NMDA receptor and results in "excitotoxic" neuronal damage. Subsequent studies by Zeevalk and Nicklas provided further evidence that neuronal damage following metabolic impairment is mediated via NMDA receptor activation and excitotoxic mechanisms *(35,36)*.

In vivo, systemic 3-NPA does not result in an elevation of extracellular glutamate levels in the CNS *(4,37)*, yet much of the neuronal damage is thought to occur by excitotoxic mechanisms. Evidence includes neuron morphology *(3)*, the relative sparing of NADPH-diaphorase neurons *(4)*, protection against striatal 3-NPA toxicity by prior decortication *(4,37)*, and evidence of poly(ADP-ribosyl)ation in striatal neurons *(38)*. Moreover, systemic 3-NPA exacerbates the toxicity of intrastriatal NMDA administration *(39)*. The indirect excitotoxic hypothesis helps to explain the similarity between neuronal damage induced by repeated systemic 3-NPA administration and acute excitotoxic insults such as kainic acid toxicity and ischemia (*see* ref. *[3]*). Moreover, the striatal damage produced by systemic 3-NPA resembles that produced by the intrastriatal injection of excitotoxins *(6)*.

In addition to the evidence of excitotoxic/necrotic neuron death, there are also biochemical and morphological results suggestive of apoptosis in striatum following systemic 3-NPA *(38,40)*.

It is not known if NMDA antagonists protect against neuronal toxicity following repeated systemic administration of 3-NPA. However, (+)-5-methyl-10,11-dihydro-5*H*-dibenzo[*a,d*]cylohepten-5,10-imine hydrogen maleate (MK-801, dizocilpine maleate, a noncompetitive NMDA antagonist) is ineffective against intrastriatal administration of 3-NPA, presumably due to the strength of the metabolic impairment resulting in the involvement of non-NMDA mechanisms *(4)*. MK-801 does attenuate lesion size following intrastriatal, but not intrahippocampal, malonate injection *(14)*.

In vitro, there is also evidence in support of the indirect excitotoxic hypothesis for 3-NPA toxicity. 3-NPA exacerbates NMDA toxicity, particularly when Mg^{2+} is removed from the extracellular medium *(41)*. MK-801 is neuroprotective in cultured neurons *(16,42,43)*, although it may delay rather than prevent the neuron death *(42)*. Even in non-neural cells that lack NMDA receptors, 3-NPA toxicity is thought to result from similar mechanisms including failure of the energy-dependent plasma membrane Na^+/K^+ pump, resulting in swelling and ultimately lysis of the cell *(44)*. However, in one study NMDA antagonists did not protect against 3-NPA toxicity in cultured striatal and cortical neurons, and the morphology of the cell death was indicative of apoptosis *(45)*.

The above results indicate that although there is considerable support for the indirect excitotoxic hypothesis, it does not fully account for 3-NPA toxicity in vivo or in vitro. The density and subtypes of NMDA receptors *(46)* and expression of glutamate transporters *(47)* do not explain the striatal vulnerability. Antagonism of NMDA receptors, using MK-801, does not block the toxicity of intrastriatal 3-NPA injection *(6)*. The striatal lesion resulting from systemic 3-NPA administration consists of a necrotic core in which both neurons and glia are absent *(48)*. In cultured cerebellar granule neurons, NMDA antagonists delay but do not prevent 3-NPA toxicity *(42)*. NMDA antagonists also do not reduce 3-NPA-induced neuron death in cultured striatal and cortical neurons *(45)*.

Recent results shed some light on these discrepancies. In primary cultures of hippocampal neurons, 3-NPA (5 m*M*) induces both a rapid necrotic and delayed apoptotic neuronal death *(49)*. Increasing levels of extracellular glutamate favor necrosis. Antagonism of NMDA receptors (MK-801, APV) or non-NMDA receptors (1,2,3,4-tetrahydro-6-nitro-2,3-dioxo-benzo(f)-quinoxaline-7-sulfonamide [NBQX]) prevents the necrotic cell death, but many of the neurons then die via apoptosis. The 3-NPA-induced apoptosis can be blocked by cycloheximide, but not by NMDA (or non-NMDA) antagonists (Fig 2). These results demonstrate that 3-NPA can lead to rapid necrotic death involving NMDA receptors, or slow apoptotic death independent of NMDA receptors. However, they do not fully explain the regional damage produced by systemic 3-NPA administration.

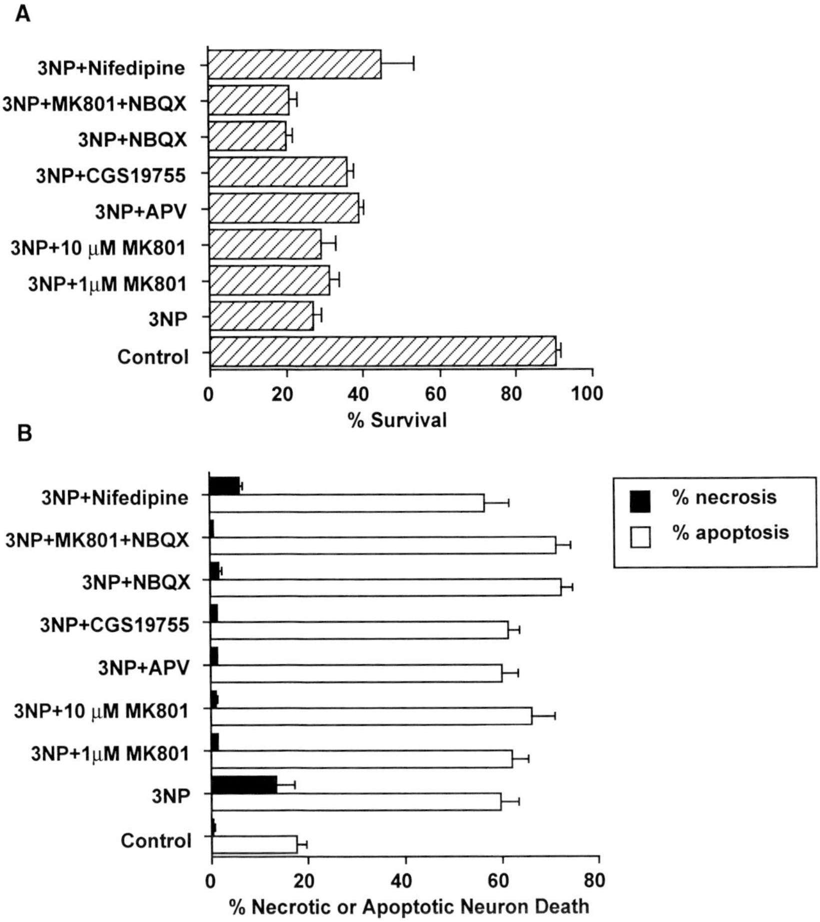

Fig. 2. Effects of glutamate receptor and L-type calcium channel blockade on 3-NPA-induced cell death. Primary cultures of hippocampal neurons were treated with 3-NPA in the presence of various antagonists. The drug concentrations were 5 m*M* 3-NPA, 100 μ*M* APV, 10 μ*M* CGS19755, 50 μ*M* NBQX, 1 and 10 μ*M* MK-801, and 50 μ*M* nifedipine. The percentages of necrotic cells **(A)**, percentages of apoptotic cells **(B)**, and neuronal survival were quantified at the 48-h time point. The results are the means ± SEM ($n = 6$). (*$p < 0.05$; **$p < 0.01$, as compared to cultures treated with 3-NPA alone.)

HOW DOES 3-NPA CAUSE APOPTOSIS?

Although progressive mitochondrial impairment would ultimately kill all cells, the fact that cycloheximide protects against 3-NPA-induced apoptosis indicates that activation of apoptotic cell death pathways accelerates the cell death. 3-NPA produces a gradual decline in neuronal ATP levels *(49)*. The affinity of the Na^+, K^+-ATPase for ATP is in the nanomolar range, and a large decline in cellular ATP levels is necessary for enzyme activity to be affected. In contrast, kinases are much more vulnerable to a loss of ATP. Thus, a decline in ATP would affect cellular kinase activity prior to inhibiting ion pumps. Stauropsorine, a broad-spectrum kinase inhibitor, effectively induces apoptosis in both neural and non-neural cells *(50)*. 3-NPA-induced loss of ATP is likely to impair kinase activity and trigger apoptosis prior to the impairment of the Na^+, K^+-ATPase and activation of excitotoxic mechanisms. In support of this hypothesis, the pattern of cytoskeletal disruption induced by 3-NPA (in the presence of MK-801) is very similar to that caused by staurosporine (unpublished results).

ROLE OF OXIDATIVE STRESS

Oxidative stress can result in necrosis or apoptosis, depending on the severity of the insult. Following 3-NPA administration, oxidative stress could occur as a direct result of inhibition of the electron transport chain or secondary to excitotoxicity *(51–54)*. In vitro, it is difficult to detect increased oxidative stress in neurons treated with 3-NPA (Fig. 3). In contrast, there is extensive in vivo evidence of increased free radical formation in rats treated with 3-NPA *(55,56)*, and 3-NPA toxicity is attenuated in copper/zinc superoxide dismutase transgenic mice *(57)*. However, the free radical spin trap alpha-phenyl-*n*-tert-butyl-nitrone (PBN) does not protect against 3-NPA toxicity *(56,58)*, although other antioxidant treatments such as 5,5-dimethyl–1-pyrroline-*n*-oxide (DMPO) may be effective *(56)*.

Together, these results suggest that the increase in oxidative stress detected in vivo is not the result of inhibition of the electron transport chain, but may be secondary to excitotoxic insult or result from an inflammatory response secondary to the neuronal damage. Many studies demonstrate that proinflammatory cytokines exacerbate ischemic damage *(59,60)*, in part through oxidative mechanisms such as increased expression of nitric oxide synthase. The possible role of cytokines in 3-NPA-induced neuronal damage has not been examined, although there is evidence that 3-NPA induces an inflammatory response. 3-NPA-induced striatal lesions are characterized by a necrotic core with neutrophil infiltration *(18)*, expression of inducible nitric oxide synthase *(61)*, and immunoreactivity of serum/immune complement factors (C3b/C4B4) *(18)*. Observations from our laboratory further indicate that levels of tumor necrosis factor-α (TNF-α), a proinflammatory cytokine, are elevated following 3-NPA-induced striatal degeneration (Fig. 4). This is consistent with the hypothesis that much of the oxidative

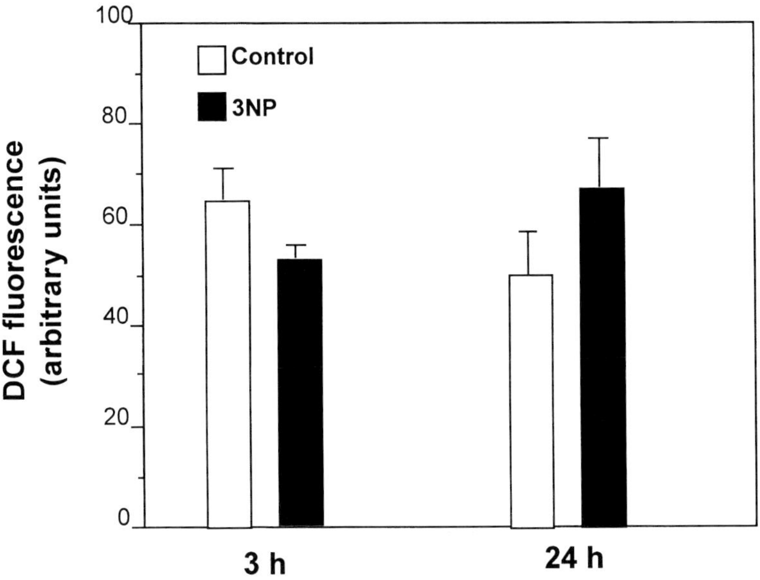

Fig. 3. Minimal oxidative stress following 3-NPA administration in vitro. Primary neuronal cultures were treated with 5 m*M* 3-NPA for 3 h or 24 h. Prior to treatment, cells were incubated for 50 min in the presence of 50 m*M* 2,7-dichlorofluorescin diacetate (DCF; Molecular Probes, Eugene, OR), and then washed three times (2 mL per wash) in Hank's balanced salt solution containing 10 m*M* HEPES and 10 m*M* glucose. Cells were imaged using a confocal laser scanning microscope. The results indicate that 3-NPA does not result in a significant increase in reactive oxygen species detected by the DCF assay.

stress detected in vivo following 3-NPA administration results from inflammatory mechanisms secondary to the initial neuronal damage.

ROLE OF DOPAMINE

As outlined previously, 3-NPA administration can result in rapid necrotic or slow apoptotic death, both in vivo and in vitro. Ionotropic glutamate receptors, particularly NMDA receptors, mediate the necrotic but not the apoptotic death. However, these results do not explain the preferential vulnerability of striatum to neuronal damage following systemic 3-NPA administration. The striatum is densely innervated by dopaminergic fibers that originate in the substantia nigra. Several lines of evidence indicate that dopamine may exacerbate excitotoxicity in striatum. Dopamine levels are increased in the striatum during and following forebrain ischemia *(62,63)*

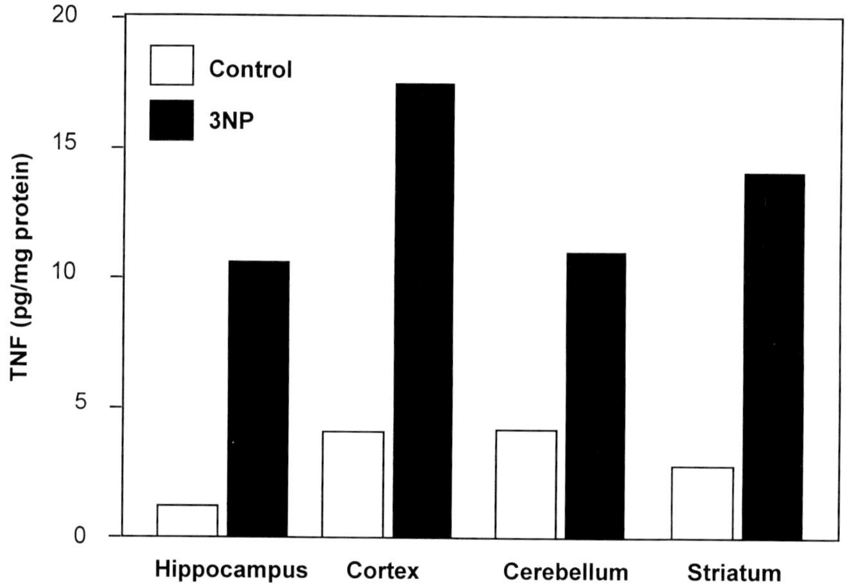

Fig. 4. 3-NPA elevates TNF-α levels in vivo. TNF-α levels were determined using an enzyme-linked immunosorbent assay (ELISA) kit specific for rat TNF-α (BioSource). In a control and a rat treated with 3-NPA (20 mg/kg/d for 5 d), TNF-α levels were determined in the hippocampus, cortex, cerebellum, and striatum. The results demonstrate that the elevation in TNF-α is not restricted to expected vulnerable brain regions (hippocampus and striatum). A similar profile of control and elevated TNF-α levels is observed following transient global ischemia *(59)*.

and substantia nigra lesions markedly attenuate the striatal damage resulting from ischeima *(64)*. Moreover, dopamine depletion attenuates striatal damage induced by both NMDA and non-NMDA receptor agonists *(65–67)*.

To determine if the dopaminergic nigrostriatal pathway contributes to the vulnerability of striatum to metabolic impairment, we used 6-hydroxydopamine to lesion substantia nigra and deafferent the ascending dopaminergic pathway. Dopamine depletion significantly attenuated lesion volume in the striatum following systemic 3-NPA administration (Table 1) *(68)*.

SUMMARY AND CONCLUSIONS

3-NPA is becoming a widely used tool for investigations of impaired energy metabolism and mitochondrial dysfunction in neurodegenerative disorders such as Huntington's disease, as well as following acute insults such as ischemia. In rats, systemic administration of 3-NPA results in preferen-

Table 1
Substantia Nigra Lesions Protect Against Striatal 3-NPA Toxicity

Treatment	Lesion size (6-OHDA-treated)	Lesion size Control
6-OHDA	6.98 ± 1.7	15.4 ± 1.5
Sham control	18.1 ± 3.2	20.4 ± 4.5

In four animals with unilateral 6-hydroxydopamine (6-OHDA) lesions of substantia nigra and four sham-lesion controls, 3-NPA was administered (20 mg/kg/d) until the rats developed motor impairment. One control was excluded from analysis because it did not develop motor impairment or striatal lesions after 8 d of treatment. The remaining animals developed motor impairment over 4–8 d. Tyrosine hydroxylase immunoreactivity confirmed the specificity of the 6-OHDA lesions. Striatal lesion volume (mm^3, mean ± SEM) was estimated from consecutive cresyl violet stained sections. The results illustrate that the 6-OHDA lesion of the nigrostriatal pathway reduced lesion size by 55% as compared to the contralateral side.

tial damage to striatum, with less consistent damage to hippocampus and neocortex. Although there is considerable evidence in support of the hypothesis that indirect excitotoxic mechanisms contribute to the neurodegeneration induced by 3-NPA, this hypothesis does not fully account for 3-NPA toxicity. In primary neuronal cultures, NMDA receptor antagonists delay, but do not prevent, cell death induced by 3-NPA and they shift the cell death mechanism from necrosis to apoptosis. The apoptotic cascade is thought to result from impaired kinase activity, rather than failure of ion pumps and excitotoxic mechanisms. In vivo, it is difficult to account for the preferential vulnerability of striatum based on the indirect excitotoxic hypothesis. However, recent results suggest that dopamine can exacerbate excitotoxic insult and lesions of the dopaminergic nigrostriatal pathway attenuate the striatal damage induced by systemic 3-NPA administration. In vitro, 3-NPA does not result in a large increase in free radial production, yet antioxidants can provide protection against 3-NPA toxicity in vivo. This likely reflects increased oxidative stress associated with an inflammatory response, secondary to the initial 3-NPA-induced neurodegeneration.

Together, these results suggest that a multifaceted approach may be necessary to effectively intervene against neurodegeneration resulting from metabolic impairment. Glutamate receptor antagonists can prevent the necrotic but not apoptotic death. Interference with dopamine receptors or metabolism may also be beneficial. The apoptotic death could be minimized by caspase inhibition, although this has not been examined with regard to 3-NPA in studies to date. Antioxidants and antiinflammatory agents may not prevent 3-NPA toxicity directly, but may minimize the subsequent neuronal damage caused by an inflammatory response.

ACKNOWLEDGMENTS

Research from the author's laboratory was supported by NIH Grants P50 AG05144 and P50 AG10836.

REFERENCES

1. He F, Zhang S, Qian F, Zhang C. Delayed dystonia with striatal CT lucencies induced by a mycotoxin (3-nitropropionic acid). Neurology 1995;45: 2178–2183.
2. Liu X, Luo X, Hu W. Studies on the epidemiology and etiology of moldy sugarcane poisoning in China. Biomed Environ Sci 1992;5:161–177.
3. Hamilton BF, Gould DH. Nature and distribution of brain lesions in rats intoxicated with 3-nitropropionic acid: a type of hypoxic (energy deficient) brain damage. Acta Neuropathol (Berl) 1987;72:286–297.
4. Beal MF, Brouillet E, Jenkins BG, et al. Neurochemical and histologic characterization of striatal excitotoxic lesions produced by the mitochondrial toxin 3-nitropropionic acid. J Neurosci 1993;13:4181–4192.
5. Miller PJ, Zaborszky L. 3-Nitropropionic acid neurotoxicity: visualization by silver staining and implications for use as an animal model of Huntington's disease. Exp Neurol 1997;146:212–229.
6. Beal MF. Neurochemistry and toxin models in Huntington's disease. Curr Opin Neurol 1994;7:542–547.
7. Borlongan CV, Koutouzis TK, Sanberg PR. 3-Nitropropionic acid animal model and Huntington's disease. Neurosci Biobehav Rev 1997;21:289–293.
8. Guyot MC, Hantraye P, Dolan R, Palfi S, Maziere M, Brouillet E. Quantifiable bradykinesia, gait abnormalities and Huntington's disease-like striatal lesions in rats chronically treated with 3-nitropropionic acid. Neuroscience 1997; 79:45–56.
9. Pulsinelli WA, Brierly JB, Plum F. Temporal profile of neuronal damage in a model of transient forebrain ischemia. Ann Neurol 1982;11:491–498.
10. Wieloch T. Hypoglycemia-induced neuronal damage prevented by an *N*-methyl-D-aspartate antagonist. Science 1985;230:681–683.
11. Gould DH, Wilson MP, Hamar DW. Brain enzyme and clinical alterations induced in rats and mice by nitroaliphatic toxicants. Toxicol Lett 1985; 27:83–89.
12. Pike CJ, Walencewicz AJ, Glabe CG, Cotman CW. In vitro aging of beta-amyloid protein causes peptide aggregation and neurotoxicity. Brain Res 1991;563:311–314.
13. Irizarry MC, Soriano F, McNamara M, et al. Abeta deposition is associated with neuropil changes, but not with overt neuronal loss in the human amyloid

precursor protein V717F (PDAPP) transgenic mouse. J Neurosci 1997; 17:7053–7059.
14. Pang Z, Umberger GH, Geddes JW. Neuronal loss and cytoskeletal disruption following intrahippocampal administration of the metabolic inhibitor malonate: lack of protection by MK–801. J Neurochem 1996;66:474–484.
15. Brouillet E, Guyot MC, Mittoux V, et al. Partial inhibition of brain succinate dehydrogenase by 3-nitropropionic acid is sufficient to initiate striatal degeneration in rat. J Neurochem 1998;70:794–805.
16. Fink SL, Ho DY, Sapolsky RM. Energy and glutamate dependency of 3-nitropropionic acid neurotoxicity in culture. Exp Neurol 1996;138:298–304.
17. Hamilton BF, Gould DH. Correlation of morphologic brain lesions with physiologic alterations and blood–brain barrier impairment in 3-nitropropionic acid toxicity in rats. Acta Neuropathol (Berl) 1987;74:67–74.
18. Nishino H, Shimano Y, Kumazaki M, Sakurai T. Chronically administered 3-nitropropionic acid induces striatal lesions attributed to dysfunction of the blood–brain barrier. Neurosci Lett 1995;186:161–164.
19. Sato S, Gobbel GT, Li Y, et al. Blood–brain barrier disruption, HSP70 expression and apoptosis due to 3-nitropropionic acid, a mitochondrial toxin. Acta Neurochir Suppl (Wien) 1997;70:237–239.
20. Olney JW. Excitotoxin-mediated neuron death in youth and old age. Prog Brain Res 1990;86:37–51.
21. Choi DW. Excitotoxic cell death. J Neurobiol 1992;23:1261–1276.
22. Westbrook GL. Glutamate receptors and excitotoxicity. Res Publ Assoc Res Nerv Ment Dis 1993;71:35–50.
23. Majno G, Joris I. Apoptosis, oncosis, and necrosis. An overview of cell death. Am J Pathol 1995;146:3–15.
24. Ankarcrona M, Dypbukt JM, Bonfoco E, et al. Glutamate-induced neuronal death: a succession of necrosis or apoptosis depending on mitochondrial function. Neuron 1995;15:961–973.
25. Ferrer I, Martin F, Serrano T, et al. Both apoptosis and necrosis occur following intrastriatal administration of excitotoxins. Acta Neuropathol 1995;90: 504–510.
26. Gwag BJ, Koh JY, DeMaro JA, Ying HS, Jacquin M, Choi DW. Slowly triggered excitotoxicity occurs by necrosis in cortical cultures. Neuroscience 1997;77:393–401.
27. Portera-Cailliau C, Price DL, Martin LJ. Non-NMDA and NMDA receptor-mediated excitotoxic neuronal deaths in adult brain are morphologically distinct: further evidence for an apoptosis-necrosis continuum. J Comp Neurol 1997;378:88–104.
28. Sohn S, Kim EY, Gwag BJ. Glutamate neurotoxicity in mouse cortical neurons: atypical necrosis with DNA ladders and chromatin condensation. Neurosci Lett 1998;240:147–150.
29. Leist M, Nicotera P. Apoptosis, excitotoxicity, and neuropathology. Exp Cell Res 1998;239:183–201.
30. Olney JW. Excitotoxic amino acids and neuropsychiatric disorders. Annu Rev Pharmacol Toxicol 1990;30:47–71.

31. Behrens MI, Koh JY, Muller MC, Choi DW. NADPH diaphorase-containing striatal or cortical neurons are resistant to apoptosis. Neurobiol Dis 1996;3:72–75.
32. Siesjo BK, Bengtsson F. Calcium fluxes, calcium antagonists, and calcium-related pathology in brain ischemia, hypoglycemia, and spreading depression. J Cereb Blood Flow Metab 1989;9:127–140.
33. Choi DW. Calcium-mediated neurotoxicity: relationship to specific channel types and role in ischemic damage. Trends Neurosci 1988;11:465–469.
34. Novelli A, Reilly JA, Lysko PG, Henneberry RC. Glutamate becomes neurotoxic via the *N*-methyl-D-aspartate receptor when intracellular energy levels are reduced. Brain Res 1988;451:205–212.
35. Zeevalk GD, Nicklas WJ. Mechanisms underlying initiation of excitotoxicity associated with metabolic inhibition. J Pharmacol Exp Ther 1991;257:870–878.
36. Zeevalk GD, Nicklas WJ. Evidence that the loss of the voltage-dependent Mg^{2+} block at the *N*-methyl-D-aspartate receptor underlies receptor activation during inhibition of neuronal metabolism. J Neurochem 1992;59:1211–1220.
37. Fu Y, He F, Zhang S, Huang J, Zhang J, Jiao X. 3-Nitropropionic acid produces indirect excitotoxic damage to rat striatum. Neurotoxicol Teratol 1995;17:333–339.
38. Sugino T, Nozaki K, Tokime T, Hashimoto N, Kikuchi H. 3-Nitropropionic acid induces poly(ADP-ribosyl)ation and apoptosis related gene expression in the striatum in vivo. Neurosci Lett 1997;237:121–124.
39. Simpson JR, Isacson O. Mitochondrial impairment reduces the threshold for in vivo NMDA-mediated neuronal death in the striatum. Exp Neurol 1993;121:57–64.
40. Sato S, Gobbel GT, Honkaniemi J, et al. Apoptosis in the striatum of rats following intraperitoneal injection of 3-nitropropionic acid. Brain Res 1997;745:343–347.
41. Greene JG, Sheu SS, Gross RA, Greenamyre JT. 3-Nitropropionic acid exacerbates *N*-methyl-D-aspartate toxicity in striatal culture by multiple mechanisms. Neuroscience 1998;84:503–510.
42. Weller M, Paul SM. 3-Nitropropionic acid is an indirect excitotoxin to cultured cerebellar granule neurons. Eur J Pharmacol 1993;248:223–228.
43. Zeevalk GD, Derr-Yellin E, Nicklas WJ. NMDA receptor involvement in toxicity to dopamine neurons in vitro caused by the succinate dehydrogenase inhibitor 3-nitropropionic acid. J Neurochem 1995;64:455–458.
44. Pass MA, Carlisle CH, Reuhl KR. 3-Nitropropionic acid toxicity in cultured murine embryonal carcinoma cells. Natural Toxins 1994;2:386–394.
45. Behrens MI, Koh J, Canzoniero LM, Sensi SL, Csernansky CA, Choi DW. 3-Nitropropionic acid induces apoptosis in cultured striatal and cortical neurons. NeuroReport 1995;6:545–548.
46. Buller AL, Larson HC, Schneider BE, Beaton JA, Morrisett RA, Monaghan DT. The molecular basis of NMDA receptor subtypes: native receptor diversity is predicted by subunit composition. J Neurosci 1994;14:5471–5484.

47. Rothstein JD, Martin L, Levey AI, et al. Localization of neuronal and glial glutamate transporters. Neuron 1994;13:713–725.
48. Wullner U, Young AB, Penney JB, Beal MF. 3-Nitropropionic acid toxicity in the striatum. J Neurochem 1994;63:1772–1781.
49. Pang Z, Geddes JW. Mechanisms of cell death induced by the mitochondrial toxin 3-nitropropionic acid: acute excitotoxic necrosis and delayed apoptosis. J Neurosci 1997;17:3064–3073.
50. Bertrand R, Solary E, O'Connor P, Kohn KW, Pommier Y. Induction of a common pathway of apoptosis by staurosporine. Exp Cell Res 1994; 211:314–321.
51. Cheng Y, Sun AY. Oxidative mechanisms involved in kainate-induced cytotoxicity in cortical neurons. Neurochem Res 1994;19:1557–1564.
52. Schulz JB, Henshaw DR, Siwek D, et al. Involvement of free radicals in excitotoxicity in vivo. J Neurochem 1995;64:2239–2247.
53. Lafon-Cazal M, Pietri S, Culcasi M, Bockaert J. NMDA-dependent superoxide production and neurotoxicity. Nature 1993;364:535–537.
54. Reynolds IJ, Hastings TG. Glutamate induces the production of reactive oxygen species in cultured forebrain neurons following NMDA receptor activation. J Neurosci 1995;15:3318–3327.
55. Fu YT, He FS, Zhang SL, Zhang JS. Lipid peroxidation in rats intoxicated with 3-nitropropionic acid. Toxicon 1995;33:327–331.
56. Schulz JB, Henshaw DR, MacGarvey U, Beal MF. Involvement of oxidative stress in 3-nitropropionic acid neurotoxicity. Neurochem Int 1996;29:167–171.
57. Beal MF, Ferrante RJ, Henshaw R, et al. 3-Nitropropionic acid neurotoxicity is attenuated in copper/zinc superoxide dismutase transgenic mice. J Neurochem 1995;65:919–922.
58. Nakao N, Brundin P. Effects of alpha-phenyl-tert-butyl nitrone on neuronal survival and motor function following intrastriatal injections of quinolinate or 3-nitropropionic acid. Neuroscience 1997;76:749–761.
59. Saito K, Suyama K, Nishida K, Sei Y, Basile AS. Early increases in TNF-alpha, IL-6 and IL-1 beta levels following transient cerebral ischemia in gerbil brain. Neurosci Lett 1996;206:149–152.
60. Kogure K, Yamasaki Y, Matsuo Y, Kato H, Onodera H. Inflammation of the brain after ischemia. Acta Neurochir Suppl (Wien) 1996;66:40–43.
61. Nishino H, Fujimoto I, Shimano Y, Hida H, Kumazaki M, Fukuda A. 3-Nitropropionic acid produces striatum selective lesions accompanied by iNOS expression. J Chem Neuroanat 1996;10:209–212.
62. Globus MY, Ginsberg MD, Harik SI, Busto R, Dietrich WD. Role of dopamine in ischemic striatal injury: metabolic evidence. Neurology 1987;37:1712–1719.
63. Slivka A, Brannan TS, Weinberger J, Knott PJ, Cohen G. Increase in extracellular dopamine in the striatum during cerebral ischemia: a study utilizing cerebral microdialysis. J Neurochem 1988;50:1714–1718.
64. Globus MY, Ginsberg MD, Dietrich WD, Busto R, Scheinberg P. Substantia nigra lesion protects against ischemic damage in the striatum. Neurosci Lett 1987;80:251–256.

65. Buisson A, Pateau V, Plotkine M, Boulu RG. Nigrostriatal pathway modulates striatum vulnerability to quinolinic acid. Neurosci Lett 1991;131:257–259.
66. Chapman AG, Durmuller N, Lees GJ, Meldrum BS. Excitotoxicity of NMDA and kainic acid is modulated by nigrostriatal dopaminergic fibres. Neurosci Lett 1989;107:256–260.
67. Filloux F, Wamsley JK. Dopaminergic modulation of excitotoxicity in rat striatum: evidence from nigrostriatal lesions. Synapse 1991;8:281–288.
68. Maragos WF, Jakel R, Pang Z, Geddes JW. 6-Hydroxydopamine injections into the nigrostriatal pathway attenuate striatal malonate and 3-nitropropionic acid lesions. Exp Neurol 1998;154:637–644.

8

Gender-Related Difference of the Effect of 3-Nitropropionic Acid on Striatal Artery

Keiya Nakajima, Yasunobu Shimano, Kunio Torii, and Hitoo Nishino

INTRODUCTION

Systemic intoxication of rats with 3-nitropropionic acid (3-NPA), a mycotoxin that inhibits succinate dehydrogenase in the mitochondrial respiratory chain, induces the striatum-selective lesions and motor symptoms (hyperkinesia, rolling, paddling, hypotonia, recumbency, etc.) reminiscent of Huntington's disease *(1–4)*. The lesions are localized in the centrolateral striatum and not in other parts of the brain. Inside the lesions, glial fibrillary acidic protein (GFAP) positive astroglia and microtubule-associated protein (MAP)-2-positive neurons are lost and the extravasation of immunoglobulin G (IgG) (the blood–brain barrier dysfunction) is detected *(5–8)*. The mechanisms for the striatum selective lesions have been proposed as follows : the enhancement of excitotoxicity (decrease in the threshold for excitotoxicity) due to the deficiency of high-energy ATP *(9–11)*, the dopamine (DA) toxicity by excessive release and turnover *(6,12)*, the high vulnerability of the lateral striatal artery (lSTR artery) accompanying the dysfunction of the BBB *(13)*, or the involvement of high 3-NPA uptake activity by an acid transporter (glial glutamate transporter). The motor disturbances and histological (striatal) damages by 3-NPA were often detected in adult male rats but not in adult females or younger rats *(6)*. Thus, the 3-NPA intoxication is a good model to investigate the pathophysiology of neuronal/glial cell death, neurodegenerative disorders, autoimmune diseases, stroke, or that of any disorders in which gender difference is often observed. Here, we review the recent finding of the gender differences of 3-NPA intoxication.

From: *Mitochondrial Inhibitors and Neurodegenerative Disorders*
Edited by: P. R. Sanberg, H. Nishino, and C. V. Borlongan © Humana Press Inc., Totowa, NJ

Table 1
Incidence of Motor Symptoms (Hyperkinesia, Rolling, Paddling, Recumbency etc.)

Animals	Percent motor symptoms
Male	
Control	50
Castrated	67
Castrated + E_2	10
Female	
Control	0
OVX	40
OVX + tamo	0
OVX + testo	67

E_2, estradiol (60 µg/rat, s.c.) treatment twice a week for 4 wk; OVX, ovariectomy; tamo, tamoxifen (2 mg/kg, p.o.) treatment every day for 4 wk; testo, testosterone propionate (500 µg/rat, s.c.) treatment twice a week for 4 wk.

GENDER DIFFERENCE IN THE INCIDENCE OF MOTOR SYMPTOMS

Gender difference in the incidence of motor symptoms and histological damages have not to date been investigated for chronic intoxication models of 3-NPA. However, in an acute intoxication (20 mg/kg, s.c., once a day for 2 consecutive days), there is a marked gender difference in the incidence of motor symptoms and histological damages *(6,13)*. Within hours after the second administration of 3-NPA, 50% of the male rats exhibited motor symptoms. Castration alone did not have a large effect, but estradiol treatment after castration decreased the incidence of motor symptoms (10%). Female rats exhibited no motor abnormalities at all, but 40% of ovariectomized females developed motor symptoms. Estrogen or tamoxifen (an estrogen receptor antagonist) treatment suppressed the motor symptoms completely, while testosterone treatment enhanced (67%) the incidence of motor symptoms in ovariectomized females (Table 1) *(13)*.

VARIATION IN HORMONAL LEVELS AFTER OVARIECTOMY AND TAMOXIFEN TREATMENT

Serum sex hormone levels in females were: estradiol, approx 130 pg/mL; testosterone, approx 110 pg/mL, measured using enzyme immunoassay. After ovariectomy, the estradiol level decreased to less than 20% of control but the

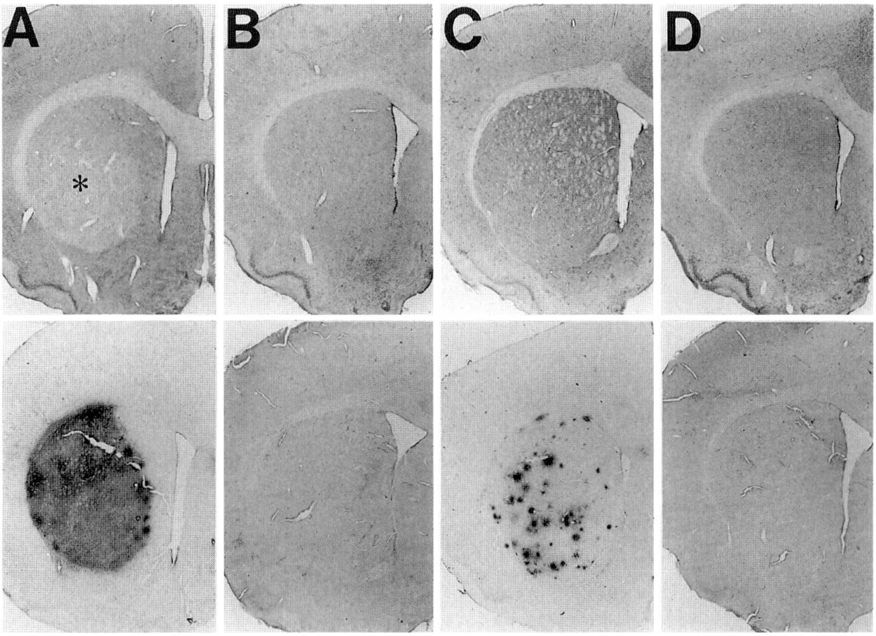

Fig. 1. Gender difference in striatal damage (*) 3 h after two administrations of 3-NPA. **(A)** male; **(B)** female; **(C)** ovariectomized female; **(D)** tamoxifen treated ovariectomized female. *Upper*, hematoxylin eosin staining; *lower*, IgG immunostaining.

testosterone level increased up to threefold that of control. Tamoxifen treatment did not affect the level of estradiol much, but suppressed the testosterone increase in ovariectomized rats. Weight of the uterus was correlated positively with the estrogen levels and inversely with the incidence of motor symptoms *(13)*.

GENDER DIFFERENCE IN HISTOLOGICAL DAMAGE

Figure 1 shows histological damage in a different group of animals after two administrations of 3-NPA. Fifty percent of male rats had motor symptoms and striatum-selective lesions with extravasation of IgG but females had no motor abnormalities or visible histological damages. Forty percent of ovariectomized females exhibited motor symptoms and striatal lesion. Ovariectomized females treated with estradiol or tamoxifen had no damages.

In parallel with the extravasation of IgG, the immunoreactivity against the factor VIII-related antigen (a marker protein of endothelial cells) increased in the lSTR artery, and GFAP-positive astroglias disappeared from the parenchyma of the centrolateral part of the striatum (Fig. 2).

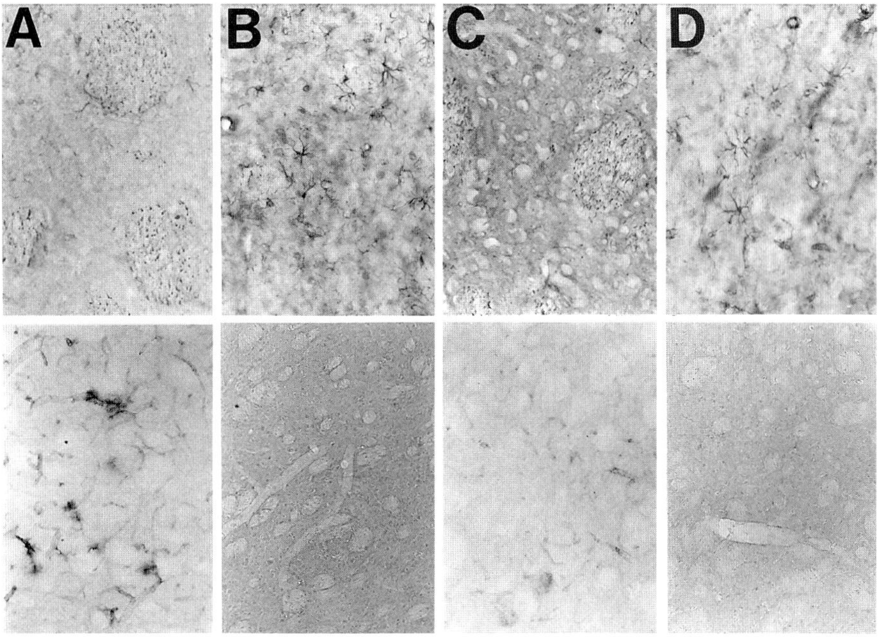

Fig. 2. Gender difference in striatal damage 3 h after two administrations of 3-NPA. **(A)** male; **(B)** female; **(C)** ovariectomized female; **(D)** tamoxifen treated ovariectomized female. *Upper*, GFAP immunostaining; *lower*, factor VIII immunostaining.

DISCUSSION

Acute systemic intoxication with 3-NPA induced striatum-selective lesions in rats with motor symptoms reminiscent of Huntington's disease *(1–4)*. The striatal lesions were characterized by damage to the lSTR artery leading to the disruption of the blood-brain barrier *(5–8)*. There was a marked gender difference in the incidence of motor symptoms and the extent of striatal damage *(6,13)*. Males were very vulnerable while females were resistant. Estrogens protected against whereas testosterone exacerbated both motor and histological damage *(13)*.

Estrogen receptors are distributed in neurons, glias, endothelial cells, and epithelial cells in the brain *(14)*. Estrogen may function as a neuroprotectant against oxidative injury and excitotoxicity *(15,16)*. Glial processes could be enhanced by the estrogen treatment *(17)*, and the release of endothelium-derived relaxing factor (EDRF) is enhanced *(18–20)* by estrogen replacement in ovariectomized females. These studies suggest that estrogens would provide protection for neurons, astroglias, and the vascular system. Our

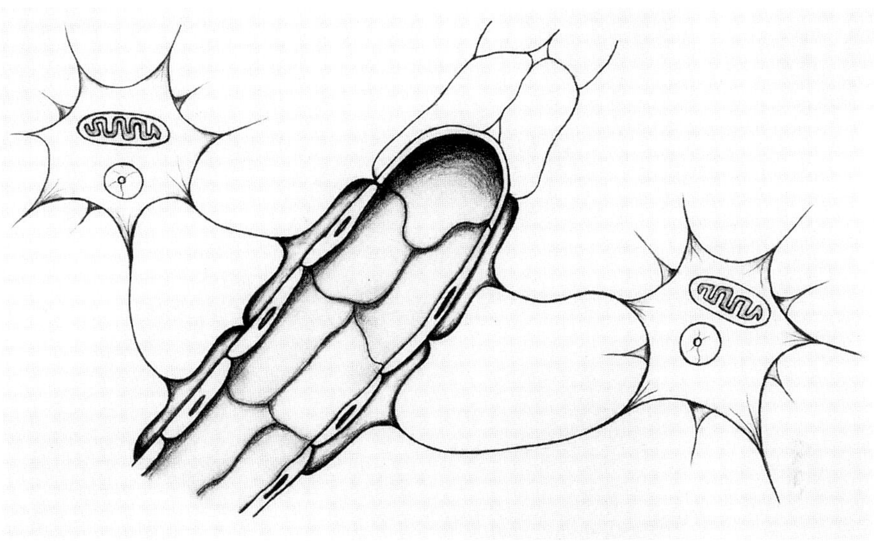

Fig. 3. Endothelial cells and end-feet of astroglias are major components of the blood–brain barrier. It is not yet known which of the two is the first target of 3-NPA.

observations reveal that the treatment of ovariectomized females with estradiol or tamoxifen (an antagonist of estrogen receptor) was protective against behavioral and histological damages, suggesting that the estrogen receptor is not involved in the protective action of estrogen in 3-NPA toxicity. With regard to the action of estrogens on endothelial cells or astroglias, to date it is not known which is the first target of 3-NPA (Fig. 3). To elucidate the mechanism, further in vivo as well as in vitro experiments are required. Mechanisms of action of estrogen in terms of the regulation of inflammatory cytokines (interleukin-1 [IL-1], interferon-γ [INFγ], tumor necrosis factor-α[TNF-α]), intracellular Ca^{2+} overload, free radical scavenging system, stress fibers and actin filaments in endothelial cells, or the GFAP in astroglias, are worth considering. These studies may offer significant information for the real mechanism of 3-NPA toxicity.

A marked gender difference in the vulnerability of the lSTR artery to 3-NPA intoxication, and the protective action of estrogens for this vulnerability, provide us crucial clues to understand the pathophysiology of stroke, neurodegenerative diseases, autoimmune diseases, etc. The lSTR artery is a perforate artery that is most often affected by stroke (putaminal bleeding) *(21)*, and there is a significant gender difference in the incidence of stroke in the brain (men are two to three times more often affected than women) *(22)*. Thus, this model may be useful to develop future therapeutic approaches for these disorders.

REFERENCES

1. Brouillet M, Hantraye P, Ferrante RT, et al. Chronic mitochondrial energy impairment produces selective striatal degeneration and abnormal choreiform movements in primates. Proc Natl Acad Sci USA 1995;92:7105–7109.
2. Shimano Y, Kumazaki M, Sakurai T, et al. Chronically administered 3-nitropropionic acid produces selective lesions in the striatum and reduces muscle tonus. Obesity Res 1995;3(Suppl 5):779s–784s.
3. Borlongan CV, Koutouzis TK, Freeman TB, et al. Behavioral pathology induced by repeated systemic injections of 3-nitropropionic acid mimics the motoric symptoms of Huntington's disease. Brain Res. 1995;679:254–257.
4. Palfi S, Ferrante RJ, Brouillet E, et al. Chronic 3-nitropropionic acid treatment in baboons replicates the cognitive and motor deficits of Huntington's disease. J Neurosci 1996;16:3019–3025.
5. Nishino H, Shimano Y, Kumazaki M, et al. Chronically administered 3-nitropropionic acid induces striatal lesions attributed to dysfunction of the blood–brain barrier. Neurosci Lett 1995;186:161–164.
6. Nishino H, Kumazaki M, Fukuda A, et al. Acute 3-nitropropionic acid intoxication induces striatal astrocytic cell death and dysfunction of the blood–brain barrier: involvement of dopamine toxicity. Neurosci Res 1997;27:343–355.
7. Hamilton BF, Gould DH. Nature and distribution of brain lesions in rats intoxicated with 3-nitropropionic acid: a type of hypoxic (energy deficient) brain damage. Acta Neuropathol 1987;72:286–297.
8. Hamilton BF, Gould DH. Correlation of morphologic brain lesions with physiologic alterations and blood–brain barrier impairment in 3-nitropropionic acid toxicity in rats. Acta Neuropathol 1987;74:67–74.
9. Beal MF. Does impairment of energy metabolism result in excitotoxic neuronal death in neurodegenerative illness? Ann Neurol 1992;31:119–130.
10. Beal MF, Brouillet E, Jenkins BG, et al. Neurochemical and histologic characterization of striatal excitotoxic lesions produced by the mitochondrial toxin 3-nitropropionic acid. J Neurosci 1993;13:4181–4192.
11. Simpson JR, Isacson O. Mitochondrial impairment reduces the threshold for in vivo NMDA-mediated neuronal death in the striatum. Exp Neurol 1993; 121:57–64.
12. Pei G, Ebendal T. Specific lesions in the extrapyramidal system of the rat brain induced by 3-nitropropionic acid (3-NPA). Exp Neurol 1995;132:105–115.
13. Nishino H, Nakajima K, Kumazaki M, et al. Estrogen protects against while testosterone exacerbates vulnerability of the lateral striatal artery to chemical hypoxia by 3-nitropropionic acid. Neurosci Res 1998;30:303–312.
14. Langub MC, Watson RE Jr. Estrogen receptor-immunoreactive glia, endothelia, and ependyma in guinea pig preoptic area and median eminence: electron microscopy. Endocrinology 1992;130:364–372.
15. Behl C, Widmann M, Trapp T, Holsboer F. 17-β Estradiol protects neurons from oxidative stress-induced cell death *in vitro*. Biochem Biophys Res Commun 1995;216:473–482.

16. Goodman Y, Bruce AJ, Cheng B, Mattson MP. Estrogens attenuate and corticosterone exacerbates excitotoxicity, oxidative injury, and amyloid β-peptide toxicity in hippocampal neurons. J Neurochem 1996;66:1836–1844.
17. Tranque PA, Suarez I, Olmos G, Fernandez B, Garcia-Segura LM. Estradiol-induced redistribution of glial fibrillary acidic protein immunoreactivity in the brain. Brain Res 1987;406:348–351.
18. Williams JK, Adams MR, Klopfenstein HS. Estrogen modulates responses of atherosclerotic coronary arteries. Circulation 1990;81:1680–1687.
19. Gisclard V, Miller VM, Vanhoutte PM. Effects of 17-beta-estradiol on endothelium-dependent responses in the rabbit. J Pharmacol Exp Ther 1988;244:19–22.
20. Miller VM, Vanhautte PM. Progesterone and modulation of endothelium-dependent responses in canine coronary arteries. Am J Physiol 1991;261: R1022–R1027.
21. Kase CS, et al. Intracerebral hemorrhage. In: Barnett HJM, et al., eds. Stroke, 2nd edit. Churchill Livingstone, New York, 1992, pp. 561–616.
22. Hier DB, et al. Hypertensive putaminal hemorrhage. Ann Neurol 1977;11: 152–159.

9

Variable Susceptibility to Neurotoxicity of Systemic 3-Nitropropionic Acid

Tajrena Alexi, Richard L. M. Faull, and Paul E. Hughes

INTRODUCTION

Between 1972 and 1989 nearly 900 people in China accidentally contracted food poisoning from the consumption of mildewed sugarcane *(1,2)*. As a consequence, 88 people died and many of the survivors developed encephalopathy and dystonia. The victims were mainly children and it is unclear whether children show an increased susceptibility to the poisoning or whether they simply consumed larger amounts of the sweet sugarcane as compared to the adults. Most victims developed an acute illness characterized by nausea, abdominal pain, seizures, convulsions, and temporary coma (up to 20 d). Some of the patients fully recovered with no permanent damage. Others developed neurological damage in the basal ganglia and a dystonic movement disorder. The dystonia included symptoms such as spasms, jerklike movements, facial grimacing, and speech disturbances.

The poisoning was found to be caused by the fungus *Arthrinium*, which had contaminated the sugarcane. This fungus synthesizes the compound 3-nitropropionic acid (3-NPA), which was found to cause poisoning symptoms and neurotoxicity in experimental animals that were similar to the case studies of the Chinese patients *(3,4)*. 3-NPA was found to inhibit ATP synthesis by inactivating succinate dehydrogenase (SDH), a mitochondrial enzyme active in the tricarboxylic acid cycle *(5)*. The disruption of mitochondrial energy metabolism by 3-NPA leads to widespread loss of ATP throughout the entire body. However, for unknown reasons the brain is the most vulnerable organ and 3-NPA poisoning can lead to permanent neurological damage, particularly in the basal ganglia. Interestingly, other mitochondrial poisons such as cyanide and carbon monoxide also cause selective damage to the basal ganglia despite widespread collapse of energy metabolism (for review *see* refs. *[6,7]*).

From: *Mitochondrial Inhibitors and Neurodegenerative Disorders*
Edited by: P. R. Sanberg, H. Nishino, and C. V. Borlongan © Humana Press Inc., Totowa, NJ

3-NPA AS A MODEL OF HUNTINGTON'S DISEASE

The physiological damage caused by 3-NPA metabolic compromise resembles the genetic disorder Huntington's disease (HD), which is characterized by progressive degeneration of the basal ganglia and choreic motor dysfunction *(8–10)*. Although the primary movement dysfunction resulting from 3-NPA poisoning in humans is dystonia and not chorea, the two movement disorders are closely related and both result from physiological alterations within the basal ganglia. Exposure to 3-NPA in experimental nonhuman primates leads to both chorea and dystonia *(11,12)*. Postmortem and in vivo clinical measures in HD brain support a role for metabolic compromise in HD (for review *see* ref. *[6]*). 3-NPA is currently used as an animal model of HD *(9,10,13)*.

VARIABLE SUSCEPTIBILITY TO 3-NPA

As evidenced by the case studies of the sugarcane poisoning outbreaks in China, some individuals are susceptible to permanent neurological damage while others recover unscathed. It is not clear what accounts for these differences, as it is difficult to draw correlations in uncontrolled human poisoning outbreaks. However, controlled animal studies also show variable susceptibility to 3-NPA toxicity, yet the reasons remain elusive *(14,17)*. Although it is clear that there is an age-related susceptibility to 3-NPA toxicity in experimental animals, there is still variability within each age group as well as between age groups *(18,20)*. Of note is that in experimental animal studies it is younger animals that are more resilient. This is in contrast to the Chinese poisoning incidents whereby children were the main victims. It is strongly believed, however, that in the case of the accidental food poisonings children were the main victims because of overindulgence in the sweet sugarcane as compared with less indulgent adults.

We have studied the variable toxic effects of systemic injection of 3-NPA in laboratory rats. Young adult (10-wk-old) Sprague–Dawley male rats were given a single injection of 3-NPA intraperitoneally (i.p.) and controls received vehicle (phosphate-buffered saline). Based on pilot studies, a single dose of 30 mg/kg, i.p. was used. This dose caused mortality in 21.7% of rats by 16 h after injection. Other than moderate lethargy, surviving rats did not display any observable pathological behaviors (such as stereotypies or generalized seizures). By 24 h, activity levels and behavioral patterns of rats returned to normal.

After 7 d we assessed the effects of 3-NPA on the activity of SDH, the biochemical target of 3-NPA. We measured endogenous SDH activity *in situ* in the brain by histochemical detection of oxidative enzymes. Fresh frozen sections were incubated for 1 h at 37°C in the substrate for SDH, succinate (0.05 *M*), and the chromogen nitroblue tetrazolium (0.5 mg/mL), which changes color in proportion to the activity of the enzyme as previously reported *(16)*. Negative controls were conducted by adding 3-NPA to or removing succinate from the incubation mix. Quantitative imaging was used to obtain densitometric measurements of chromagen staining recorded as relative optical densities (MicroComp Integrated Image Analysis System, Southern Micro Instruments, Atlanta, GA). Statistical analysis was carried out by a multivariant analysis of variance (ANOVA) with a post-hoc Student Newman–Keuls test for statistical differences (total $n = 24$).

We found that doses of 30 mg/kg, i.p. caused a severe loss of SDH activity throughout the brain at 7 d after injection. In the striatum, SDH activity declined to approximately one quarter of control values. Of particular interest was that in a subset of rats (24.6%) postmortem analysis revealed a well-circumscribed region within the central striatum (bilaterally) where SDH activity declined dramatically. SDH activity approached undetectable levels within this area and appeared as a white circle set against the purple staining of the rest of the brain. We refer to this region as the "depleted pocket" (Fig. 1A,B).

Although all animals receiving 30 mg/kg 3-NPA showed severe losses in SDH activity, most of the animals did not develop neurological damage as assessed by histochemical staining with hematoxylin/eosin at 7 d. It was only in the subset of animals that had the depleted pockets of SDH activity within the striatum that had developed neurological lesions. In fact, neurological damage was vastly concentrated within the pockets of SDH depletion. Because only the subset of animals with depleted pockets had developed lesions, we classified them as "vulnerable," while the remainder of the rats that recovered from 3-NPA exposure were classified as "resilient" (Fig. 1A–C). Prior to histological evaluation, there were no obvious behavioral indicators as to which individual rats were vulnerable. This variable vulnerability resembles the outcomes of the clinical case studies of the sugarcane poisoning outbreaks in China. It is also consistent with published experimental studies on systemic 3-NPA administration in Sprague–Dawley *(18–20)* and Wistar *(15,17)* rat strains. A more chronic dosing regimen has been found to greatly reduce this variability, with most rats eventually showing striatal lesions *(10,14,21)*.

NEUROLOGICAL TOXICITY OF 3-NPA

We then sought to determine which came first: was the decline in SDH activity simply due to the loss of cells or did the decline in SDH activity

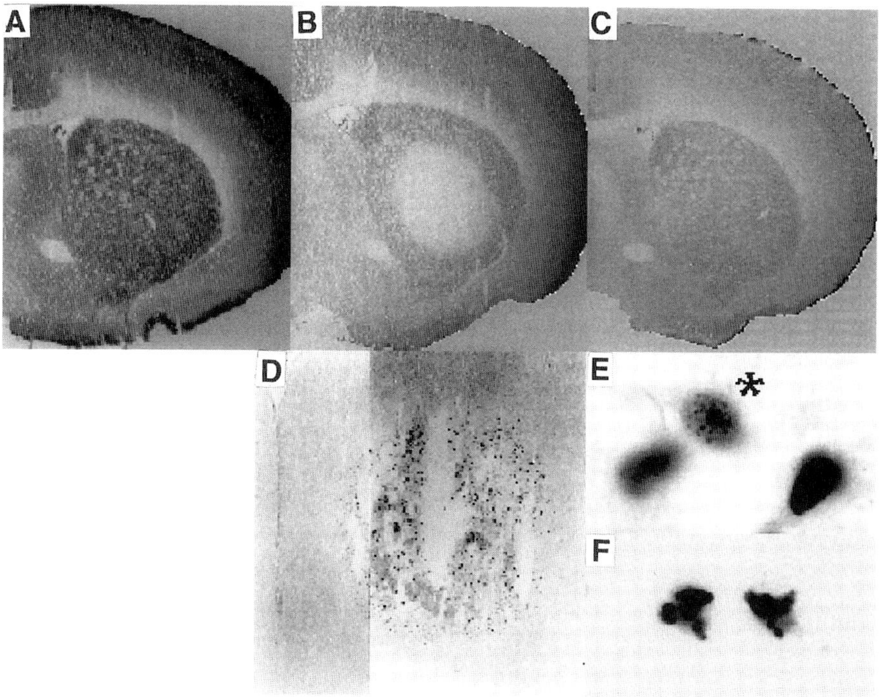

Fig. 1. Histochemical staining for SDH activity and TUNEL staining in rat striatum after acute systemic 3-NPA. Sprague–Dawley rat striatum after systemic 3-NPA (30 mg/kg, i.p.) injection. SDH activity in control **(A)**, vulnerable rat at 3 d **(B)**, and resilient rat at 7 d **(C)** and TUNEL induction in vulnerable rat at 3 d **(D)** after injection. Vulnerable rat striatum shows a pocket of fully depleted SDH activity **(B)** in the same area where TUNEL is induced **(D)**, whereas resilient rats do not develop this region **(C)**. High-power magnification of TUNEL cells in vulnerable rat striatum at 3 d **(E)** and 7 d **(F)** after 3-NPA injection. Asterisks in **(E)** point to a cell showing condensed chromatin (apoptotic) bodies while the other cells show diffuse (necrotic) staining. The cells in **(F)** show apoptotic bodies that have budded off from the nucleus.

cause the death of cells? We looked at a more detailed time course of 1, 3, 7, and 10 d (total $n = 84$) to determine whether the SDH activity declined prior to cell death or vice versa. We found something very unexpected. First, the pocket of SDH depletion in the striatum in vulnerable rats that we had seen at 7 d was evident as early as 1 d after 3-NPA injection, thus allowing early identification of those animals that were vulnerable to lesioning. Second, the maximal depletion of SDH activity was similar between vulnerable and resilient rats, and both groups had full recovery of SDH activity levels back to control levels by 10 d after 3-NPA injection (except within the area of the pocket in vulnerable animals). However, the temporal pattern of inhibition of SDH activity vastly differed between vulnerable and resilient animals (Fig. 2A). We found that in vulnerable rats the maximal decline of SDH activity throughout the brain occurred immediately at 1 d after 3-NPA injection. In resilient rats the decline was gradual and took 7 d to reach maximal depletion. The absolute maximal declines in SDH activity for both groups was about one quarter of control levels, with control levels being similar for both groups. But although maximal declines remained similar for vulnerable and resilient animals, only vulnerable animals had developed neurological lesions—both within the pocket and within the striatal area surrounding the pocket. This suggested that it wasn't the severity of the inhibition of SDH activity by 3-NPA that caused the neurological damage, but rather the speed at which the decline occurred.

We next wished to identify individual dead and dying cells more accurately than our histochemical hematoxylin/eosin assessment of neurological damage. Because dead and dying cells contain fragmented DNA, we used terminal deoxytransferase-mediated dUTP-biotin nick end labeling (TUNEL) to stain them. The TUNEL procedure was performed by a modification *(16)* of the method of Gavrieli et al. *(22)*. In brief, fresh frozen sections were fixed with 4% paraformaldehyde in phosphate buffer (pH 7.4) for 15 min, treated with 1% hydrogen peroxide in absolute methanol for 5 min, and incubated in terminal deoxytransferase (TdT; 1:100) with biotin-14-dATP (1:100) in TdT buffer (Life Technologies, New Zealand) for 1 h at 37°C in a humidified chamber. Unincorporated biotinylated dATP was washed away in 2× sodium chloride/ sodium citrate buffer (SSC) for 15 min, and salts were removed by a rapid dip in distilled water. Sections were then developed by standard immunocytochemical procedures using ExtrAvidin Peroxidase and 3,3'-diaminobenzidine (Sigma-Aldrich, Australia). Positive controls were conducted by pretreating tissue sections with DNase I, which creates extensive DNA fragmentation and TUNEL-positive nuclei. Negative controls were sections incubated in the absence of TdT which gave no labeling.

Control rats had essentially no TUNEL staining in the striatum (0.01–0.06 cells/mm^2). In vulnerable rats, 3-NPA (30 mg/kg, i.p.) massively induced TUNEL within the striatal SDH depleted pocket at 1, 3, and 7 but not 10 d after injection (Fig. 1B,D and Fig. 2B). Maximal labeling in the striatal pocket was at 3 d after 3-NPA injection (Table 1). TUNEL was also significantly induced in the region surrounding the striatal pocket at 1, 3, and 7 but not 10 d after 3-NPA, and was maximal at 1 d. The timing of the maximal

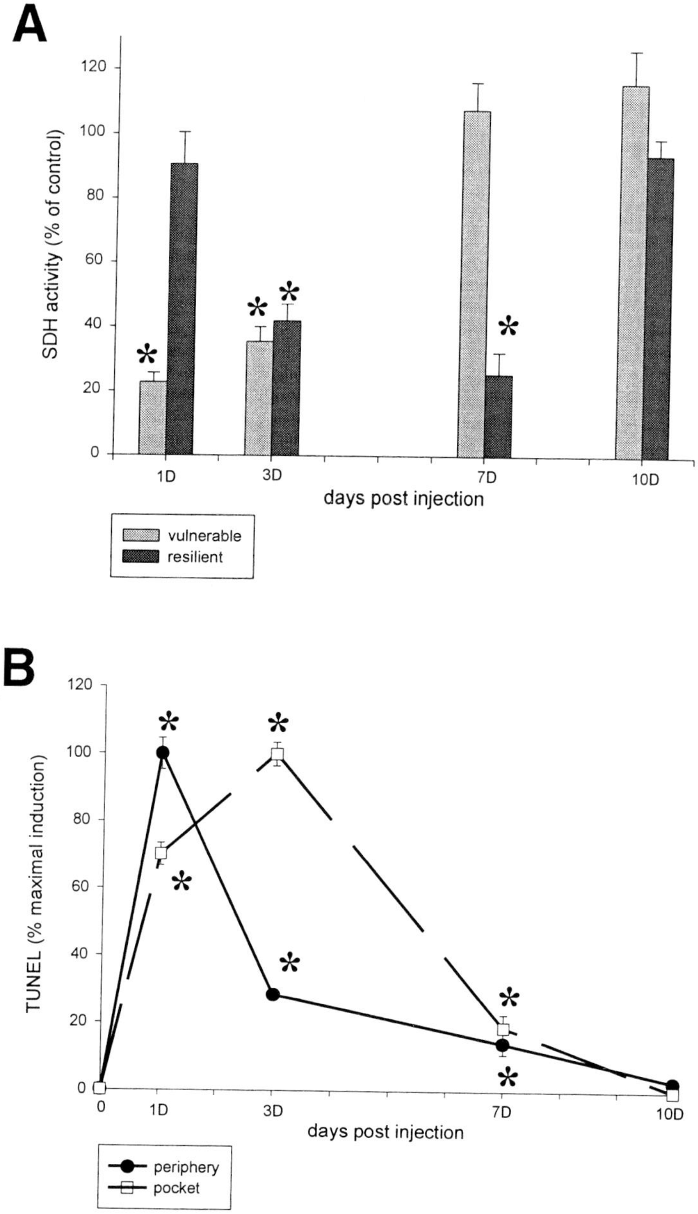

Fig. 2. Comparison of the time course of inhibition of SDH activity and TUNEL induction by 3-NPA in rat striatum. Sprague–Dawley rat striatum after systemic 3-NPA (30 mg/kg, i.p.) injection. **(A)** Relative optical densities of blue diformazan deposits representing SDH activity are presented as percentage of control striatum (means ± SEM). Measurements were taken from the region surrounding the SDH-

number of TUNEL-positive cells at 3 d was delayed relative to that of maximal inhibition of SDH activity which occurred at 1 d. A detailed examination of TUNEL-stained cells revealed that approximately half of them contained condensed chromatin bodies that were either within the nucleus (asterisks in Fig. 1E) or had been released from the nucleus (Fig. 1F). These features are characteristic of an apoptotic process of cell death, which is characterized by internucleosomal DNA fragmentation, chromatin clumping, and cell shrinkage followed by nuclear fragmentation and budding off of apoptotic bodies, and then rapid clearance of cell debris by phagocytosis *(23,24)*. The majority of the remaining cells displayed diffuse TUNEL staining, which is more suggestive of a necrotic process of random DNA fragmentation and cell death (other cells in Fig. 1E). An induction of TUNEL has also been associated with postmortem HD brain *(25)*.

Because TUNEL induction was associated with decreases in SDH activity in vulnerable rats, we expected to see an induction of TUNEL in resilient rats because both groups of animals had similar maximal declines in SDH activity. Remarkably, resilient rats did not show any induction of TUNEL out to 10 d in any area of the brain. In vulnerable rats TUNEL was induced within the first 24 h after 3-NPA injection, which was the same timing as the maximal decrease in SDH activity. In resilient rats there was still no induction of TUNEL even 72 h after the time of maximal decline in SDH activity (i.e., maximal decline in SDH activity was at 7 d after 3-NPA injection and by 10 d there was still no induction of TUNEL). In other words, the fast decline in SDH activity in vulnerable animals was associated with an induction in the number of TUNEL stained dead and dying cells, whereas the gradual decline of SDH activity in resilient animals was not.

At 1 and 3 d after 3-NPA injection there was a large number of apparently healthy cells within the striatal pocket that stained normally for hematoxylin/esoin and were negative for TUNEL. These cells coexisted among

---

depleted pocket in vulnerable rats (*gray bars*) and from a comparable area in resilient rats (*black bars*). Vulnerable rats showed a more rapid 1 d temporal decline whereas resilient rats showed a gradual 7-d decline after 3-NPA. Both groups of animals showed full recovery of SDH activity by 10 d. **(B)** TUNEL induction within the pocket (*squares*) and the surrounding striatal periphery (*circles*) in vulnerable rats. TUNEL cell counts are represented as percentage of maximal induction, i.e., we divided all time points by the time point with the highest induction. Real values for maximal induction were: periphery = 2.11 ± 0.1 cells/mm^2 at 1 d; pocket = 1401.5 ± 49.1 cells/mm^2 at 3 d. Control TUNEL cell counts were 0.03 ± 0.002 cells/mm^2. No induction of TUNEL was found in resilient rats (0.02 ± 0.001 cells/mm^2). $^*p < 0.01$ ANOVA with a post-hoc Student Neuman–Keuls test.

Table 1
Summary of the Toxic Effects of Systemic 3-NPA on Male Rats

Measure	Control	Vulnerable		Resilient
		Periphery	Pocket	
Toxicity % (total *n*=69)	—	24.6%[a]		53.7%[a]
SDH activity maximal decline (% of control)	100%	22.7 ± 2.9%[b]	2.8 ± 0.3 %[c]	25.5 ± 5.8%[d]
TUNEL maximal induction (cells/mm²)	0.03 ± 0.002	2.11 ± 0.1[b]	1401.5 ± 49.1[c]	No change

Systemic 3-NPA (30 mg/kg, i.p.) was given to 10-wk-old male Sprague–Dawley rats. Controls received a single injection of vehicle (PBS). Animals showed variable susceptibility to neurotoxicity in the striatum (TUNEL). The level of depletion of SDH activity was similar in all animals except that vulnerable rats developed a pocket of depleted SDH activity within the striatum. TUNEL induction occurred only in vulnerable animals in both the pocket and in the striatal peripheral region surrounding the pocket.

[a]3-NPA caused mortality in the remaining animals (21.7%). [b]At 1 day. [c]At 3 d. [d]At 7 d.

TUNEL-positive cells within the pocket. Thus, there existed a large population of neurons that were still alive within the pocket but that had little or no SDH activity. Therefore, the loss in SDH activity apparently precedes the death of neurons, at least for many cells.

The variability in susceptibility to neurotoxicity due to 3-NPA in our Sprague–Dawley rats could be explained by a number of factors. One factor that might explain the differences in the temporal pattern of inhibition of SDH activity is the blood–brain barrier. Suppose vulnerable animals had a "weaker" blood–brain barrier that allowed higher concentrations of 3-NPA to pass from the peripheral blood supply into the brain. This rush of 3-NPA into the brain would likely constitute a severe challenge. This might cause a fairly rapid decline in SDH activity. In turn, resilient rats might have a "stronger" blood–brain barrier that allows 3-NPA to only trickle into the brain, thus resulting in a longer time course of inhibition of SDH activity. A dysfunction of the blood–brain barrier has in fact been found to occur with systemic 3-NPA administration *(15,26)*. Nishino and colleagues *(15,26)* have proposed that the specific biophysical circumstances of the lateral striatal artery suggest that it is a weak link in the blood–brain barrier. Based on those reports we can speculate that 3-NPA may itself cause a dysfunction of the blood–brain barrier which would only become manifest in animals that previously had a vulnerable blood–brain barrier to begin with. Thus, only animals with vulnerable blood–brain barriers would succumb to 3-NPA poisoning. However, the total brain concentration of 3-NPA is not the only factor involved in the neurological outcome because an equivalent inhibition of SDH activity in vulnerable and resilient animals leads to lesioning only in vulnerable animals. Perhaps the more gradual inhibition seen in resilient rats might have allowed for recruitment of compensatory metabolic mechanisms, thus preventing the toxicity of 3-NPA in the striatum.

Another factor that might be involved in the variable vulnerability of rats is differences in physiological responses to the mechanism of toxicity by 3-NPA. 3-NPA (and its functional mimic malonic acid) is known to incur a secondary excitotoxic mechanism along the lines of glutamate receptor activation which leads to an irreversible cytotoxic intracellular calcium overload (for review *see* refs. *[6,13,27]*). An intriguing study by Schauwecker and Steward *(28)* demonstrated that certain mouse strains are more resilient to the neurotoxic effects of systemic injection of an excitotoxin. Thus, the physiological reasons behind this interstrain difference in excitotoxin vulnerability might also apply to our intrastrain variability in metabolic impairment-induced secondary excitotoxicity. In addition, current research has now identified a number of other processes that are involved in the mechanism of 3-NPA toxicity (for review *see* ref. *[13]*). Such factors include the following: one, oxidative stress—which produces cytotoxic free radicals—and antioxidant systems *(27,29)*; two, cellular ionic integrity, especially the function of the ATP-dependent Na^+, Ca^{2+} antiporter *(17,30)*, which, if not maintained, can lead to neuronal depolarization *(31)* and toxic intracellular calcium overload *(17,30)*; three, striatal dopamine transmission, which can lead to the production of free radicals and intracellular calcium overload *(15)*; and four, cellular biochemical metabolism, especially of the major striatal cell phenotype, γ-aminobutyric acid (GABA) neurons, as GABAergic neurons appear to be especially vulnerable to 3-NPA-induced inhibition of the tricarboxylic acid cycle *(32)*. Thus, differences in animals in any of these or any combination of these could be implicated as causative factors in the variable vulnerability to 3-NPA that we see in our Sprague–Dawley rats.

SUMMARY AND CONCLUSIONS

Our studies show that there is a variable susceptibility to neurotoxicity of systemic 3-NPA in Sprague–Dawley rats. We have learned that a rapid (1 d) decline in brain SDH activity induces neurological damage in the striatum whereas a gradual (7 d) decline does not. This distinction occurred despite the fact that the absolute levels of the declines in SDH activity were almost identical. The rapid decline seen in vulnerable rats was associated with a striatal pocket of fully depleted SDH activity and neurological lesioning, which was associated both with the pocket and with the peripheral region surrounding the pocket. The fast decline in SDH activity in vulnerable rats is followed by an induction of TUNEL and cell death whereas the slower decline in resilient rats is not.

The existence of animals that are resilient to 3-NPA suggests that there are mediating factors that prevent or bypass the cell death seen in vulnerable animals. Alternatively, vulnerable animals might have physiological deficits not found in resilient animals. The identification of the specific factors conferring either resilience or vulnerability to metabolic poisons could provide strategies for therapeutic intervention in conditions of metabolic failure such as ischemia and stroke, and possibly neurodegenerative disorders such as HD that may also involve metabolic compromise.

ACKNOWLEDGMENTS

We thank Dr. David Batchelor for help with preparation of the figures. This work was partially supported by the National Institutes of Health (TA), the Health Research Council of New Zealand (PEH, RLMF), and the New Zealand Neurological Foundation (RLMF).

REFERENCES

1. Ludolph AC, He F, Spencer PS, et al. 3-Nitropropionic acid—exogenous animal neurotoxin and possible human striatal toxin. Can J Neurol Sci 1991; 18:492–498.
2. He F, Shoulin Z, Qian F, et al.. Delayed dystonia with striatal CT lucencies induced by a mycotoxin (3-nitropropionic acid). Neurology 1995;45:2178–2183.
3. Hu W. The isolation and structure identification of a toxic substance, 3-nitropropionic acid, produced by *Arthrinium* from mildewed sugar cane. Clin J Prev Med 1986;20:321–323.
4. Liu X. Studies on the mycology and mycotoxins in an outbreak of deteriorated sugar cane poisoning. Clin J Prev Med 1989;23:345–348.
5. Alston TA, Mela L, Bright HJ. 3-Nitropropionate, the toxic substance of *Indigofera*, is a suicide inactivator of succinate dehydrogenase. Proc Natl Acad Sci USA 1977;74:3767–3771.
6. Greene J G, Greenamyre. JT. Bioenergetics and excitotoxicity: the weak excitotoxic hypothesis. In: Olanow CW, Jenner P, Youdim M, eds. Neurodegeneration and Neuroprotection in Parkinson's Disease. Academic Press, London, 1996, pp. 125–142.
7. Dawson R Jr, Beal MF, Bondy SC, et al.. Excitotoxins, aging, and environmental neurotoxins: implications for understanding human neurodegenerative diseases. Toxicol Applied Pharmacol 1995;134:1–17.

8. Brouillet E, Hantraye P. Effects of chronic MPTP and 3-nitropropionic acid in nonhuman primates. Curr Opin Neurol 1995;8:469–473.
9. Borlongan CV, Koutouzis TK, Sanberg PR. 3-Nitropropionic acid animal model and Huntington's disease. Neurosci Biobehav Rev 1997;21:289–293.
10. Borlongan CV, Nishino H, Sanberg PR. Systemic, but not intraparenchymal, administration of 3-nitropropionic acid mimics the neuropathology of Huntington's disease: a speculative explanation. Neurosci Res 1997;28:185–189.
11. Brouillet E, Hantraye P, Ferrante RJ, et al. Chronic mitochondrial energy impairment produces selective striatal degeneration and abnormal choreiform movements in primates. Proc Natl Acad Sci USA 1995;92:7105–7109.
12. Palfi S, Ferrante RJ, Brouillet E, et al. Chronic 3-nitropropionic acid treatment in baboons replicates the cognitive and motor deficits of Huntington's disease. J Neurosci 1996;16:3019–3025.
13. Alexi T, Hughes PE, Faull RLM, et al. 3-Nitropropionic acid's lethal triplet: cooperative pathways of neurodegeneration. NeuroReport 1998;9:R57–R64.
14. Guyot M.-C, Hantraye P, Dolan R, et al. Quantifiable bradykinesia, gait abnormalities and Huntington's disease-like striatal lesions in rats chronically treated with 3-nitropropionic acid. Neuroscience 1997;79:45–56.
15. Nishino H, Kumazaki M, Fukuda A, et al. Acute 3-nitropropionic acid intoxication induces striatal astrocytic cell death and dysfunction of the blood–brain barrier: involvement of dopamine toxicity. Neurosci Res 1997;27:343–355.
16. Alexi T, Hughes PE, Knusel B, et al. Metabolic compromise with systemic 3-nitropropionic acid produces striatal apoptosis in Sprague Dawley rats but not BALB/c ByJ mice. Exp Neurol 1998;153:74–93.
17. Fukuda A, Deshpande SB, Shimano Y, et al. Astrocytes are more vulnerable than neurons to cellular Ca^{2+} overload induced by a mitochondrial toxin, 3-nitropropionic acid. Neuroscience 1998;87:497–507.
18. Beal MF, Brouillet E, Jenkins BG, et al. Neurochemical and histologic characterization of striatal excitotoxic lesions produced by the mitochondrial toxin 3-nitropropionic acid. J Neurosci 1993;13:4181–4192.
19. Bossi SR, Simpson JR, Isacson O. Age dependence of striatal neuronal death caused by mitochondrial dysfunction. NeuroReport 1993;4:73–76.
20. Brouillet, E, Jenkins BG, Hyman BT, et al. Age-dependent vulnerability of the striatum to the mitochondrial toxin 3-nitropropionic acid. J Neurochem 1993;60:356–359.
21. Borlongan CV, Koutouzis TK, Freeman TB, et al. Behavioral pathology induced by repeated systemic injections of 3-nitropropionic acid mimics the motoric symptoms of Huntington's disease. Brain Res 1995;697:254–257.
22. Gavrieli Y, Sherman Y, Ben-Sasson SA. Identification of programmed cell death *in situ* via specific labeling of nuclear DNA fragmentation. J Cell Biol 1992;119:493–501.
23. Charriaut-Marlangue C, Ben-Ari Y. A cautionary note on the use of the TUNEL stain to determine apoptosis. NeuroReport 1995;7:61–64.
24. Hughes PE, Alexi T, Schreiber SS. A role for the tumour suppressor gene p53 in regulating neuronal apoptosis. NeuroReport 1997;8:v–xii.

25. Dragunow M, Faull RLM, Lawlor P, et al. *In situ* evidence for DNA fragmentation in Huntington's disease striatum and Alzheimer's disease temporal lobes. NeuroReport 1995;6:1053–1057.
26. Nishino H, Shimano Y, Kumazaki M, et al. Chronically administered 3-nitropropionic acid induces striatal lesions attributed to dysfunction of the blood–brain barrier. Neurosci Letts 1995;186:161–164.
27. Beal MF. Aging, energy, and oxidative stress in neurodegenerative diseases. Ann Neurol 1995;38:357–366.
28. Schauwecker PE, Steward O. Genetic determinants of susceptibility to excitotoxic cell death: implications for gene targeting approaches. Proc Natl Acad Sci USA 1997;94:4103–4108.
29. Schulz J B, Matthews RT, Jenkins BG, et al. Blockade of neuronal nitric oxide synthase protects against excitotoxicity in vivo. J Neurosci 1995;15: 8419–8429.
30. Deshpande SB, Fukuda A, Nishino H. 3-Nitropropionic acid increases the intracellular Ca^{2+} in cultured astrocytes by reverse operation of the Na^+-Ca^{2+} exchanger. Exp Neurol 1997;145:38–45.
31. Riepe M, Hori N, Ludolph AC, et al. Inhibition of energy metabolism by 3-nitropropionic acid activates ATP-sensitive potassium channels. Brain Res 1992; 586:61–66.
32. Hassel B, Sonnewald U. Selective inhibition of the tricarboxylic acid cycle of GABAergic neurons with 3-nitropropionic acid in vivo. J Neurochem 1995; 65:1184–1191.

10

The 3-Nitropropionic Acid Model of Huntington's Disease

Do Alterations in the Expression of Metabolic mRNAs Predict the Development of Striatal Pathology?

Keith J. Page, Alicia Meldrum, and Stephen B. Dunnett

INTRODUCTION

Metabolic compromise has been found to precede both neuronal loss and the appearance of motoric and cognitive deficits in Huntington's disease (HD) patients. Studies using both magnetic resonance imaging (MRI) and positron emission tomography (PET) to assess the metabolic status of the central nervous system (CNS) have demonstrated anatomically restricted deficits in energy metabolism in the caudate putamen and frontal cortex of both confirmed HD patients and also those genetically at risk of developing the disease *(1–6)*. These findings have been confirmed biochemically in regional assessments of oxidative damage and metabolic dysfunctions in the HD brain *(7)*. Taken together, these findings strongly suggest that a preclinical determinant of HD is the specific inhibition of energy metabolism in those brain regions most affected during the course of the pathology. Precisely how metabolic inhibition leads to neurodegeneration in the HD brain is currently not fully understood. However, analysis of the consequences of metabolic dysfunction in the CNS has opened avenues for researchers interested in establishing the pathological basis for not only Huntington's disease but also Alzheimer's and Parkinson's diseases and schizophrenia *(8,9)*.

In animals, CNS energy metabolism can be inhibited in a number of ways, and in the context of HD, two main approaches have been used. Both make use of the competitive or noncompetitive inhibition of succinate dehydrogenase (SDH), an enzyme critical to both the Krebs cycle and the mitochondrial electron transport *(10–12)*. Direct intrastriatal infusions of malonic acid (malonate) produce lesions that in many ways are very similar to those seen

From: *Mitochondrial Inhibitors and Neurodegenerative Disorders*
Edited by: P. R. Sanberg, H. Nishino, and C. V. Borlongan © Humana Press Inc., Totowa, NJ

in advanced HD; medium spiny projection neurons are targeted while interneurons expressing neuronal nitric oxide synthase (nNOS) and NADPH-diaphorase (NADPHd) are largely spared. The neurotoxic effects are age dependent, with older animals displaying a lower toxicity threshold than younger ones, and the morphology of neurons undergoing malonate toxicity is very similar to that seen in HD *(13–15)*. Intrastriatal infusion of malonate in old animals provides a reasonable approximation of the end point of the disease, modeling effectively the anatomical pattern of the latter stages of striatal pathology (Chapter 14). In this respect, malonate toxicity has its greatest value in allowing us to correlate more accurately the motoric and cognitive deficits seen in animals bearing these lesions, with damage to specific neural systems, particularly the direct and indirect striatonigral projections, which are known to be differentially affected in HD 16. However, as malonate toxicity is an acute treatment with a highly restricted anatomical range, as a model of progressive neurodegenerative nature of HD, when considered in isolation, its usefulness is limited.

HD is a genetic disorder, which has its neuropathological basis in an unstable poly-CAG expansion in exon 1 of the *huntingtin* gene *(17)*. The consequence of this mutation is the translation of a novel protein with a lengthened polyglutamine tract and this mutation is believed to confer a neurotoxic "gain of function" to the novel Huntingtin protein *(18)*. Both the wild-type and the mutant Huntingtin proteins are ubiquitously expressed in the CNS as well as in the majority of other tissues *(19–23)*. HD pathology undergoes a very precise pattern of neurodegenerative progression, one that does not allow the drawing of a simple relationship between expression of mutant huntingtin in the CNS with the development of the pathology *(24)*. Although recent immunohistochemical studies have identified a correlation between expression of wild-type huntingtin in the rat CNS with the compartmentalized striatal origins of HD neurodegeneration *(25,26)*.

To model HD pathology in experimental animals without using genetic manipulation of the *huntingtin* gene, researchers have adopted an approach that, although along similar biochemical lines to malonate toxicity, avoids using acute and immediately neurotoxic manipulations of the CNS. 3-Nitropropionic acid (3-NPA), like malonate, is an irreversible SDH inhibitor, yet when chronically administered, produces replicable and well-characterized CNS lesions of which the first to appear are localized in the dorsolateral striatum *(27–30)*. The mechanism for the neurotoxic actions of 3-NPA are not fully understood, but may involve interactions between metabolically inhibited neurons and high levels of endogenous glutamate and dopamine

(27,31–33). Alternatively, 3-NPA-induced neurotoxicity may depend upon the compromise of the blood–brain barrier following metabolic inhibition and death of astrocytes, leading to localized striatal hemorrhage *(34,35)*. There is also evidence to support roles for nitric oxide and the generation of hydroxyl free radicals in 3-NPA-induced toxicity *(36)* and further evidence supporting apoptosis as a neurodegenerative mechanism for 3-NPA in in vitro model systems as well as in HD *(37–41)*.

Systemically administered 3-NPA has been widely adopted as an approach to replicating experimentally some of the features of striatal pathology and the consequent motoric impairments in rats and primates *(8,28–30,42,43)*. In our studies to date *(51)*, we have used 3-NPA toxicity to analyze the consequences of metabolic inhibition and resultant striatal pathology on the expression cytochrome oxidase (COX-II and COX-IV) and SDH mRNA and on the expression of astrocytic glial fibrillary protein (GFAP) mRNA.

We predicted that subtoxic 3-NPA would lead to the dose-dependent regulation of COX-II, COX-IV, and SDH mRNA expression and activity in the CNS. If the unique pattern of 3-NPA-induced striatal toxicity was a reflection of the increased susceptibility of striatal neurons, then we may have expected to see specific alterations in striatal COX and SDH activities and in the expression of COX-II, COX-IV, and SDH mRNA.

MATERIALS AND METHODS

Thirty-two male Lister Hooded rats (300 g at the start of experiment) were used in this study; they were housed in groups of four under normal light/dark conditions with food and water freely available. We established three experimental groups: group 1 received daily injections of water (n = 5); group 2 received 3-NPA (n = 9, 1 mg/kg, i.p., in water pH 7.4, 1/d); group 3 received 3-NPA (n = 9, 20 mg/kg, i.p., 1/d) for 14 consecutive days. Body weights and appearance were recorded. Individuals in the 20 mg/kg 3-NPA-treated groups began to lose weight after the fifth injection and mean body weight at sacrifice was approx 80% of starting weight. From d 5, these animals were maintained on a diet of soggy rat chow and received daily injections of 2 mL of glucose/saline intraperitoneally approx 6 h after the 3-NPA injection. At sacrifice, all the rats in the 20 mg/kg treated group displayed hindlimb paralysis and recumbency.

Six hours after the final injection on d 14, the rats were sacrificed by exposure to elevated concentrations of CO_2 followed by cervical dislocation. Brains were rapidly removed, snap-frozen in prechilled isopentane, and stored at –70°C for subsequent analysis of COX-II, COX-IV, SDH, and

Table 1
Oligonucleotide Design Details

Target mRNA	Bases	GenBank	Reference
Rat COX-II	531-575	M64496	Cao et al., 1991 *(47)*
Rat COX-IV	442-486	X14209	Goto et al., 1989 *(46)*
Rat GFAP	324-368	L27219	Chen and Liem, 1993 *(48)*
Rat SDH -1	91-136	H33531	Lee et al., 1995 *(49)*
Rat SDH-2	152-276	H32609	Lee et al., 1995 *(49)*

Oligonucleotide probes were designed using Oligo Primer Analysis software for Macintosh (Rychlik 1989–1992). Appropriate oligonucleotide sequences were typically obtained from the 5'-untranslated regions and contained 55%:45% GC:AT with minimal potential for hairpins or alignment autohybridizations.

GFAP mRNA expression by radioactive *in situ* hybridization histochemistry (ISHH), and COX and SDH activity by enzyme histochemistry. The procedure for the ISHH has been described in detail elsewhere, as have the methods for assessing COX and SDH using enzyme histochemistry *(44,45)*. We designed 45-mer sense and antisense oligonucleotide probes for rat COX-II (mitochondrial subunit) and COX-IV (nuclear subunit) and rat GFAP mRNA using published full-length sequences *(46–48)*. Two pairs of probes against a partial sequence of rat mRNA that displayed high homology to human SDH mRNA were designed and used in parallel confirmatory hybridizations *(49)*. Oligonucleotide probes were synthesized commercially and purified by polyacrylamide gel electrophoresis (Genosys, Cambridge, UK; *see* Table 1). First, we mapped the expression of COX-II, COX-IV, and SDH mRNA in 14 μm sagittal brain sections from two naive rats (identical strain, sex, and age to those used experimentally).

RESULTS

The expression of COX-II, COX-IV, and SDH mRNA in control rats is presented in Fig. 1. COX-II, COX-IV, and SDH mRNA were expressed ubiquitously in the CNS. Analysis of emulsion-dipped sections confirmed the neuronal localization of COX-II, COX-IV, and SDH mRNA (not shown). Hybridizations using complementary "sense" oligonucleotide probes gave no autoradiographic signal when performed in parallel to the antisense hybridizations. There was little variability in the level of mRNA expression across CNS loci, with the exception that white matter regions such as the internal capsule and corpus callosum displayed no autoradiographic signal for any of the target mRNAs.

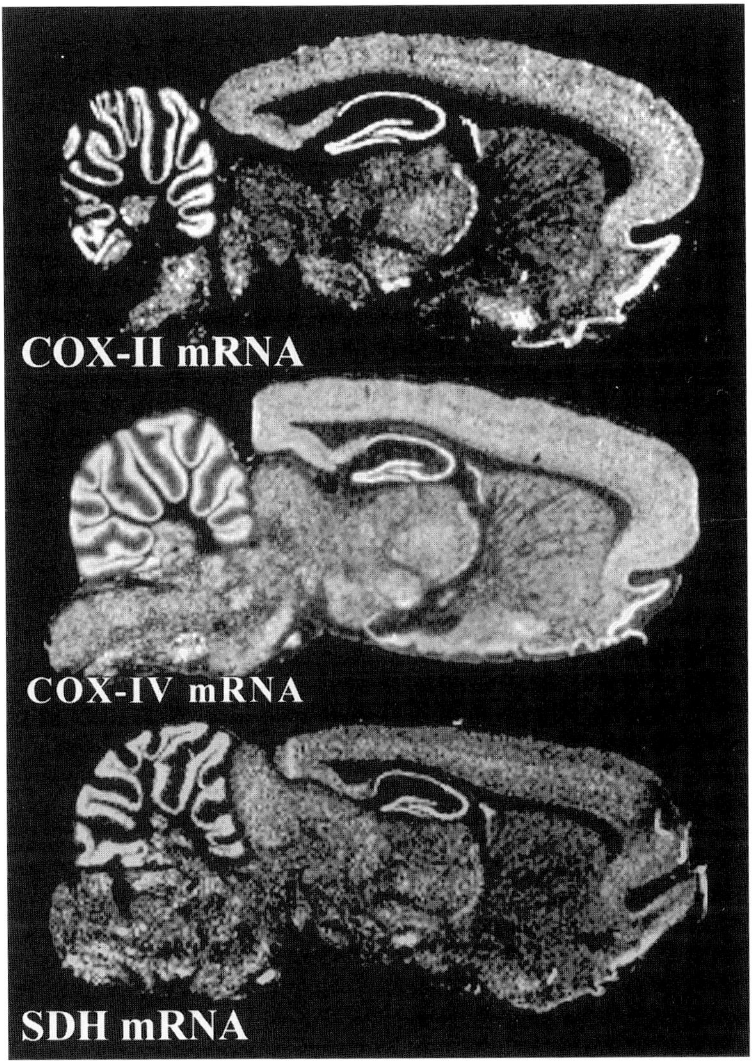

Fig. 1. Dark-field autoradiograms showing the expression of COX-II mRNA, COX-IV mRNA, and SDH mRNA in sagittal sections of rat brain. All three target mRNAs are ubiquitously and strongly expressed in gray matter. Expression in white matter, the corpus callosum, and internal capsule, for example, is typically low. "Sense" hybridizations gave no autoradiographic signal (*not shown*).

Analysis of Nissl-stained 14-μm sections confirmed the presence of focal bilateral lesions in the dorsolateral striatum in the 20 mg/kg treated group and in none of the other groups. We measured the optical density (OD) of

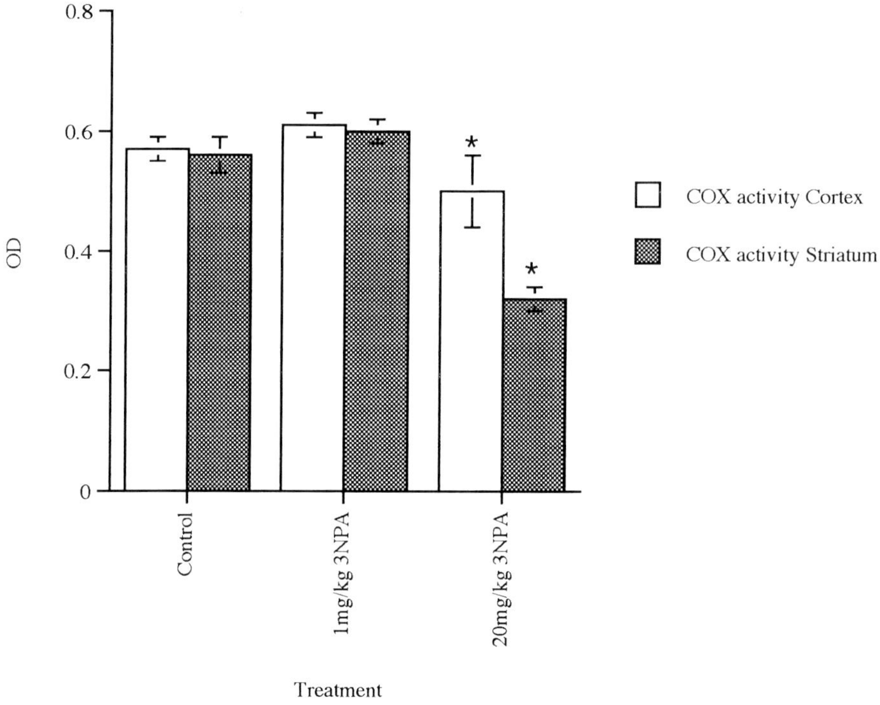

Fig. 2. Optical density of sections stained histochemically to demonstrate the activity of COX. COX activity is significantly reduced in the neocortex and striatum of rats treated with 20 mg/kg 3-NPA ($*p < 0.05$), but unchanged in rats treated with 1 mg/kg 3-NPA.

COX and SDH staining and of COX-II, COX-IV, and SDH mRNA expression in ventromedial striatum and neocortex. ODs were measured using NIH Image software (Version 1.5). Ventromedial (VM) striatum was chosen because of its proximity to the dorsolateral striatal lesions. In the sections in which striatal lesions were present, they never included the VM striatal region where measurements were made in equivalent nonlesioned rats.

Unfixed 14-μm sections stained histochemically for COX and SDH activity revealed marked effects. COX activity was reduced in nonlesioned VM striatum and neocortex in the 20 mg/kg 3-NPA-treated groups. COX activity in the 1 mg/kg 3-NPA-treated group was not significantly different from COX activity in the control rats (*see* Fig. 2). SDH activity was reduced in a dose-dependent manner across all experimental conditions; SDH activity was greatest in controls, less in 1 mg/kg 3-NPA-treated and least of all in the 20 mg/kg 3-NPA-treated rats (*see* Fig. 3).

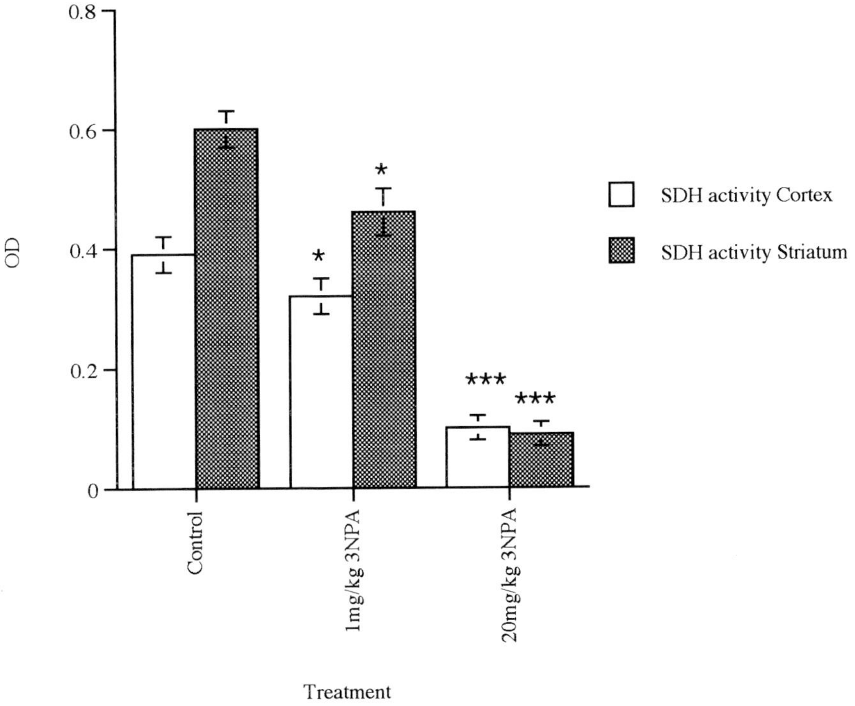

Fig. 3. Optical density of sections stained histochemically to demonstrate the activity of SDH. SDH activity is significantly reduced in neocortex and striatum of rats treated with 20 mg/kg of 3-NPA (***$p < 0.0001$) and also in those treated with 1 mg/kg of 3-NPA (*$p < 0.05$), thus indicating that although the lesions induced by systemic administration of 3-NPA are localized to the striatum, the pharmacological effects of 3-NPA (SDH inhibition) affect CNS loci outside of the basal ganglia to approxiamtely the same degree.

Quantitation of the COX-II and COX-IV mRNA expression in VM striatum and neocortex revealed no effects on COX-II or COX-IV mRNA expression in any group other than the 20 mg/kg 3-NPA treated rats (*see* Figs. 4, 5, and 6). In the 20 mg/kg 3-NPA-treated rats, expression of COX-IV mRNA was significantly decreased in the VM striatum. The presence of bilateral lesions was evident by an absence of COX-II, -IV, and SDH mRNA expression in dorsolateral striatum (*see* Fig. 4). There were no significant differences in the expression of SDH mRNA in any of the 3-NPA-treated groups (*see* Figs. 4 and 7).

The expression of GFAP mRNA was unchanged in the rats that had been systemically treated with 1 mg/kg 3-NPA; however, in the lesioned, 20 mg/kg

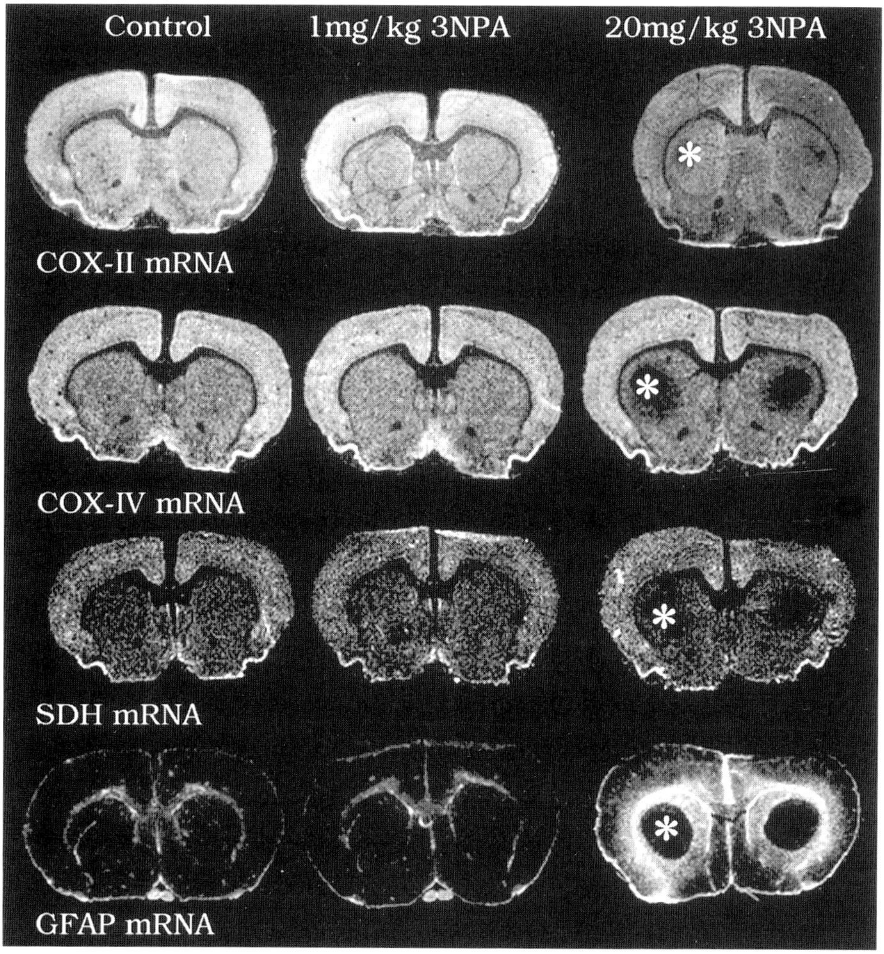

Fig. 4. Dark-field autoradiograms showing the expression of COX-II, COX-IV, SDH, and GFAP mRNA in coronal sections through the rat striatum, from subjects treated systemically with either 1 mg/kg or 20 mg/kg of 3-NPA or in control subjects. The presence of focal and bilateral striatal lesions is indicated by asterisks in the sections from 20 mg/kg 3-NPA-treated rats. There is a loss of signal within the lesion core for all target mRNAs, indicating loss of striatal neurons in the cases of COX-II, COX-IV, and SDH mRNA. In the sections hybridized to show GFAP mRNA, there is a clear and massive increase in message encircling the lesion core but no signal within it, suggesting perhaps that unlike directly excitotoxic lesions in which astrocytes rapidly invade the area of maximum neurodegeneration, 3-NPA-induced lesions are particularly damaging to the reactive astrocytes themselves.

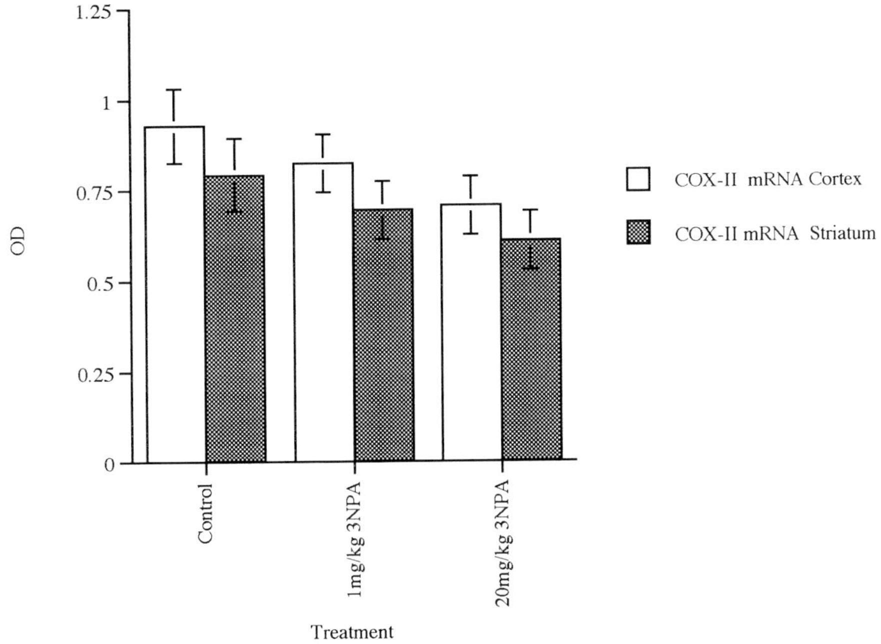

Fig. 5. Quantitation of the expression of COX-II mRNA in neocortex and striatum of rats treated with either 1 or 20 mg/kg 3-NPA or in untreated controls. There were no statistically significant differences in the expression of COX-II mRNA in these subjects, but a tendency for COX-II mRNA to decrease in both neocortex and striatum with increasing dose of 3-NPA.

3-NPA-treated rats, there were pronounced increases in the expression of GFAP mRNA in both VM striatum and overlying neocortex, reflecting astrocytosis in response to the striatal lesion. It was interesting to observe that GFAP mRNA expression was completely absent from the core of the lesion (*see* Fig. 4), a feature of systemic 3-NPA-induced striatal toxicity not seen in striatal lesions induced by the intracerebral injection of quinolinic acid or other glutamate receptor agonists (Thian, Page, and Dunnett, unpublished observations).

DISCUSSION

These experiments have demonstrated that although systemic injections of 3-NPA produce dose-dependent decreases in COX and SDH activity in both striatal and extrastriatal CNS loci, these changes in the activity of critical metabolic enzymes typically do not occur by regulation in the expres-

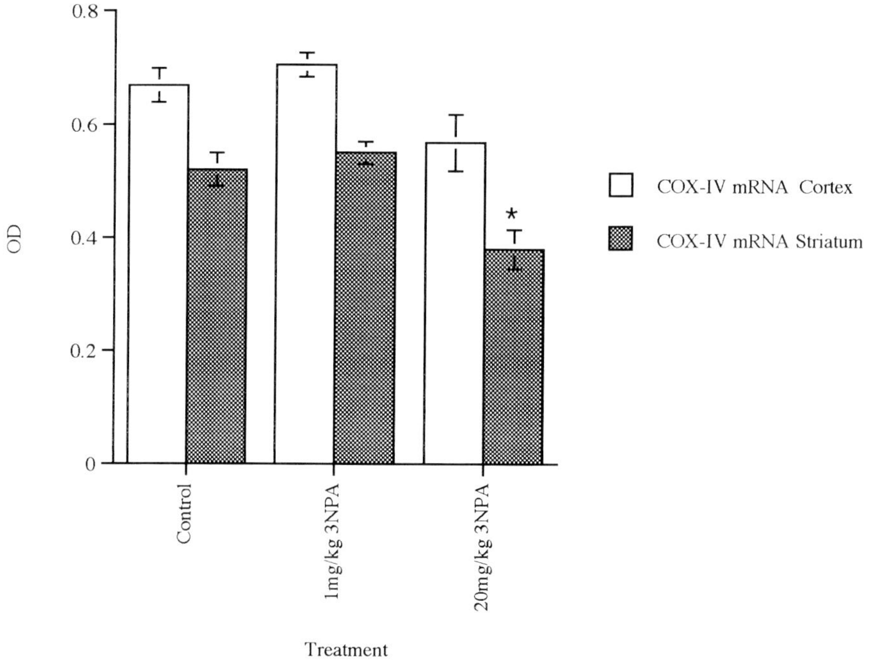

Fig. 6. Quantitation of the expression of COX-IV mRNA. There is a significant decrease in COX-IV expression in the striatum of 20 mg/kg 3-NPA treated rats (*$p < 0.05$) and a tendency for COX-IV mRNA to be decreased in the neocortex of the same group.

sion of the mRNAs encoding them. Systemic administration of 3-NPA at 20 mg/kg was found to produce severe weight loss in animals that preceded hindlimb paralysis and recumbency. It was, however, possible to maintain the viability of the most affected subjects using twice daily injections of glucose/saline and allowing free access to hydrated food. Bilateral striatal lesions were evident only in the group treated with the highest dose of 3-NPA (20 mg/kg). The striatal toxicity of 3-NPA at 20 mg/kg was accompanied in surviving regions of the striatum by reductions in the expression of COX-IV mRNA. There were no significant changes in COX-II, COX-IV, or SDH mRNA expression at either of the lower doses. The finding that SDH and COX activities are reduced in both striatal and nonstriatal CNS loci suggests that the specific patterns of neurotoxicity induced by systemic 3-NPA are not simple reflections of preferential SDH inhibition in these nuclei, but rather some other characteristic of vulnerable striatal neurons or their neural/glial environment.

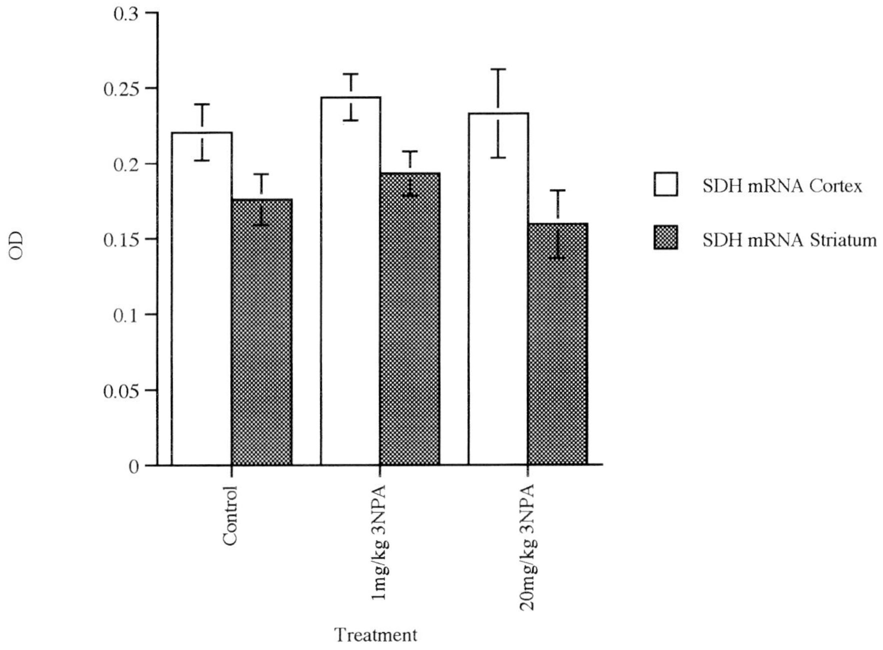

Fig. 7. Quantitation of the expression of SDH mRNA. There were no significant differences between differentially treated groups and no readily identifiable trends.

The astrocytic response in the 20 mg/kg 3-NPA-treated rats was markedly dissimilar to that seen following the induction of similar sized lesions by the direct injection of quinolinic acid. Reactive astrocytes, rather than invading the core of the lesion, as we have observed several days after striatal injections of quinolinic acid, congregated densely at the interface between lesioned and intact striatum acid (Thian, Page, and Dunnett, unpublished observations). It has been suggested that one of the primary modes of action of systemic 3-NPA in in vivo systems is to attack astrocytes supporting the blood–brain barrier, thereby promoting its breakdown and initiating a hemorrhage-like neurotoxic event. Ultrastructural changes in astrocyte end-feet in the striatum have been observed as early as 1 h following the second daily injection of 3-NPA (20 mg/kg i.p.), a time point at which there were no ultrastructural abnormalities detected in cortical sections *(35)*. In the same series of experiments, Nishino and colleagues convincingly demonstrate a greater vulnerability of astrocytes to 3-NPA *(34,35)*. It is unclear, however, why striatal astrocytes should exhibit enhanced vulnerability to systemic 3-NPA, as it is clear from our studies that the striatum exhibits no greater or lesser inhibition of SDH activity following the systemic administration of 3-NPA.

In the present study it was clear that systemic 3-NPA did not alter greatly the expression of metabolic mRNAs. We predicted that we would see the greatest regulation in the expression of SDH mRNA, reflecting the specific inhibition of SDH activity. Indeed in the 20 mg/kg 3-NPA-treated rats, we report a global decrease in SDH activity to approx 15% of that seen in control rat brain, without significant up- or down-regulation of SDH mRNA. It is possible that the regulation of COX and SDH mRNAs following 3-NPA treatment is transient and at the time point different from the one at which we sacrificed our subjects. At 6 h after the final injection, SDH and COX mRNA may have returned to normal levels of expression. Studies are underway to assess the time course of the regulation of metabolic gene expression after a single high dose (40 mg/kg) of 3-NPA.

Our studies have revealed several important features of striatal 3-NPA toxicity. First, regulation of both COX and SDH mRNA expression was surprisingly not seen in brain sections where the activities of both enzymes had been reduced. Both COX and SDH activity are decreased globally following 3-NPA administration; therefore the pattern of enzyme inhibition does not match the pattern of neurodegeneration. Lastly, the astrocytic responses to 3-NPA lesions are markedly different from those seen following quinolinic acid-induced lesions, supporting the hypothesis that damage to the striatal astrocyte population may be a central feature of 3-NPA-induced toxicity.

The results of our studies and others allow us to make certain observations on the usefulness of 3-NPA as a model of the striatal pathology seen in HD. Neuroanatomically, 3-NPA-induced toxicity has many similarities to the late-stage HD striatum; medium spiny projection neurons rather than magnocellular interneurons are targeted and the morphological features of surviving NADPH-containing neurons are similar for both pathologies *(27,30,50)*. However, unlike the progressive neurodegeneration seen in the HD striatum, considerable evidence is accumulating to suggest that 3-NPA-induced degeneration appears to be an *acute* neurotoxic event rather than a *progressive* degeneration. Thus, the anatomical consequences of 3-NPA-induced toxicity may be superficially similar those seen in the late-stage HD brain, but the mechanism through which this neuropathology is reached is likely to be rather different. 3-NPA-induced toxicity has been related to a unique feature of the striatal neuronal environment, the relatively high density of glutamatergic and/or dopaminergic inputs to medium spiny neurons. Perhaps it is more likely that 3-NPA-induced toxicity is a reflection of a unique feature of astrocytes, and why striatal astrocytes would be preferentially targeted will be an important direction for future research.

REFERENCES

1. Maziotta JC, Phelps ME, Pahl JJ, et al. Reduced cerebral glucose metabolism in asymptomatic subjects at risk for Huntington's disease. N Engl J Med 1987;316:357–362.
2. Kuhl DE, Phelps ME, Markham CH, Metter EJ, Riege WH, Winter P. Cerebral metabolism and atrophy in Huntington's disease determined by 18FDG and computerised tomography. Ann Neurol 1982;12:425–434.
3. Sax DS, Powser R, Kim A, Bhatia R, Cupples LA, Myers RH. Evidence of cortical metabolic dysfunction in early Huntington's disease by single-photon-emission computed tomography. Mov Disord 1996;11:671–677.
4. Martin WRW, Clark C, Ammann W, Stoessl AJ, Shytbel W, Hayden MR. Cortical glucose metabolism in Huntington's disease. Neurology 1992; 42:223–229.
5. Antonini A, Leenders KL, Spiegel R, et al. Striatal glucose metabolism and dopamine D2 receptor binding in asymptomatic gene carriers and patients with Huntington's disease. Brain 1996;119:2085–2095.
6. Harms L, Meierkord H, Timm G, Pfieffer L, Ludolph AC. Decreased *N*-acetylaspartate/choline ratio in the frontal lobe of patients with Huntington's disease; a proton magnetic resonance spectroscopy study. J Neurol Neurosurg Psychiatry 1997;62:27–30.
7. Browne SE, Bowling AC, MacGarvery U, et al. Oxidative damage and metabolic dysfunction in Huntington's disease—selective vulnerability of the basal ganglia. Ann Neurol 1997;41:646–653.
8. Beal MF, Hyman BT, Koroshetz W. Do defects in mitochondrial energy metabolism underlie the pathology of neurodegenerative disease? Trends Neurosci 1993;16:125–131.
9. Coyle JT, Puttfarcken P. Oxidative stress glutamate and neurodegenerative disorders. Science 1993;262:689–695.
10. Alston TA, Mela L, Bright HF. 3-Nitropropionic acid the toxic substance of *Indigofera* is a suicide inactivator of succinate dehydrogenase. Proc Natl Acad Sci USA 1977;74:3767–3771.
11. Coles CJ, Edmondson DE and Singer TP. Inactivation of succinate dehydrogenase by 3-nitropropionic acid. J Biol Chem 1979;254:5161–5167.
12. Webb JL. Enzyme and Metabolic Inhibitors, 2nd edit. Academic Press, New York, 1966.
13. Henshaw R, Jenkins BG, Schulz JB, et al. Malonate produces striatal lesions by indirect NMDA-receptor activation. Brain Res 1994;647:161–166
14. Beal MF, Brouillet E, Jenkins B, Henshaw R, Rosen B, Hyman BT. Age-dependent striatal excitotoxic lesions produced by the endogenous mitochondrial inhibitor malonate. J Neurochem 1993;61:1147–1150.

15. Greene JG, Greenamyre JT. Characterization of the excitotoxic potential of the reversible succinate dehydrogenase inhibitor malonate. J Neurochem 1995;64:430–436.
16. Richfield EK, Maguire-Zeiss KA, Vonkemanne HE, Voorn P. Preferential loss of preproenkephalin versus preprotachykinin neurons from the striatum of Huntington's disease patients. Ann Neurol 1995;38:852–860.
17. Huntington's Disease Collaborative Research Group. A novel gene containing a trinucleotide repeat that is expanded and unstable on Huntington's disease chromosomes. Cell 1993;72:971–983.
18. Albin RL and Tagle DA. Genetics and molecular biology of Huntington's disease. Trends Neurosci 1995;18:11–14.
19. Strong TV, Tagle DA, Valdes JM, et al. Widespread expression of the human and rat huntington's disease gene in brain and nonneural tissues. Nat Genet 1993;5:259–265.
20. diFiglia M, Sapp E, Chase K, et al. Huntingtin is a cytoplasmic protein associated with vesicles in human and rat brain neurons. Neuron 1995;14:1075–1081.
21. Sharpe AH, Loev SJ, Schilling G, et al. Widespread expression of Huntington's disease gene (IT15) product. Neuron 1995;14:1065–1074.
22. Wood JD, MacMillan JC, Harper PS, Lowenstein PR, Jones AL. Partial characterisation of murine *huntingtin* and apparent variations in the subcellular localisation of Huntingtin in human mouse and rat brain. Hum Mol Genet 1996;5:481–487.
23. Bhide PG, Day M, Sapp E, et al. Expression of normal and mutant huntingtin in the developing brain. J Neurosci 1996;16:5523–5535.
24. Vonsattel J-P, Myers RH, Stevens TJ, Ferrante RJ, Bird ED, Richardson EP. Neuropathological classification of Huntington's disease. Neuropathol Exp Neurol 1985;44:559–577.
25. Hedreen JC, Folstein SE. Early loss of neostriatal striosome neurons in Huntington's disease. J Neuropathol Exp Neurol 1995;54:105–120.
26. Kosinski CM, Cha J-H, Young AB, et al. Huntingtin immunoreactivity in the rat neostriatum: differential accumulation in projection and interneurons. Exp Neurol 1997;144:239–247.
27. Beal MF, Brouillet E, Jenkins BG, et al. Neurochemical and histologic characteristion of striatal excitotoxic lesions produced by the mitochondrial toxin 3-nitropropionic acid. J Neurosci 1993;13:4181–4192.
28. Brouillet E, Hantraye P, Ferrante RJ, et al. Chronic mitochondrial energy impairment produces selective striatal degeneration and abnormal choreiform movements in primates. Proc Natl Acad Sci USA 1995;92:7105–7109.
29. Borlongan CV, Koutouzis TK, Freeman TB, Cahill DW, Sanberg PR. Behavioural pathology induced by repeated systemic injections of 3-nitropropionic acid mimics the motoric symptoms of Huntington's disease. Brain Res 1995; 697:254–257.
30. Brouillet E, Jenkins BG, Hyman BT, et al. Age-dependent vulnerability of the striatum to the mitochondrial toxin 3-nitropropionic acid. J Neurochem 1993;60:356–359.

31. Simpson JR, Isacson O. Mitochondrial impairment reduces the threshold for in vivo NMDA-mediated neuronal death in the striatum. Exp Neurol 1993; 121:57–64.
32. Zeevalk GD, Derr-Yellin E, Nicklas WJ. NMDA receptor involvement in toxicity to dopamine neurons in vitro caused by the succinate dehydrogenase inhibitor 3-nitropropionic acid. J Neurochem 1995;64:455–458.
33. Bowyer JF, Clausing P, Schmued L, et al. Paternally administered 3-nitropropionic acid and amphetamine can combine to produce damage to terminals and cell bodies in the striatum. Brain Res 1996;712:221–229.
34. Nishino H, Shimano Y, Kumazaki M, Sakurai T. Chronically administered 3-nitropropionic acid induces striatal lesions attributed to dysfunction of the blood–brain barrier. Neurosci Lett 1995;186:161–164.
35. Nishino H, Kumazaki M, Fukada A, et al. Acute 3-nitropropionic acid intoxication induces astrocytic cell death and dysfunction of the blood–brain barrier: involvement of dopamine toxicity. Neurosci Res 1997;27:343–355.
36. Beal MF, Ferrante RJ, Henshaw R, et al. 3-Nitropropionic acid neurotoxicity is attenuated in copper/zinc superoxide dismutase transgenic mice. J Neurochem 1995;65:919–922.
37. Zeitlin S, Liu J-P, Chapman DL, Papaioannou VE, Efstratiadis A. Increased apoptosis and early embryonic lethality in mice nullizygous for the Huntington's disease gene homologue. Nat Genet 1995;11:155–163.
38. Portera-Cailliau C, Hedreen JC, Price DL, Koliatsos VE. Evidence for apoptotic cell death in Huntington disease and excitotoxic animal models. J Neurosci 1995;15:3774–3787.
39. Behrens MI, Koh J, Canzoniero LMT, Sensi SL, Csernansky CA, Choi DW. 3-Nitropropionic acid induces apoptosis in cultured striatal and cortical neurons. NeuroReport 1995;6:545–548.
40. Sato S, Gobbel GT, Honkaneimi J, et al. Apoptosis in the striatum of rats following intraperitoneal injection of 3-nitropropionic acid. Brain Res 1997;745:343–347.
41. Pang Z, Geddes JW. Mechanisms of cell death induced by the mitochondrial toxin 3-nitropropionic acid: acute necrosis and delayed apoptosis. J Neurosci 1997;19:3064–3073.
42. Palfi S, Ferrante RJ, Brouillet E, et al. Chronic 3-nitropropionic acid treatment in baboons replicates the cognitive and motor deficits of Huntington's disease. J Neurosci 1996;16:3019–3025.
43. Borlongan CV, Koutouzis TK, Sanberg PR. 3-Nitropropionic acid animal model of Huntington's disease. Neurosci Biobehav Rev 1997;21:289–293.
44. Sirinathsinghji DJS, Dunnett SB. Imaging gene expression in neural grafts. In: Sharif NA, ed. Molecular Imaging in the Neuroscience: A Practical Approach. IRL Press, Oxford, 1993, pp. 43–68.
45. Baker DM, Santer RM. Development of a quantitative histochemical method for determination of succinate dehydrogenase activity in autonomic neurons and its application to the study of aging in the autonomic nervous system. J Histochem Cytochem 1990;38:525–531.

46. Goto Y, Amuro N, Okazaki T. Nucleotide sequence of cDNA for rat brain and liver cytochrome *c* oxidase subunit IV. Nucleic Acids Res 1989;17:2851.
47. Cao J, Revzin A and Ferguson-Miller S. Conversion of a mitochondrial gene for mammalian cytochrome *c* oxidase subunit II into its universal codon equivalent and expression *in vivo* and *in vitro*. Biochemistry 1991;30:2642–2650.
48. Chen W-J, Liem RRKH GenBank Accession Number L27219. 1993.
49. Lee NH, Weistock KG, Kirkness EF, et al. Comparative expressed-sequence-tag analysis of differential gene expression profiles in PC-12 cells before and after nerve growth factor treatment. Proc Natl Acad Sci USA 1995; 92:8303–8307.
50. Wullner U, Young AB, Penney JB, Beal MF. 3-Nitropropionic acid toxicity in the striatum. J Neurochem 1994;63:1772–1781.
51. Page KJ, Dunnett SB, and Everitt BJ. 3-Nitropropionic acid-induced changes in the expression of metabolic and astrocytic mRNAs. NeuroReport 1998;9: 2881–2886.

11

Mechanisms of Action of 3-Nitropropionic Acid

Dopamine Overflow and Vulnerability of the Lateral Striatal Artery

Michiko Kumazaki, Chucharin Ungsuparkorn, Shripad B. Deshpande, Atsuo Fukuda, and Hitoo Nishino

INTRODUCTION

Neurons in the CA1 hippocampus, striatum, or reticular nucleus of the thalamus are most often damaged during brain ischemia. An abundant glutamatergic innervation that drives excitotoxity and the vulnerability of γ-aminobutyric acid-ergic (GABAergic) neurons to such stress has been suggested for the neuronal cell death of these areas *(1–4)*. Intoxication with 3-nitropropionic acid (3-NPA), a mycotoxin that inhibits succinate dehydrogenase irreversibly, leads to high-energy (ATP) deficiency *(5–7)* and production of free radicals *(8,9)*, thus results in cellular hypoxia and cellular damage. Systemic intoxication with 3-NPA induces striatum-selective lesions with motor symptoms, reminiscent of Huntington's disease *(10–13)*. Enhancement of excitotoxicity *(5–7)* and breakdown of the blood–brain barrier *(14–17)* have been suggested but mechanisms are not yet well understood. 3-NPA-induced striatal lesions are located in the lateral part of the striatum associated with dysfunction of the blood–brain barrier *(17)*. This suggests that specific mechanisms are operating in this area of the brain. In this chapter, we propose that the dopamine (DA) toxicity due to DA overflow and the vulnerability of the lateral striatal (lSTR) artery resulting from damage to endothelial cells and end-feet of astrocytes contribute to development of striatum-selective lesions following systemic 3-NPA intoxication. These characteristics may be major predisposing factors for neurodegenerative diseases, autoimmune diseases, or stroke that often affect the striatum.

From: *Mitochondrial Inhibitors and Neurodegenerative Disorders*
Edited by: P. R. Sanberg, H. Nishino, and C. V. Borlongan © Humana Press Inc., Totowa, NJ

Table 1
Research on 3-NPA Toxicity

Authors	Topic	Journal, Year, Volume, Pages
He F, et al. *(18)*	Three case reports	Chin Med J 1987; 67:395–396
Hamilton BF, Gould DH *(15)*	Blood–brain barrier impairment	Acta Neuropathol 1987;74:67–74
Beal MF, et al. *(6)*	Enhancement of excitotoxicity	J Neurosci 1993;13:4181–4192
Simpson JR, Isacson O *(7)*	*N*-Methyl-d-aspartate-mediated neuronal cell death	Exp Neurol 1993;121:57–64
Beal MF, et al. *(8)*	Cu/Zn superoxide dismutase	J Neurochem 1995;65:919–922
Schulz JB, et al. *(9)*	nNOS involvement	J Neurosci 1995;15:8419–8429
Nishino H, et al. *(17)*	Astrocytic cell death/blood–brain barrier dysfunction	Neurosci Lett 1995;186:161–164
Nishino H, et al. *(17)*	DA toxicity/Ca^{2+} overload/astrocytic cell death	Neurosci Res 1997;27:343–355

3-NPA-INDUCED NEURONAL AND GLIAL CELL DEATH

A clinical case of 3-NPA intoxication was first reported 12 yr ago *(18)*. These patients presented with the motor symptoms resembling Huntington's disease. The report drew extensive attention, especially in light of attempts to understand the pathophysiology of Huntington's disease and to develop an animal model. Table 1 summarizes the research in this area during the past 12 yr, which can be grouped along two major lines: one concerned with neuronal cell death and the other with glial/endothelial cell death. The current status of research in the latter field is reviewed here.

THE LSTR ARTERY IS HIGHLY VULNERABLE TO 3-NPA INTOXICATION

The lSTR artery, a perforate artery that arises from the middle cerebral artery, is very vulnerable to 3-NPA intoxication. In the following sections we discuss this vulnerability with the following perspectives: dysfunction of the blood–brain barrier, astrocytic cell damage, and endothelial cell damage.

Extravasation of Immunoglobulin G (Blood–Brain Barrier Dysfunction)

The damage to the blood–brain barrier in the striatum was addressed in very early reports of 3-NPA intoxication *(14,15)*, which stated that the damage to the blood–brain barrier was involved mainly and this would modulate the neuronal cell death at later stages. Recently, we detected damage to the

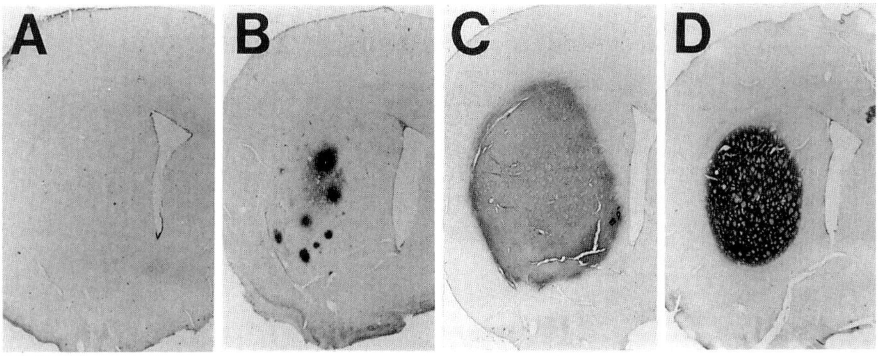

Fig. 1. Extravasation of IgG (blood–brain barrier dysfunction). **(A)** 24 h after one administration of 3-NPA. **(B,C)** 2 h and 5 h after two administrations of 3-NPA, respectively, in rats with motor symptoms. **(D)** 1 wk after two administrations in rats with no motor symptom.

blood–brain barrier by the extravasation of immunoglobulin G (IgG) in both acute and subacute intoxications in adult male rats *(16,17)*. In subacute intoxication even with a lower dose of 3-NPA, the extravasation of IgG was observed in the striatum after a long period (1 mo) of intoxication *(16)*. However, in an acute high-dose intoxication *(17)*, the IgG extravasation appeared at first with dotted patterns around the branches of the lSTR artery within hours in animals that exhibited motor symptoms (rolling, paddling, extension, recumbence, etc.), and extended to the entire lateral striatum thereafter (Fig. 1).

ASTROCYTIC CELL DEATH

One striking feature of 3-NPA intoxication is that the striatal astrocytes are damaged selectively in the brain *(17)*. Figure 2 shows typical morphological damage to striatal astrocytes. No visible changes were detected after the first administration (20 mg/kg, s.c.) of 3-NPA, but within hours after the second administration on the next day, half of the intoxicated animals exhibited motor symptoms. Glial fibrillary acidic protein (GFAP) positive astroglias disappeared from the parenchyma of the striatum and unusual (dotted) shaped astrocytes remained inside the fiber bundles. Before disappearance, GFAP-positive astrocytes in the centrolateral striatum become swollen and bushy *(17)*, then burst out. At this point, damage to neurons is not yet considerable (not shown) *(17)*. The other half of the animals, which showed no abnormal motor symptoms, were perfused for histological

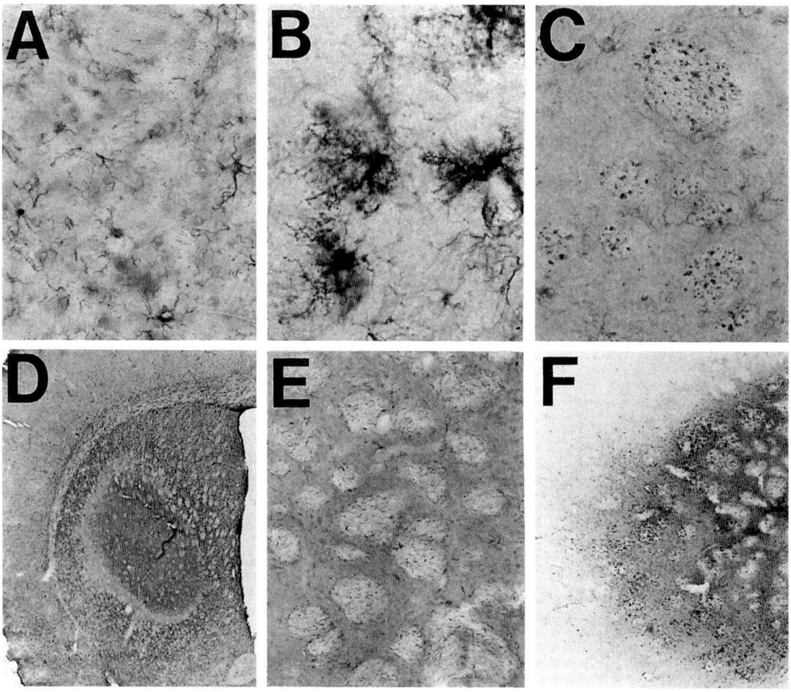

Fig. 2. Astroglial cell damage. **(A)** 24 h after one administration of 3-NPA. **(B,C)** 2.5 h after two administrations of 3-NPA. GFAP-positive cells became bushy/ edematous, and disappeared from the parenchyma of the striatum. They remained in fiber bundles and around vessels. **(D)** A typical striatal lesion (astroglial cell loss) 1 wk after two administrations of 3-NPA. **(E)** Higher magnification of the striatal lesion. **(A–E)** GFAP immunostaining. **(F)** Invasion of macrophages/ microglias 1 wk after two administrations of 3-NPA (OX-41 immunostaining).

examination after a week. One third of these demonstrated typical striatal lesions. Inside the lesion, GFAP-positive astrocytes and microtubule-associated protein-2 (MAP-2)-positive neurons were lost, and invasion of neutrophils, microglia, and macrophages, and the extravasation of IgG were detected.

Endothelial Cell Damage

Along with the damage to the striatal astrocytes, endothelial cells of the lSTR artery and its branches were disintegrated *(19)*. Figure 3 represents typical examples of such damage. Within hours after the second administration of 3-NPA, immunoreactivity for factor VIII-related antigen (a marker protein of endothelial cells) increased in the centrolateral striatum where the

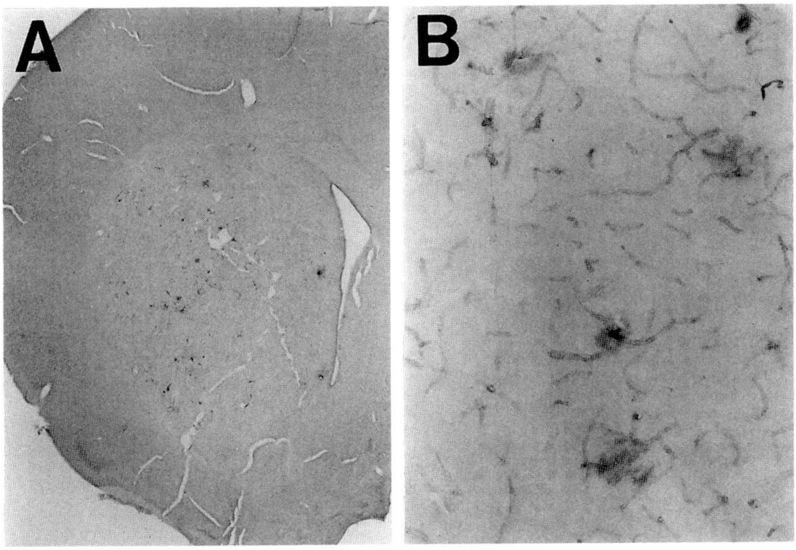

Fig. 3. Endothelial cell damage (factor VIII immunostaining). **(A)** Immunoreaction of factor VIII is enhanced only in the striatum (3 h after two administrations of 3-NPA). **(B)** Higher magnification of the striatum.

lSTR artery feeds but not in other parts of the brain. Under higher magnification, the immunoreaction was especially strong around the bifurcations of smaller branches of the lSTR artery, and often extended out of the lumen.

DA OVERFLOW AND DA TOXICITY

DAergic neurons in the substantia nigra are easily damaged by 3-NPA intoxication *(20)*. Extracellular levels of striatal DA and its metabolite dihydroxyphenylacetic acid (DOPAC) were measured using microdialysis in unanesthetized animals. DA and its metabolite increased within hours after the first administration (20 mg/kg, s.c.) of 3-NPA, but they returned to nearly basal levels after 12 h *(17)*. When 3-NPA was administered again on the subsequent day, DA and DOPAC levels increased sharply and remained at higher levels. These animals also exhibited motor symptoms *(17)*. Pretreatment with quinpirole (a D_2-receptor agonist) attenuated both motor symptoms and IgG extravasation whereas spiperone (a D_2-receptor antagonist) increased the incidence of motor symptoms and the extent of IgG extravasation *(17)*. Striatal DA depletion by prior 6-OHDA lesion in the unilateral nigrostriatal DA pathway attenuated the IgG extravasation following spiperone treatment *(17)*. Figure 4 shows the effect of extracellular

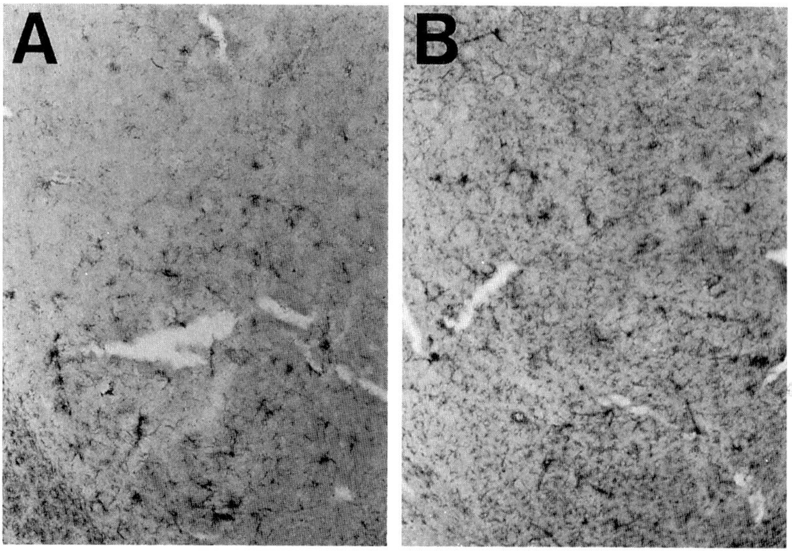

Fig. 4. Effect of extracellular DA tonus on astrocytic cell death 3 h after intrastriatal injection of 3-NPA (17 m*M*, 10 µL) in normal **(A)** and DA-depleted **(B)** striatum. GFAP-positive astroglias became more or less bushy/edematous in both injection sites, and some have already disappeared in **(A)**.

DA tonus/level on astrocytic cell death following intrastriatal (local) injection of 3-NPA. Astrocytic deformation (bushy edematous structure) or death is less extensive in DA-depleted striatum.

DISCUSSION

3-NPA inhibits succinate dehydrogenase irreversibly, thus disturbing the complex II of the mitochondrial electron transport system. This leads to the failure of proton exclusion from the inner membrane of the mitochondria and then disruption of high-energy (ATP) production. Deficiency in high-energy ATP results in disturbance of ATP-dependent pumps, leading to retention of Na^+, Ca^{2+}, Cl^-, and water within cells *(17)*, and depolarization of membrane potential. This may enhance excitotoxicity and the formation of oxidative free radicals. Glutamatergic excitotoxicity thus is strongly proposed to be involved in neuronal cell death in 3-NPA intoxication *(6)*.

Another intriguing feature of 3-NPA toxicity is the dysfunction of the blood–brain barrier in the centrolateral part of the striatum, which is supplied by the lSTR artery. Dysfunction of the blood–brain barrier implies disturbance of the endothelial cells and the surrounding astroglias *(19)*. Systemically administered 3-NPA circulates throughout the body and reaches endothelial cells in all organs, including the brain, but produces lesions or damage only to the lSTR artery. The underlying mechanisms of the greater vulnerability of endothelial cells of the lSTR artery are not known but can be explained on the basis of the following factors: it is a perforate artery and has high intraluminal pressure, large diameter to wall thickness ratio, and acute angled bifurcations. Thus, in the resting state, the wall of the lSTR artery is subjected to greater tension (Laplace law). The turbulent flow around the branchings of the lSTR artery suggest the existence of a stressful condition *(17)*. Accordingly, under any additional stressful states such as 3-NPA intoxication, ischemia, or hypertension, the endothelial cells of the lSTR artery would be damaged very easily. Once the endothelial cells are damaged, 3-NPA may enter the extracellular space of the brain more readily. The astrocytic end-feet, the first barrier that covers the basement membrane of the capillaries, would take up 3-NPA via their acid transporter because 3-NPA is an acid having a similar structure to glutamate. Our preliminary data suggested that intrastriatally (locally) injected 3-NPA solution with lower pH (6.0) was more toxic to astroglias than either a neutral or alkaline solution (pH 7.5–8.0). This may suggest the involvement of an astroglial acid glutamate transporter in uptake of 3-NPA and production of striatum selective lesions (Nishino H, Fukuda A, Deshpande SB, *unpublished data*).

Approximately 50% of the rats exhibited motor symptoms with striatum-selective lesions after two administrations of 3-NPA. However, other areas such as the cerebral and cerebellar cortex were histologically resistant, and no damage to the blood–brain barrier was detected even after three administrations of 3-NPA. This very sharp all-or-none difference in the vulnerability of the blood–brain barrier in the striatum compared to that in other areas cannot be explained solely by the difference in intraluminal (endothelial) characteristics. Extravascular characteristics in the striatum, such as an abundant release of glutamate from the corticostriatal glutamatergic terminals or an excessive release of DA *(17)* from the nigrostriatal DAergic terminals may exacerbate the vulnerability of the lSTR artery, striatal neurons, and glia. The former possibility *(6)* has been well studied in relation to excitotoxic neuronal cell death and the latter possibility has been reported here and elsewhere *(17)* in relation to glial cell death.

Thus far, a large amount of data have been reported for the striatum-selective mechanism in 3-NPA intoxication, but its exact nature has not yet been clearly delineated, involving as it does neuronal/glial damage, intraluminal/extravascular characteristics, glutamatergic/DAergic toxicity, and so on. Future studies on the distribution of high-affinity binding of or high-affinity receptors for 3-NPA, the different characteristics of endothelial cells of the lSTR artery compared with those of other vessels, the use of other chemical agents that inhibit other steps of the electron transport system or oxidative phosphorylation, or those aimed at determining the role of oxidative free radicals and of high energy deficiency may offer clues to elucidate the exact mechanisms of 3-NPA toxicity. However, the 3-NPA intoxication model is a useful tool for understanding the pathophysiology of and developing a therapeutic approach for neural/glial cell death, neurodegenerative disorders, autoimmune diseases, stroke, and others.

REFERENCES

1. Choi DW. Glutamate neurotoxicity and diseases of the nervous system. Neuron 1988;1:623–634.
2. DiFiglia M. Excitotoxic injury of the neostriatum: a model for Huntington's disease. Trends Neurosci 1990;13:286–289.
3. Meldrum B, Garthwaite J. Excitatory amino acid neurotoxicity and neurodegenerative disease. Trends Pharmacol Sci 1990;11:379–387.
4. Nishino H, Czurko A, Fukuda A, et al. Pathophysiological process after transient ischemia of the middle cerebral artery in the rat. Brain Res Bull 1994;35:51–56.
5. Beal MF. Does impairment of energy metabolism result in excitotoxic neuronal death in neurodegenerative illness? Ann Neurol 1992;31:119–130.
6. Beal MF, Brouillet E, Jenkins BG, et al. Neurochemical and histologic characterization of striatal excitotoxic lesions produced by the mitochondrial toxin 3-nitropropionic acid. J Neurosci 1993;13:4181–4192.
7. Simpson JR, Isacson O. Mitochondrial impairment reduces the threshold for in vivo NMDA-mediated neuronal death in the striatum. Exp Neurol 1993; 121:57–64.
8. Beal MF, Ferrante RJ, Henshaw R, et al. 3-Nitropropionic acid neurotoxicity is attenuated in copper/zinc superoxide dismutase transgenic mice. J Neurochem 1995;65:919–922.
9. Schulz JB, Matthews RT, Jenkins BG, et al. Blockade of neuronal nitric oxide synthase protects against excitotoxicity in vivo. J Neurosci 1995;15:8419–8429.
10. Brouillet M, Hantraye P, Ferrante RT, et al. Chronic mitochondrial energy impairment produces selective striatal degeneration and abnormal choreiform movements in primates. Proc Natl Acad Sci USA 1995;92:7105–7109.
11. Shimano Y, Kumazaki M, Sakurai T, et al. Chronically administered 3-nitropropionic acid produces selective lesions in the striatum and reduces muscle tonus. Obesity Res 1995;3(Suppl 5):779s–784s.
12. Borlongan CV, Koutouzis TK, Freeman TB, et al. Behavioral pathology induced by repeated systemic injections of 3-nitropropionic acid mimics the motoric symptoms of Huntington's disease. Brain Res 1995;679:254–257.

13. Palfi S, Ferrante RJ, Brouillet E, et al. Chronic 3-nitropropionic acid treatment in baboons replicates the cognitive and motor deficits of Huntington's disease. J Neurosci 1996;16:3019–3025.
14. Hamilton BF, Gould DH. Nature and distribution of brain lesions in rats intoxicated with 3-nitropropionic acid: a type of hypoxic (energy deficient) brain damage. Acta Neuropathol 1987;72:286–297.
15. Hamilton F, Gould DH. Correlation of morphologic brain lesions with physiologic alterations and blood–brain barrier impairment in 3-nitropropionic acid toxicity in rats. Acta Neuropathol 1987;74:67–74.
16. Nishino H, Shimano Y, Kumazaki M, et al. Chronically administered 3-nitropropionic acid induces striatal lesions attributed to dysfunction of the blood–brain barrier. Neurosci Lett 1995;186:161–164.
17. Nishino H, Kumazaki M, Fukuda A, et al. Acute 3-nitropropionic acid intoxication induces striatal astrocytic cell death and dysfunction of the blood–brain barrier: involvement of dopamine toxicity. Neurosci Res 1997;27:343–355.
18. He F, Zhang S, Zhang C, et al. Extrapyramidal lesions induced by mildewed sugar cane poisoning. Three case reports. Chin Med J 1987;67:395–396.
19. Nishino H, Nakajima K, Kumazaki M, et al. Estrogen protects against while testosterone exacerbates vulnerability of the lateral striatal artery to chemical hypoxia by 3-nitropropionic acid. Neurosci Res 1998;30:303–312.
20. Pei G, Ebendal T. Specific lesions in the extrapyramidal system of the rat brain induced by 3-nitropropionic acid (3-NPA). Exp Neurol 1995;132:105–115.

12
Mitochondrial Inhibition and Neuronal Death in Huntington's Disease

María Isabel Behrens

Huntington's disease (HD) is a neurodegenerative disease that belongs to the group of disorders caused by CAG triplet repeat expansions, which leads to expansions of polyglutamine in a protein of unknown function named huntingtin *(1)*. HD causes progressive dementia and chorea and there is no effective therapy available at present. Its pathological characteristic is the selective degeneration of subsets of neurons, primarily those in the striatum and neocortex. Medium spiny neurons in the striatum are extensively destroyed, with preservation of the small subpopulation of neurons containing NADPH-diaphorase or acetylcholinesterase (AChE). The mechanism of neuronal death in HD is unknown. Recent studies in transgenic animals expressing a construct of huntingtin protein encoding the N-terminal end including the glutamine repeat, as well as examination of autopsy material from HD patients, have revealed the presence of intranuclear inclusions of ubiquinated huntingtin peptides containing polyglutamine repeats *(2–4,7,8)*. Intranuclear inclusions are observed in other CAG repeat diseases, such as spinocerebellar atrophy and dentatorubralpallidoluysian atrophy *(3,4,9–11)*.

One hypothesis to explain the mechanism of neuronal death observed in HD is impairment of energy metabolism *(12–14,17)*. There are several lines of evidence supporting this speculative hypothesis from studies in HD patients and animal models. As anecdotal evidence, HD patients have high caloric needs, perhaps beyond that explained by their movement disorder, that causes marked reduction of weight during the active phase of the disease. Magnetic resonance spectroscopy studies of HD patients show an increase in lactate concentration in certain brain regions *(18)*. Furthermore, recent reports show decreased mitochondrial enzyme activities (complex

From: *Mitochondrial Inhibitors and Neurodegenerative Disorders*
Edited by: P. R. Sanberg, H. Nishino, and C. V. Borlongan © Humana Press Inc., Totowa, NJ

II–III and to a lesser degree complex IV) in the caudate and putamen of HD patients *(19,20)*.

The hypothesis of energy impairment is also strongly supported by animal studies in which neuropathological, neurochemical, and behavioral features of HD can be reproduced by systemic administration or local injection of metabolic toxins into the neostriatum *(15,16,21–28)*. The most well known animal model of HD is that generated by the administration of the complex II mitochondrial inhibitor 3-nitropropionic acid (3-NPA). The systemic or intrastriatal administration of 3-NPA to mice or primates induces striatal degeneration that mimics the neuropathological features of HD *(15,21–25,28)*. In addition, accidental ingestion of mildewed sugarcane containing 3-NPA by children in China was shown to induce acute encephalopathy followed by delayed dystonia *(29–31)*.

The specific mechanisms by which loss of mitochondrial function and consequent cellular energy depletion might lead to neuronal death are not fully delineated. One possibility is triggering of excitotoxicity *(32,33)*. The inability to maintain cellular ATP levels may lead to partial neuronal depolarization with persistent activation of glutamate receptors by ambient glutamate levels *(12,37)*, excessive influx of calcium, and eventual cell death *(34)*. Evidence in favor of this hypothesis comes from animal studies in which removal of the corticostriatal glutamate inputs or pretreatment with glutamate antagonists causes attenuation of the lesion size *(15,36)*.

While working on the effect of 3-NPA in vitro, we surprisingly observed that the induced death of cortical and striatal neurons did not follow the characteristic pattern of excitotoxicity observed in culture *(33,35,38,39)*. Instead, the exposure of mouse striatal or cortical cultures to 3-NPA produced gradual neuronal degeneration characterized by cell body shrinkage, DNA fragmentation, and protection of neuronal death by protein synthesis inhibitors, while no protection was observed with glutamate antagonists *(41)*. These characteristics of cell death are all compatible with apoptosis. Similar results were described by Pang and Geddes *(44)* in hippocampal neurons. They described two types of cell death induced by 3-NPA, acute excitotoxic necrosis and delayed apoptosis. Our results using other mitochondrial inhibitors acting at different sites of oxidative phosphorylation are also in accordance with a combination of the two forms of neuronal death. Carbonyl cyanide *p*-trifluoromethoxyphenylhydrazone (FCCP), rotenone, or oligomycin are able to induce apoptosis combined with a variable degree of excitotoxic necrosis (*40*; Behrens et al., *unpublished results*). In addition, inhibition of the glycolytic enzyme glyceraldehyde-3-phosphate dehydrogenase (GAPDH) also induces a slow decrease in ATP levels that leads to apoptosis of neurons in culture *(45)*. It appears that the prominence of glutamate receptor-mediated excitotoxic necrosis depends on the rapidity of ATP depletion. Exposure to insults producing rapid energy depletion such as high concentrations of cyanide (20 m*M* for 20 min) or oxygen–glucose deprivation (45 min; *[46]*) likely induce substantial glutamate release and consequently prominent excitotoxicity, whereas insults producing gradual energy depletion may induce less fulminant excitotoxicity and permit the triggering of programmed cell death. Recent studies in vivo are in accordance with our in vitro data showing 3-NPA-induced neuronal apoptosis. Systemic administration of 3-NPA to rats induced striatal apoptosis detected by dUTP-biotin nick-end labeling (TUNEL) staining *(28)*.

The results described suggest that the toxicity of 3-NPA and other mitochondrial inhibitors might involve both an excitotoxic component and an apoptosis component. In this respect, the original descriptions of the lesions induced by 3-NPA injections to rat brains showed both neuronal cell swelling and neuronal cell shrinkage *(47)*. The mode of cell death that predominates in vivo might depend on the velocity at which the toxin acts, a chronic low dose toxicity might lead to apoptosis *(22)*, whereas a fast acting toxin, which might be more feasible to reproduce experimentally, might lead to excitotoxic necrosis.

Further support for a role of apoptosis in the mechanism of cell death induced by energy impairment comes from studies in which mitochondrial inhibition by 3-NPA as well as several other apoptosis-inducing insults and GAPDH inhibition can mimic in vitro a pattern of neuronal selective damage seen in HD *(48)*, in which the subpopulation of neurons containing NADPH diaphorase or AChE are spared relative to the general striatal neuronal population *(42,43,45)*. Therefore, the striatal interneurons that are spared in HD are resistant not only to NADH excitotoxicity but also to apoptosis. It is interesting to note in this respect that these neurons do not show intranuclear inclusions *(3)*.

Altogether, the results in vitro allow one to suggest the interesting possibility that the mechanism by which energy impairment leads to neuronal death in HD might be not only through excitotoxicty but also through apoptosis.

An increasing number of studies of human postmortem brains from HD patients have found evidence of DNA fragmentation suggestive of apoptosis *(49–52)*. It must be recognized that, however, the method of detection of apoptosis, by *in situ* end-labeling of DNA strand breaks (TUNEL staining), used in these studies is not specific for apoptosis.

Apoptosis is a form of cell death first described during normal development that is characterized by a distinctive morphology with coarse chromatin aggregation and margination, progressive loss of cell volume, and extrusion of membrane-bound cytoplasmic fragments (apoptotic bodies) *(53)*. It allows for the elimination of cells that have been produced in excess or are irreversibly damaged, and plays a role in the homeostasis of multicellular organisms, maintaining the regulatory balance between cell proliferation and cell death *(54)*. Apoptosis can contribute to human disease, including cancer, autoimmune diseases or degenerative disorders *(55)*. This form of cell death is characterized by the execution of an exquisitely coordinated and stereotyped genetic program that is conserved from worms to humans and is followed by phagocytosis without inflammation *(56–58)*. In contrast, necrosis is the cell death induced by various environmental insults that is characterized by early cell swelling; dilation of mitochondria and endoplasmic reticulum; and the eventual rupture of nuclear, organelle, and plasma membranes leading to inflammation of surrounding tissue. The excitotoxin-induced neuronal death is mostly accepted to be necrotic rather than apoptotic; however, low levels of excitotoxic injury might induce incomplete triggering of apoptosis *(35,59)*.

There is recent evidence showing that mitochondria play a central role in the cascade of cellular events during apoptosis *(59–67,69)*. In cells undergoing apoptosis mitochondrial cytochrome *c*, which functions as an electron carrier in the respiratory chain, translocates to the cytosol and leads eventually to activation of caspase 3, which then acts on intracellular substrates to execute the cell death program *(68,69,77)*. The antiapoptotic proteins Bcl2 or Bcl-XL, which reside in the outer mitochondrial membrane, prevent the critical release of cytochrome *c* from mitochondria and cell death *(65,70–73)*. The key role that mitochondria play in the apoptotic cascade of events strengthens the idea that mitochondrial dysfunction and energy impairment might lead to apoptosis.

There is little understanding at present of how an expansion of polyglutamines in huntingtin protein leads to neuronal death in HD. The onset of neurological signs in the transgenic animal model of HD is preceded by the appearance of huntingtin protein in neuronal nuclei in the form of a characteristic neuronal intranuclear inclusion *(2)*. These structures have been found in neuronal nuclei in the cerebral cortex and caudate or putamen of necropsy brain tissue from patients with HD *(7)*. The frequency of the lesions and their localization correlates with the length of the CAG repeat *(8)*. In addition, a stable neuroblastoma cell line expressing mutant huntingtin shows that although the wild-type form leads to the expected diffuse cytoplasmic localization, the expression of mutant truncated proteins containing the polyglutamine expansion leads to the formation of cytoplasmic and nuclear inclusions in a time- and polyglutamine length-dependent manner *(78)*. It has been shown that the expansion of a polyglutamine sequence within a truncated huntingtin protein leads to the formation of β-pleated sheet (amyloid-like) fibrils that are insoluble *(74)*. There seems to be a good correlation between the sites of severe neuronal loss and the presence of intranuclear inclusions, and it is possible that they interfere with nuclear function *(3)*. On the other hand, there is evidence that intranuclear and perinuclear aggregates of truncated huntingtin peptides containing polyglutamine repeats increases susceptibility to apoptotic stress *(5,6)*. Another report shows that the presence of intranuclear inclusions did not correlate with huntingtin-induced death in cultured striatal neurons transfected with mutant huntingtin; instead the presence of mutant huntingtin within the nucleus induced degeneration by an apoptotic mechanism *(75)*.

Alternatively, one attractive possibility to explain how an expansion of polyglutamines in huntingtin protein leads to neuronal death in HD is that somehow the polyglutamine expansion of huntingtin leads to a defect in energy production, given the evidence for this latter feature in HD. In this respect, polyglutamine peptides bind strongly to GAPDH as a function of glutamine repeats *(76)*. GAPDH is a key enzyme in the glycolytic pathway.

ACKNOWLEDGMENTS

I am very grateful to Dr. Dennis Choi for valuable suggestions, discussions, and help in writing this chapter. I also thank Dr. Osvaldo Alvarez for helpful revision of the manuscript.

REFERENCES

1. Huntington's Disease Collaborative Research Group. A novel gene containing a trinucleotide repeat that is unstable on Huntington's disease chromosomes. Cell 1993;72:971–983.
2. Davies SW, Turmaine M, Cozens BA, DiFiglia M, Sharp AH, Ross CA, Scherzinger E, Wanker EE, Mangiarini L, Bates GP. Formation of neuronal intranuclear inclusions underlies the neurological dysfunction in mice trangenic for the HD mutation. Cell 1997;90:537–548.
3. Davies SW, Beardsall K, Turmaine M, DiFiglia M, Aronin N, Bates GP. Are neuronal intranuclear inclusions the common neuropathology of triplet-repeat disorders with polyglutamine-repeat expansions? The Lancet 1998;351:131–133.
4. Becher MW, Kotzuk JA, Sharp AH, Davies SW, Bates GP, Price DL, Ross CA. Intranuclear neuronal inclusions in Huntington's disease and dentatorubral and pallidoluysian atrophy: correlation between the density of inclusions and IT15 CAG triplet repeat length. Neurobiol Dis 1998;4:387–397.
5. Lunkes A, Mandel JL. A cellular model that recapitulates major pathogenic steps of Huntington's disease. Hum Mol Genet 1998;7:1355–1361.

6. Hackam AS, Singaraja R, Wellington CL, Metzler M, McCutcheon K, Zhang T, Kalchman M, Hayden MR. The influence of Huntingtin protein size on nuclear localization and cellular toxicity. J Cell Biol 1998;141:1097–1105.
7. DiFiglia M, Sapp E, Chase KO, Davies SW, Bates GP, Vonsattel JP, Aronin N. Aggregation of Huntingtin in neuronal intranuclear inclusions and dystrophic neurites in brain. Science 1997;277:1990–1993.
8. Martindale D, Hackam AS, Wieczorek A, Ellerby L, Wellington CL, McCutcheon K, Singaraja R, Kazemi-Esfarjani P, Devon R, Bredesen DE et al. Length of the protein and polyglutamine tract influence localization and frequency of intracellular aggregates of huntingtin. Nat Genet 1998; 18:150–154.
9. Skinner PJ, Koshy BT, Cummings CJ, Klement IA, Helin K, Servadio A, Zoghbi HY, Orr HT. Ataxin-1 with an expanded glutamine tract alters nuclear matrix-associated structures. Nature 1997;389:971–974.
10. Igarashi S, Koide R, Shimohata T, Yamada M, Hayashi Y, Takano H, Date H, Oyake M, Sato T, Sato A, et al. Suppression of aggregate formation and apoptosis by transglutaminase inhibitors in cells expressing truncated DRPLA protein with an expanded polyglutamine stretch. Nat Genet 1998;18:111–117.
11. Paulson HL, Perez MK, Trottier Y, Trojanovski JQ, Subramony SH, Das SS, Vig P Mandel JL, Fischbeck KH, Pittman RN. Intranuclear inclusions of expanded polyglutamine protein in spinocerebellar ataxia type 3. Neuron 1997;19:333–344.
12. Beal MF. Does impairment of energy metabolism result in excitotoxic neuronal death in neurodegenerative illnesses? Ann Neurol 1992;31:119–130.
13. Beal MF. Mitochondria, free radicals, and neurodegeneration. Curr Opin Neurobiol 1996;6:661–666.
14. Beal MF. Mitochondrial dysfunction in neurodegenerative diseases. Biochem Biophys Acta 1998;1366:211–223.
15. Beal MF, Brouillet E, Jenkins B, Ferrante RJ, Kowall NW, Miller JM, Storey E, Srivastava R, Rosen B, Hyman BT. Neurochemical and histologic characterization of striatal excitotoxic lesions produced by the mitochondrial toxin 3-nitropropionic acid. J. Neurosci 1993;13:4181–4192.
16. Beal MF, Brouillet E, Jenkins B, Ferrante RJ, Henshaw R, Rosen B, Hyman BT. Age-dependent striatal excitotoxic lesions produced by the endogenous mitochondrial inhibitor malonate. J Neurochem 1993;61:1147–1150.
17. Young AB. Impairment of energy metabolism and excitotoxic cell death in Huntington's disease. Rev Neurol (Paris) 1997;153:496–498.
18. Jenkins BG, Koroshetz WJ, Beal MF, Rosen BR. Evidence for impairment of energy metabolism *in vivo* in Huntington's disease using localized 1H NMR spectroscopy. Neurology 1993;43:2689–2695.
19. Gu M, Gash MT, Mann VM, Javoy-Agid F, Cooper JM, Schapira AHV. Mitochondrial defect in Huntington's disease caudate nucleus Ann Neurol 1996;39:385–389.
20. Browne SE, Bowling AC, MacGarvey U, Baik MJ, Berger SC, Mugit MM, Bird ED, Beal MF. Oxidative damage and metabolic dysfunction in Huntington's disease: selective vulnerability of the basal ganglia. Ann Neurol 1997;41:646–653.

21. Borlongan CV, Koutouzis TK, Sanberg PR 3-Nitropropionic acid animal model and Huntington's disease. Neurosci Biobehav Rev 1997; 21:289–293.
22. Borlongan CV, Nishino H, Sanberg PR. Systemic, but not intraparenquimal, administration of 3-nitropropionic acid mimics the neuropathology of Huntington's disease: a speculative explanation. Neurosci Res 1997; 28:185–189.
23. Brouillet E, Jenkins BG, Hyman BT, Ferrante RJ, Kowall NW, Srivastava R, Roy DS, Rosen BR, Beal MF. Age dependent vulnerability of the striatum to the mitochondrial toxin 3-nitropropionic acid. J Neurochem 1993; 60:356–359.
24. Brouillet E, Hantraye P, Ferrante RJ, Dolan R, Leroy-Willig A, Kowall NW, Beal MF. Chronic mitochondrial energy impairment produces selective striatal degeneration and abnormal choreiform movements in primates. Proc Natl Acad Sci USA 1995;92:7105–7109.
25. Brouillet E, Guyot MC, Mittoux V, Altairac S, Conde F, Palfi S, Hamtraye P. Partial inhibition of brain succinate dehydrogenase by 3-nitropropionic acid is sufficient to initiate degeneration in rat. J Neurochem 1998;70:794–805.
26. Henshaw R, Jenkins BG, Schulz JB, Ferrante RJ, Kowall NW, Rosen BR, Beal MF. Malonate produces striatal lesions by indirect NMDA receptor activation. Brain Res 1994;647:161–166.
27. Greene JG, Porter RHP, Eller RV, Greenamyre T. Inhibition of succinate dehydrogenase by malonic acid produces an excitotoxic lesion in rat striatum. J Neurochem 1993;61:1151–1154.
28. Alexi T, Hughes PE, Knusel B, Tobin AJ. Metabolic compromise with systemic 3-nitropropionic acid produces striatal apoptosis in Sprague-Dawley rats but not in BALB/c ByJ mice. Exp Neurol 1998;153:74–93.
29. He F, Zhang S, Zhang C, et al. Chinese Med J 1987;67:395–396.
30. Ludolph AC, He F, Spencer PS, Hammerstad J, Sabri M. 3-Nitropropionic acid-exogenous animal neurotoxin and possible human striatal toxin. Can J Neurol Sci 1991;18:492–498.
31. Liu X, Luo X, Hu W. Studies on the epidemiology and etiology of moldy sugarcane poisoning in China. Biomed Environ Sci. 1992;5:161–177.
32. Olney JW (1986) Inciting excitotoxic cyticide among central neurons. Adv Exp Med Biol 1986;203:631–645.
33. Choi DW. Excitotoxic cell death. J Neurobiol 1992;23:1261–1276.
34. Choi DW. Calcium: still center-stage in hypoxic–ischemic neuronal death. Trends Neurosci 1995;18:58–60.
35. Choi DW. Ischemia-induced neuronal apoptosis. Curr Opin Neurobiol 1996; 6:667–672.
36. Ludolph AC, Seelig M, Ludolph A, Sabri MI, Spencer PS. ATP deficits and neurodegeneration induced by nitropropionic acid. Neurodegeneration 1992; 1:155–161.
37. Novelli A, Reilly JA, Lysko PG, Henneberry RC. Glutamate becomes neurotoxic via the *N*-methyl-D-aspartate receptor when intracellular energy levels are reduced. Brain Res 1988;451:205–212.
38. Gwag BJ, Lobner D, Koh JY, Wie MB, Choi DW. Blockade of glutamate receptors unmasks neuronal apoptosis after oxygen-glucose deprivation in vitro. Neuroscience 1995;68, 615–619.

39. Csernansky CA., Canzoniero LM, Sensi SL, Yu SP, Choi DW. Delayed application of aurintricarboxylic acid reduces glutamate-induced cortical neuronal injury. J Neurosci Res 1994;38:101–108.
40. Behrens MI, Koh J, Gwag BJ, Canzoniero LMT, Sensi SL, Choi DW. Mitochondrial toxins induce cycloheximide-sensitive neuronal death in murine cortical cultures. Soc Neurosci Abstr 1994;20:248.
41. Behrens MI, Koh J, Canzoniero LMT, Sensi SL, Csernansky CA, Choi DW. 3-Nitropropionic acid induces apoptosis in cultured striatal and cortical neurons. NeuroReport 1995;6:545–548.
42. Behrens MI, Koh J, Muller MC, Choi DW. NADPHdiaphorase-containing striatal or cortical neurons are resistant to apoptosis. Neurobiol Dis 1996;3: 72–75.
43. Behrens MI, Muller MC, Choi DW. Staurosporine-induced apoptosis in striatal cell cultures spares acetylcholinesterase-positive neurons. American Academy of Neurology Meeting, San Fancisco, March 23–30, 1996.
44. Pang Z, Geddes JW. Mechanism of cell death induced by the mitochondrial toxin 3-nitropropionic acid: acute excitotoxic necrosis and delayed apoptosis. Neuroscience 1997;17:3064–3073.
45. Sheline CT Choi DW. Neuronal death in cultured murine cortical cells is induced by inhibition of GAPDH and triosephosphate isomerase. Neurobiol Dis 1998;5:47–54.
46. Goldberg MP, Choi DW. Combined oxygen and glucose deprivation in cortical cell culture: calcium-dependent and calcium-independent mechanisms of neuronal injury. J Neurosci 1993;13:3510–3524.
47. Hamilton BB, Gould DH. Nature and distribution of brain lesions in rats intoxicated with 3-nitropropionic: a type of hypoxic (energy deficient) brain damage. Acta Neuropathol 1987;72:286–297.
48. Ferrante RJ, Kowall NW, Beal MF. Selective sparing of a class of striatal neurons in Huntington's disease. Science 1985;230:561–563.
49. Portera-Cailliau C, Hedreen JC, Price DL, Koliatsos VE. Evidence for apoptotic cell death in Huntington disease and excitotoxic animal models. J Neurosci 1995;15:3775–3787.
50. Dragunow M, Faull RL, Lawlor P, Beilharz EJ, Singleton K, Walker EB, Mee E. In situ evidence for DNA fragmentation in Huntington's disease striatum and Alzheimer's disease temporal lobes. NeuroReport 1995;6:1053–1057.
51. Thomas LB, Gates DJ, Richfield EK, O'Brien TF, Schweitzer JB, Steindler DA. DNA end labeling (Tunel) in Huntington's disease and other neuropathological conditions. Exp Neurol 1995;133:265–272.
52. Butterworth NJ, Williams L, Bullock JY, Love DR, Faull RL, Dragunow M. Trinucleotide (CAG) repeat length is positively correlated with the degree of DNA fragmentation in Huntington's disease striatum. Neuroscience 1998;87:49–53.
53. Wyllie AH, Kerr JFR, Currie AR. Cell death: the significance of apoptosis. Intl Rev Cytol 1989;68:251–306.
54. Jacobson MD, Weil M, Raff MC. Programmed cell death in animal development. Cell 1997;88:347–354.

55. Thompson CB. Apoptosis in the pathogenesis and treatment of disease. Science 1995;267:1456–1462.
56. Hengartner MO, Horvitz HR. *C. elegans* cell survival gene ced-9 encodes a functional homolog of the mammalian proto-oncogene bcl-2. Cell 1994;76:665–676.
57. Hengartner MO. Programmed cell death in invertebrates. Curr Opin Genet Dev 1996;6:34–38.
58. Deshmukh M, Johnson EM. (1997) Programmed cell death in neurons: focus on the pathway of nerve growth factor deprivation-induced death of sympathetic neurons. Mol Pharmacol 1997;51:897–906.
59. Ankarcrona M, Dypbukt JM, Bonfoco E, Zhivotovsky B, Orrenius S, Lipton SA, Nicotera P. Glutamate-induced neuronal death: a succession of necrosis or apoptosis depending on miotchondrial function. Neuron 1995;15:961–973.
60. Newmeyer DD, Farschon DM, Reed JC. Cell-free apoptosis in Xenopus egg extracts: Inhibition by *Bcl-2* and requirement for an organelle fraction enriched in mitochondria. Cell 1994;79:353–364.
61. Liu X, Kim CN, Yang J, Jemmerson R, Wang X. Induction of the apoptotic program in cell-free extracts: requirements of dATP and cytochrome *c*. Cell 1996;86:147–157.
62. Golstein P. Controlling cell death. Science 1997;275:1081–1082.
63. Kroemer G. The proto-oncogene *Bcl-2* and its role in regulating apoptosis. Nature Med 1997;3:614–620.
64. Kroemer G, Zamzani N, Susin SA. Mitochondrial control of apoptosis. Inmunol Today 1997;18:44–51.
65. Kluck RM, Bossy-Wetzel E, Greene DR, Newmeyer DD. The release of cytochrome *c* form mitochondria: a primary site for *Bcl-2* regulation of apoptosis. Science 1997;275:1132–1136.
66. Spector MS, Desnoyers S, Hoeppner DJ, Hengartner MO. Interaction between the *C. elegans* cell-death regulators CED-9 and CED-4. Nature 1997;385:653–656.
67. Bossy-Wetzel E, Newmeyer DD, Green DR. Mitochondrial cytochrome *c* release in apoptosis occurs upstream of DEVD-specific caspase activation and independently of mitochondrial transmembrane depolarization. EMBO J 1998;17:37–49.
68. Li P, Nijhawan D, Budihardjo I, Srinivasula SM, Ahmad M, Alnemri ES, Wang X. Cytochrome *c* and dATP-dependent formation of Apaf-1/Caspase-9 complex initiates an apoptotic protease cascade. Cell 1997;91:479–489.
69. Zou H, Henzel WJ, Liu X, Lutschg A, Wang X. Apaf-1, a human protein homologous to *C. elegans* CED-4, participates in cytochrome *c*-dependent activation of caspase-3. Cell 1997;90:405–413.
70. Kharbanda S, Pandey P, Schofield L, Israels S, Roncinske R, Yoshida K, Bharti A, Yuan ZM, Saxena S, Weichselbaum R, et al. Role for Bcl-x_L as an inhibitor of cytosolic cytochrome *c* accumulation in DNA damage-induced apoptosis. Proc Natl Acad Sci USA 1997;94:6939–6942.
71. Kim CN, Wang X, Huang Y, Ibrado AM, Liu L, Fang G, Bhalla K. Overexpression of Bcl-x_L inhibits Ara-C-induced mitochondrial loss of cytochrome *c* and other perturbations that activate the molecular cascade of apoptosis. Cancer Res 1997;57:3115–3120.

72. Vander Heiden MG, Chandel NS, Williamson EK, Acumacher PT, Thompson CB. Bcl-x_L regulates the membrane potential and volume homeostasis of mitochondria. Cell 1997;91:627–637.
73. Yang J, Liu X, Bhalla K, Kim CN, Ibrado AM, Cai J, Peng TI, Jones DP, Wang X. Prevention of apoptosis by *Bcl-2* release of cytochrome *c* from mitochondria blocked. Science 1997;275:1129–1132.
74. Scherzinger E, Lurz R, Turmaine M, et al. Huntingtin encoded polyglutamine expansions form amyloid-like protein aggregates in vitro and in vivo. Cell 1997;90:549–558.
75. Saudou F, Finkbeiner S, Devys D, Greenberg ME. Huntingtin acts in the nucleus to induce apoptosis but death does not correlate with the formation of intranuclear inclusions. Cell 1998;95:55–66.
76. Burke JR, Enghild JJ, Martin ME, Jou YS, Myers RM, Roses AD, Vance JM, Strittmatter WJ. Huntingtin and DRPLA proteins selectively interact with the enzyme GAPDH. Nat Med 1996;2:347–350.
77. Pettmann B, Henderson CE. Neuronal cell death. Neuron 1998;20:633–647.
78. Cooper JK, Schilling G, Peters MF, Herring WJ, Sharp AH, Kaminsky Z, Masone J, Khan FA, Delanoy M, Borchelt DR, Dawson VL, Dawson TM, Ross CA. Hum Mol Genet 1998;7:783–790.

13

Effects of Brain Mitochondrial Metabolism, Aging, and Caloric Restriction on Membrane Lipids and Proteins

An Electron Paramagnetic Resonance Investigation

S. Prasad Gabbita, John M. Carney, and Allan Butterfield

INTRODUCTION

Recent evidence has shown that an inevitable consequence of living in an aerobic environment is the continuous production of oxygen free radicals. The major organelle responsible for this generation of endogenous free radicals is the mitochondrion. Apart from its nurturing role including ATP synthesis in a cell, the mitochondrion is accountable for the most oxidants produced by cells during normal aerobic respiration. This makes intuitive sense considering that mitochondria consume greater than 80% of the available oxygen in the cellular milieu. The free radical theory of aging, as proposed by Harman *(1)*, postulates that oxygen-derived free radicals result in a cumulative damage to critical cellular components, eventually leading to many age-related disorders. An increase in the metabolic rate could lead to a substantial production of endogenous oxidants, such as superoxide ($O_2^{-\cdot}$), hydrogen peroxide (H_2O_2), hydroxyl radical ($OH^{\cdot}$), as by-products of normal oxygen metabolism in the mitochondria. Studies corroborating this suggestion have demonstrated that consequent damage in terms of the level of oxidative DNA damage is roughly related to metabolic rate in a number of mammalian species *(1–3)*. Apart from normal brain aging, it is hypothesized that there is a free radical mediated deterioration of neuronal membrane components leading to age-related neurodegenerative disorders such as Alzheimer's disease, Parkinsonism, amyotrophic lateral sclerosis, and Huntington's disease. Harman was the first to propose that the mitochondrion was involved in the aging process *(4)*. The dysfunctional mitochondrion is a cellular organelle that also has been implicated in several

From: Mitochondrial Inhibitors and Neurodegenerative Disorders
Edited by: P. R. Sanberg, H. Nishino, and C. V. Borlongan. Humana Press Inc., Totowa, NJ

neurodegenerative disease states *(5)*. Evidence suggests that biomolecular components of the mitochondria, such as mitochondrial DNA (mtDNA), electron transport chain enzymes (e.g., cytochrome oxidase), and lipid components (e.g., cardiolipin) undergo possible free radical mediated deterioration, resulting in a compromise of the associated bioenergetic processes *(4–5)*. The tightly coupled process of oxidative phosphorylation during mitochondrial respiration utilizes the electron transport chain to accomplish a four-electron reduction of O_2 to water with a simultaneous production of ATP through phosphorylation of ADP. A temporary or sustained loss of mitochondrial function and ATP production has been implicated in etiology of several neurodegenerative disorders *(5,6)*.

Although the mechanism underlying the age-related increases in mitochondrial production of $O_2^{-\cdot}$ and H_2O_2 is unknown, it has been well established that mitochondrial macromolecules undergo damage by self-generated reactive oxygen species (ROS) *(7)*. Higher levels of oxygen tension, e.g., during exercise, or increased respiratory chain substrate concentrations under induced metabolic stress tend to form higher levels of ROS *(1,2,8–11)*. These ROS can contribute to oxidative damage of mitochondrial lipids, proteins, and DNA.

That aging in general and aging of the central nervous system in particular may, in part, relate to the damage inflicted by oxygen free radicals and their intermediates has considerable experimental support *(12–17)*. Oxidative stress is the usual phrase used to identify the association of toxic free radicals with damage to cells and tissues. Ideally, several of the free radical and free radical generating processes would be neutralized by antioxidative defense mechanisms, of which there are a variety in organisms. However, a different paradigm, wherein animals are placed on restricted or controlled levels of dietary intake that provide adequate numbers of calories and appropriate amounts of vitamins, may, because of reduced mitochondrial respiration, lead to lower oxidative stress. Animals placed on caloric restriction appear to be healthier and live longer than their ad-libitum-fed counterparts *(19,20)*. One potential outcome of such changes might include a decrease in the production of free radicals *(10,11,21)*. A calorically restricted diet (DR) may also decrease the production of free radicals and concurrently limit their persistence and presumably the macromolecular damage *(22,23)*. Some of the strongest support for the proposition that caloric restriction retards age-related processes comes from Sohal and co-workers. Their investigations found that the levels of the superoxide radical and hydrogen peroxide were markedly lower in mice subjected to long-term caloric restriction than in normally fed controls *(20,21)*. In addition, a significant increase of free radical production with age seen in the control groups was blunted by caloric restriction in the experimental group. Moreover, this blunted decrease was accompanied by a lessened amount of oxidized mitochondrial protein and DNA.

In this chapter, we review our studies using electron paramagnetic resonance (EPR) spectrometry in conjunction with a site-specific spin label *(10,11)* to investigate: (1) whether succinate stimulation of mitochondria results in oxidative modification of membrane lipids; (2) whether there are changes in the conformational structure of cytoskeletal proteins caused by oxidative processes induced following heightened state 4 metabolic stimulation of mitochondrial respiration; (3) whether there are age-related differences in the effects of succinate stimulation of mitochondrial respiration; and (4) whether caloric restriction could exert its reported anti-aging effect by reducing this free radical production following succinate stimulation of mitochondrial respiration. The hypothesis being tested is that caloric restriction protects brain mitochondria and its biomolecular components and thus decreases metabolic generation of oxy-radicals, thereby modulating lipid membrane damage.

For these studies, we used a preparation consisting of a mixed population of synaptosomes and mitochondria as described in Gabbita et al. *(10,11)* Typically, the crude rat cortical brain homogenate was centrifuged at 1500*g* for 10 min at 4°C, after which the supernatant was collected and recentrifuged at 20,000*g* for 10 min. The resulting pellet, termed the P-2 pellet, was dispersed in Krebs buffer, pH 7.4 containing 1 m*M* desferrioxamine, an iron chelator. This P-2 fraction containing the mitochondria and synaptosomes was used for the lipid membrane study and protein oxidation study.

For the aging and caloric restriction study, Brown Norway rats of ages 6, 16, and 24 mo were used. The daily dietary regimen and maintenance of the rats fed ad libitum (AL) and the ones placed on a calorically restricted diet (DR) is as described in Gabbita et al. *(10,11)*.

EPR-ACTIVE COMPOUNDS USED FOR PROBING MEMBRANE LIPIDS AND CYTOSKELETAL PROTEINS

5-Nitroxyl stearate (5-NS) is a compound which belongs to the class of so-called doxyl nitroxide-substituted stearic acid analogs. 5-NS contains an oxazolidine ring, with resident nitroxide functionality, tethered covalently to a specific position on the stearic acid hydrocarbon tail. 5-NS has the EPR-detectable paramagnetic oxazolidine moiety at the 5-position of the hydrocarbon tail and exhibits an anisotropic motion, unlike the isotropic motion exhibited by nonoriented spin probes like the protein sulfhydryl-specific spin label 2,2,6,6-tetramethyl-4-maleimidopiperidine-1-oxyl (MAL-6) *(24,25)*. To determine if changes in the physical state of cortical synaptosomal lipid bilayer occurred after metabolic stimulation, the lipid-specific spin probe 5-NS was utilized. The 5-NS amphipathic spin probe intercalates into both leaflets of the lipid bilayer with its fatty acyl chain embedded in the hydrophobic bilayer and its polar head group oriented near the polar head groups of the lipid molecules at the hydrophilic surface of the bilayer. It is conceptualized that the polar head group of 5-NS is held firmly in place by the head groups of the bilayer lipids, while under normal conditions, the hydrophobic tail is free to undergo rapid anisotropic motion in the interior of the bilayer *(24,25)*. Because the nitroxide group (the electron paramagnetic resonance-active portion) is covalently bound to the alkyl chain of the probe, the motion of the nitroxide group reflects the intramembrane motion in the adjacent segment of the molecule. The oxazolidine moiety, being attached to the 5-position on the stearic acid molecule, allows the spin probe to be proximal to the lipid–water interface. This location permits the spin probe to report oxidative events occurring closer to the lipid–water bilayer rather than processes taking place deep within the membrane. A typical spectrum of 5-NS intercalated into the lipid bilayers of a mixed population of synaptosomes and mitochondrial membranes is shown in Fig. 1. The signal amplitude of the $M_I = 0$ central line EPR spectrum is designated by B_0. This spectrum reflects an average of all the labeled membranes present in the mixed population and is influenced by membrane lipid composition. In order to measure changes in the membrane fluidity, the model of 5-NS intercalation into the lipid bilayer is such that its long alkyl chain lies parallel to the alkyl chains of the membrane lipids (Fig. 1) *(24,25)*. Rapid anisotropic motion occurs about the long axis of the spin probe, which engenders new effective T tensor elements $T_{\|}'$ and $T_{\perp}'$. An order parameter S is calculated from these T-tensor values by the equation:

$$S = \frac{T_{\|}' - T_{\perp}'}{T_{\|}' - T_{\perp}'} * \frac{TrT}{TrT'}$$

$$TrT' = T_{\|}' + 2T_{\perp}'$$

where the primed values are spectroscopically measured (Fig. 2) and the unprimed values are known constants obtained from single crystal data *(26)*. Lower values of S indicate an environment of lower order and increased motion, suggesting increased fluidity.

As a means to measure the changes in conformation of cytoskeletal proteins in synaptosomes, the thiol-specific, covalently binding isotropic spin label MAL-6 was utilized (Fig. 1). The mixed population of synaptosomes

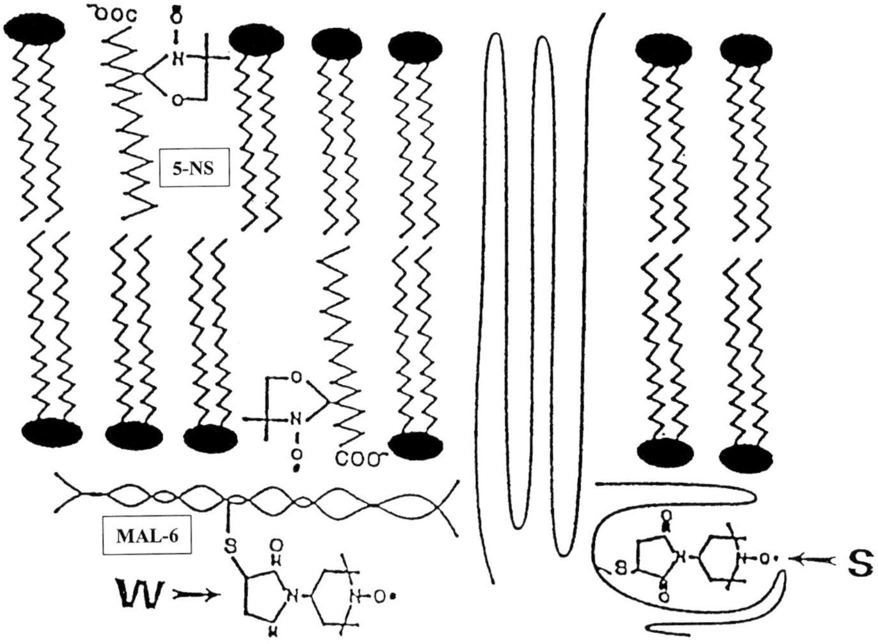

Fig.1. Localization of the EPR spin compounds, 5-NS and MAL-6, in the lipid membranes and the synaptosomal cytoskeleton, respectively. The lipid specific spin label, 5-NS, intercalates between the fatty acid chains of the lipid membrane, with the alkyl chains parallel to each other. MAL-6 binds to thiol groups of the synaptosomal cytoskeletal proteins in weakly (*W*) or strongly (*S*) immobilized sites. The ratio of the amplitudes of the weak to strongly immobilized spin label reaction sites [*W/S* ratio] is used to determine any alteration in protein conformation.

and mitochondria was treated with either 20 m*M* succinate or with equal volume of vehicle (Krebs buffer). The cytoskeletal protein modification analysis was performed by an isolation of synaptosomes via ultracentrifugation techniques using the succinate-stimulated and vehicle-treated P-2 pellet suspensions. The resulting synaptosomal preparation was lysed to expose the cytoskeleton and then labeled with MAL-6.

A typical EPR spectrum of MAL-6 covalently attached to membrane proteins in cortical synaptosomes is shown in Fig. 3. At least two distinct populations of spin label binding sites, characterized by their ability to restrict spin label motion, are observed. The weakly immobilized sites indicate covalent binding of the spin label to thiols closer to the exterior of the proteins, allowing for greater degree of motional freedom to the label. The strongly

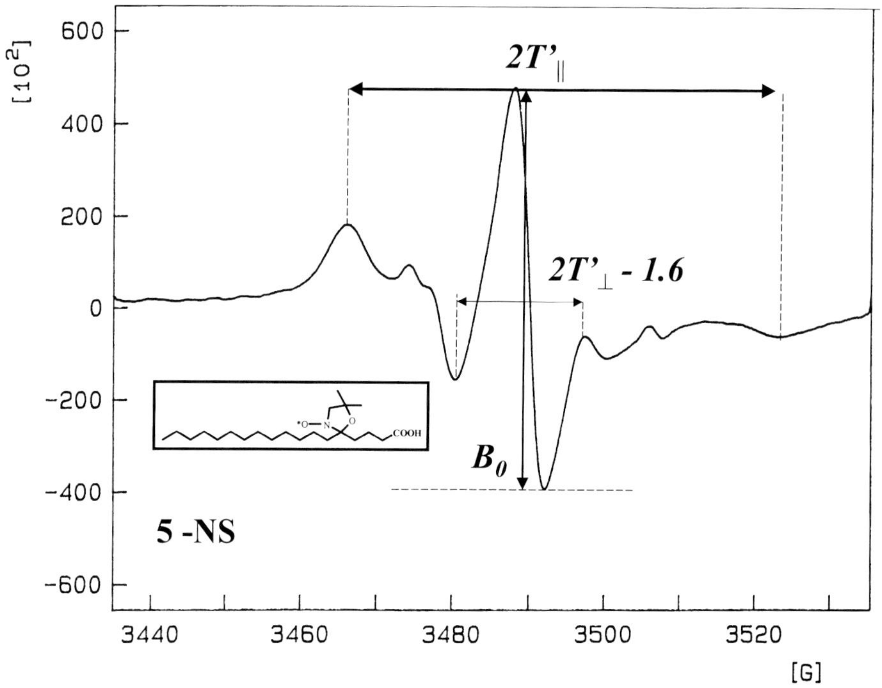

Fig. 2. Representative EPR spectrum of 5-nitroxyl stearate (5-NS). *T*-tensor measurements from which the order parameter is calculated and the signal amplitude (B_0) are indicated. The concentration of the spin probe is 1.6 μ*M* in the mitochondrial plus synaptosomal suspended in calcium-free Krebs buffer.

immobilized binding sites result from covalent linking of MAL-6 to free thiols present in sterically hindered deep pockets of the protein resulting in motional restriction of the spin label. This differential binding of the MAL-6 label results in line broadening of the low-field region of the spectrum (Fig. 3). The relevant EPR parameter measured is the ratio of the spectral amplitude of the $M_I = +1$ low-field weakly immobilized line (*W*) and that of the $M_I = +1$ low-field strongly immobilized line (*S*), which is referred to as the *W/S* ratio. Changes in the *W/S* ratio are known to be strong indicators of perturbations in the normal interaction of cytoskeletal proteins and reflect characteristic physical modifications of membrane protein structure, typically arising from chemical perturbations *(27–30)*. Previous studies in our laboratory have shown that increased protein oxidation is associated with decreased values of the *W/S* ratio of MAL-6 *(19,29,31,32)*.

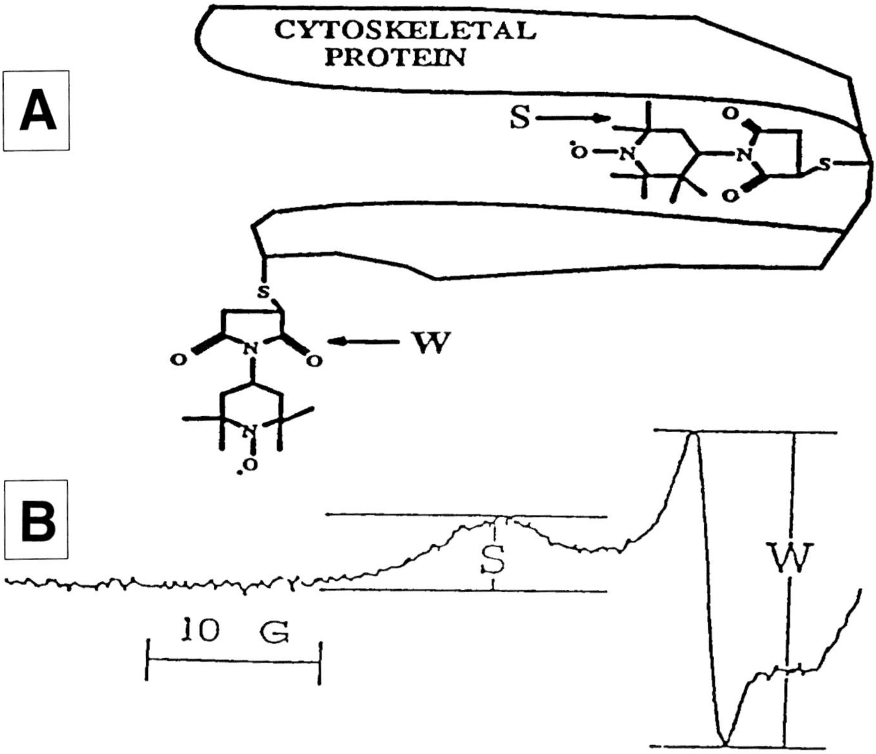

Fig. 3. Binding to cytoskeletal proteins of the synaptosomes with MAL-6. **(A)** Binding of MAL-6 to exterior proteins resulting in a weakly immobilized (*W*) or to sterically hindered protein sites deep within the pocket causing a strongly immobilized spin label (*S*). **(B)** Low-field EPR spectrum of MAL-6 labeled synaptosomal cytoskeletal proteins indicating weakly and strongly immobilized spin labeled signal amplitudes.

CHANGES IN 5-NS LABELED MEMBRANE LIPIDS FOLLOWING STIMULATION OF MITOCHONDRIAL RESPIRATION

As described earlier, a parameter of interest in the 5-NS spectrum is the amplitude (B_0) of the 5-NS signal. This value is a measure of the amplitude of the 5-NS signal and, with a constant line width, a decrease in its magnitude is indicative of direct reduction of the nitroxide head group by reactive oxygen species such as superoxide ($O_2^{-\cdot}$) and hydroxyl radical ($OH^{\cdot}$) to the corresponding EPR-silent hydroxyl amine *(10,33–39)*. As shown in Fig. 4, our results indicate that metabolic stimulation of mitochondrial respiration, using succinate as the substrate, leads to a highly significant decrease in the

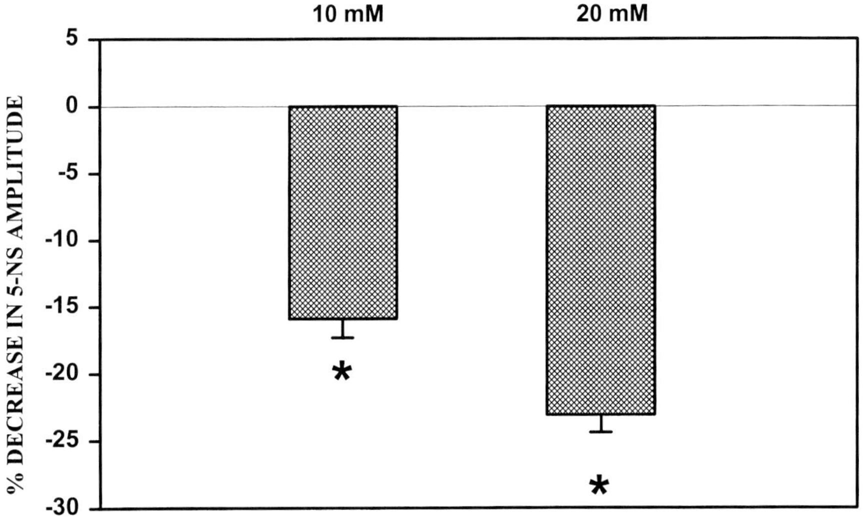

Fig. 4. Changes in the 5-NS spectrum signal amplitude (B_0) following induction of mitochondrial respiration utilizing succinate, a complex II substrate for the electron transport chain. A significant lowering in B_0 was observed ($*p < 0.0001$). This decrease was indicative of an oxy-radical flux, generated by the mitochondria, which likely can interact with the cortical membranes, initiating lipid peroxidation.

magnitude of the B_0 value of the 5-NS spin label. This decrease was indicative of an oxy-radical flux, generated by the mitochondria, which likely can interact with the cortical membranes, initiating lipid peroxidation. Simultaneously, there is a significant increase in the fluidity of the membrane lipids following succinate stimulation of the mitochondria (Fig. 5) as the damage to the lipid membranes of the mixed population of synaptosomes and mitochondria can be assessed by changes in the order or rigidity of the membrane.

The reduction of the 5-NS signal amplitude is conceivably caused due to its chemical reduction by reactive oxygen species, generated following succinate oxidation in the mitochondrial electron transport chain, to form the EPR-silent hydroxylamine. Chemical studies demonstrate that a reoxidation of the hydroxyl amine to the nitroxide is considered evidence in favor of hydroxylamine generation following reduction of nitroxides *(40)*. Potassium ferricyanide is a known oxidant of hydroxylamines to the corresponding EPR-detectable nitroxide *(36,40)*. To investigate whether the hydroxylamine is indeed generated following mitochondrial respiration, a reoxidation of the succinate-treated preparation was performed *(11)*. As shown in Fig. 6,

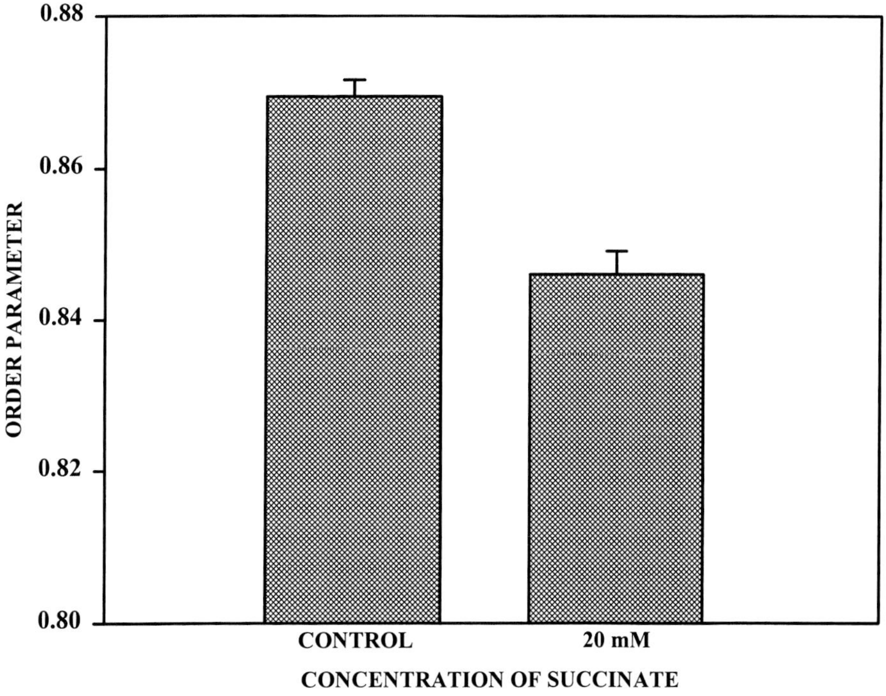

Fig. 5. Order parameter changes in cortical lipid membranes following stimulus of state 4 mitochondrial respiration with 20 m*M* succinate. A significant decrease in order, i.e., increase in membrane fluidity was observed (*$p < 0.0001$).

addition of 2 m*M* $K_3Fe(CN)_6$ leads to the 5-NS signal amplitude (B_0) being completely regenerated following oxidation of the succinate-stimulated, spin-labeled membranes. Predictably, although there was a near-complete regeneration of the 5-NS signal, the increase in membrane fluidity caused by mitochondrial respiration was unaffected by the reoxidation of the hydroxylamine (Fig. 7). Thus, our results indicated no significant attenuation of the decreased membrane order following reoxidation by $K_3Fe(CN)_6$ of the 5-NS-labeled, succinate-stimulated preparation. This was consistent with the notion that the membrane had undergone a permanent oxidative modification due to an attack by oxy-radical generation following mitochondrial respiratory stimulation. We conclude that there is a significant reduction in membrane order, i.e., an increase in membrane fluidity ensues following succinate-induced mitochondrial respiration. This observation is consistent with the conclusion that specific changes in the membrane allow the 5-NS spin probe to have an increased lateral motion in the lipid mem-

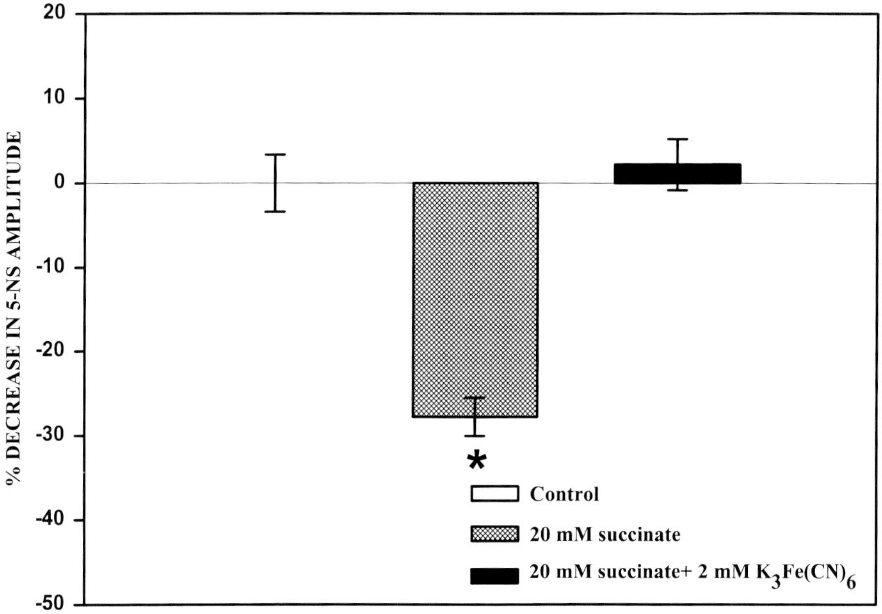

Fig. 6. A highly significant decrease in the signal amplitude, B_0, was observed with 20 m*M* succinate stimulation of mitochondrial respiration (**p* < 0.0001). As shown, the reduced 5-NS signal was completely regenerated following reoxidation with 2 m*M* $K_3Fe[CN]_6$.

brane microenvironment *(10,32)*. Figure 8 is an illustration depicting the oxidative chain of events possibly thought to occur in the membrane following ROS generation. Following the onslaught of the oxy-radicals on the fatty acid chains of the lipid molecule, the formation of a lipid hydroperoxide ensues. The creation of a peroxide moiety on the fatty acid chain results in the emergence of a hydrophilic side chain. The hydrophilic property of this fatty acid molecule compels it to move toward the aqueous phase at the membrane periphery, creating a "gap" in the membrane. The remaining fatty acid side chains now have more latitude to wiggle, and thus an increase in the fluidity of the membrane results. The intercalated 5-NS spin probe finds itself in an increasingly capacious lipid microenvironment (Fig. 8) and reflects this event by changes in *T*-tensor values, the determinants of the order parameter of the lipid membrane.

Neuronal membranes easily succumb to lipid peroxidative events caused by oxy-radical processes. The deterioration of the bilayer can result in a disruption in the ability of the membrane to prevent an elevated influx of

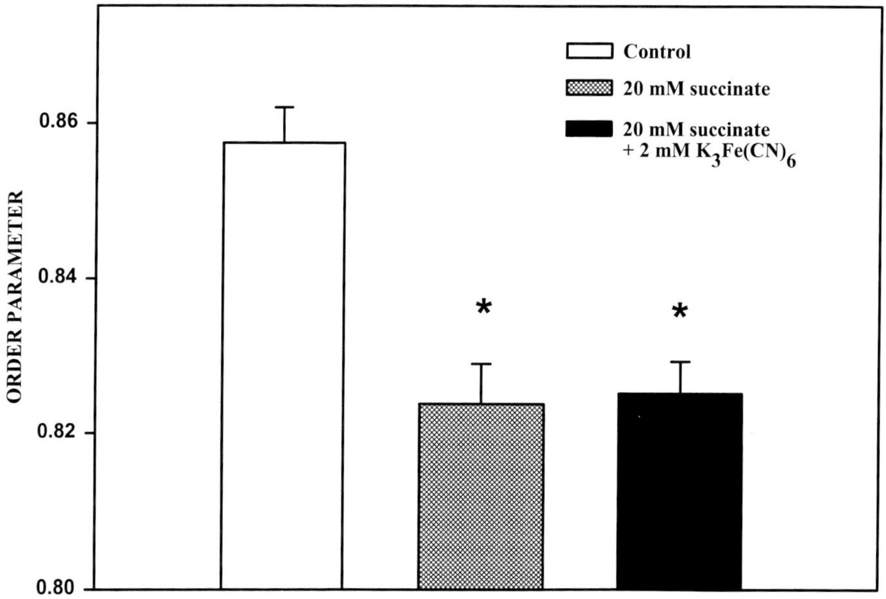

Fig. 7. The decrease in order parameter is not attenuated by reoxidation of EPR-silent hydroxylamine to corresponding EPR-detectable 5-NS. No significant difference between succinate-treated and succinate-treated followed by potassium ferricyanide reoxidized samples were observed.

calcium into the cell. This alone can potentially initiate several noxious stimuli within the cell, resulting in cell death. Damage to the mitochondrial lipid bilayer can also cause a functional decline in the sequestration of calcium in the mitochondria. This, in turn, can set off a chain of secondary free radical processes inside the cell and cause increased generation of lipid peroxidation products. Toxic aldehydic lipid peroxidation products such as malondialdehyde (MDA) and 4-hydroxy nonenal (4-HNE) can interact with proteins to form covalent adducts that alter protein conformation *(41)* and render critical cellular proteins functionally inactive *(42,43)*.

CHANGES IN MAL-6 LABELED SYNAPTOSOMAL CYTOSKELETAL PROTEIN FOLLOWING MITOCHONDRIAL STIMULATION

Metabolic stimulation of the mitochondria with succinate, a complex II substrate of the electron transport chain, was found to result in a highly significant decrease in the *W/S* ratio of subsequently isolated and spin

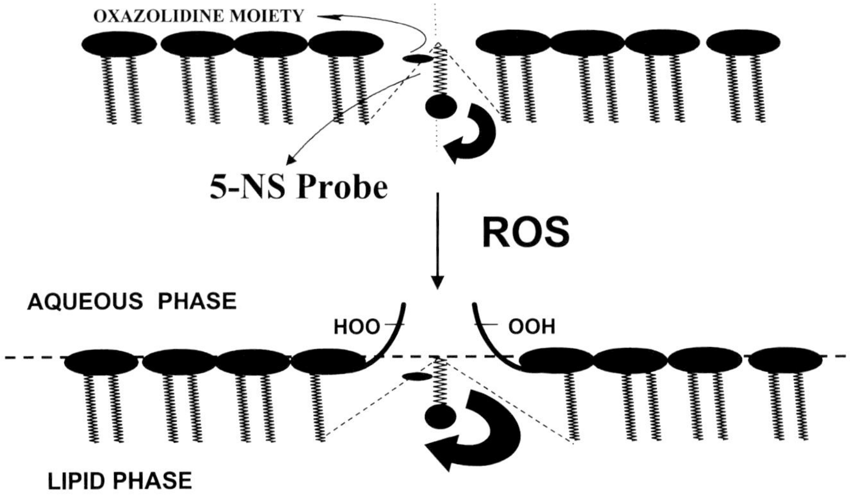

Fig. 8. A proposed model for the decrease in membrane order as assessed by 5-NS spin probe following oxy-radical attack on membrane. The lipid membrane, with the 5-NS spin probe intercalated within it, is susceptible to attack by ROS. The resultant lipid hydroperoxides, formed due to interaction with ROS, move toward the aqueous phase creating a void in the membrane. This allows for increased rotational motional freedom for the 5-NS spin probe and is reflected by an increase in the order parameter.

labeled synaptosomal membranes ($p < 0.0001$). A 30% decrease in the *W/S* ratio was found to occur in cortical homogenates from the rat brain following mitochondrial stimulation for 3 h with 20 m*M* succinate (Fig. 9). Previous studies have found that the *W/S* ratio of the MAL-6-labeled synaptosomes is significantly lower, as compared to control values, in various models of oxidative stress *(17,28,29,31,32)*. The protocols included in vivo oxidation conditions such as hyperoxia or ischemia/reperfusion injury as well as in vitro conditions with either an Fe^{2+}/ascorbate or Fe^{2+}/H_2O_2 or amyloid β-peptide. Based on these earlier findings, there was evidence of a significant oxidative modification in the cytoskeletal proteins due to an onslaught by ROS, generated following succinate-induced stimulation of mitochondrial respiration. The typical dynamic range of the *W/S* ratio in various membrane systems is between 2 and 8. Thus, a decrease in this parameter by even a single unit is indicative of an enormous oxidative modification of synaptosomal cytoskeletal proteins.

Oxidation of proteins can be a result of direct interaction with ROS such as OH· radicals or possibly by interaction with toxic aldehydic products

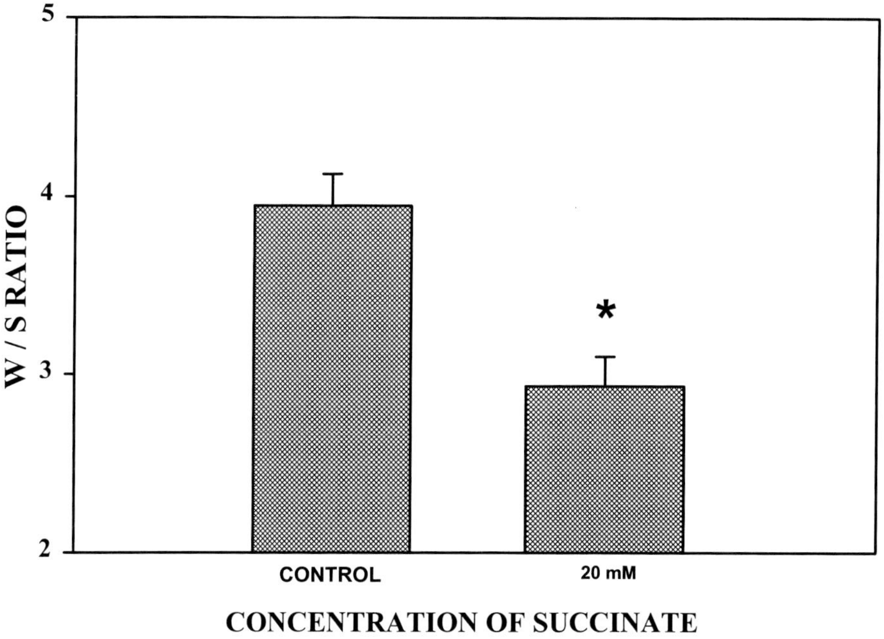

Fig. 9. A significant decrease in *W/S* ratio was observed following stimulation for 3 h of mitochondrial respiration with 20 m*M* succinate (*$p < 0.0001$). This is consistent with oxidative modification of cytoskeletal protein conformations.

(41–43). The result of this increased oxidative inactivation of proteins is an overall inability of the cell to maintain ionic homeostasis. This triggers several mechanisms leading to eventual cell death. Many neurodegenerative disease conditions including Alzheimer's disease (AD), amyotrophic lateral sclerosis, and Parkinson's disease have implicated ROS in their pathogenesis. Indeed, investigations have shown that brain autopsy samples from AD patients show significantly higher oxidized protein levels as compared to brain tissue from normal healthy elderly control *(44,45)*.

EFFECTS OF AGE, DIET, AND SUCCINATE ON OXY-RADICAL PRODUCTION

A preparation of a mixed population of cortical synaptosomes and mitochondria obtained from animals of differing ages led to a highly significant decrease in 5-NS signal amplitude following an increasing dose of succinate (Fig. 10). A significant decrease in 5-NS spectral amplitude was observed between control and 10 m*M* succinate and 10 m*M* vs 20 m*M* ($p < 0.05$) in the

6-mo-old animals (Fig. 10 A). Similarly, in the 16-mo-old animals, significant decreases in B_0 in 10 m*M* and 20 m*M* succinate treatments when compared to controls as well as when compared with each other were found (Fig. 10B). There is also a significant decrease in B_0 between control vs 10 m*M* and 20 m*M* succinate in the 25-mo-old animals (Fig. 10C). Thus, a clear consequence of stimulation of succinate dehydrogenase in a preparation from animals cortices irrespective of age is to decrease the $M_1 = 0$ central resonance line signal amplitude (B_0) of the 5-NS spin probe.

The effect of age-related changes in 5-NS signal amplitude (B_0) of spin-labeled lipid membranes following succinate-induced mitochondrial respiratory stimulation also was examined. A highly significant decrease in 5-NS signal amplitude was observed with increasing age of the animal following metabolic stimuli. The most significant differences were found between the 16- and 25-mo-old animals. Interestingly, at the 10 m*M* succinate stimulation level the generation of oxy-radicals, as evidenced by a decrease in 5-NS amplitude in 6-mo-old animals, was similar in magnitude to that of the 25-mo-old animals. The capacity to generate oxy-radicals increased with increasing concentrations of succinate as evidenced by a loss in 5-NS amplitude following respiratory stimulation of mitochondria. At 20 m*M* succinate stimulation, a highly significant age effect was observed ($p < 0.001$). Succinate stimulation of mitochondria resulted in greater membrane damage in 25-mo-old animals compared to younger age groups, suggesting increased mitochondrial leakage with age. The results in Fig. 10 are therefore consistent with a peroxidative attack on the lipid membrane bilayer. Figure 10 also shows the effect of the animal's age on mitochondria in terms of its propensity to generate oxy-radicals on metabolic stimulation. As indicated, the results demonstrate a significant overall age effect in the decrease of 5-NS signal amplitude following succinate-induced state 4 mitochondrial respiration.

Caloric restriction has been suggested to play a major role in modulation of the aging process *(20,46–48)*. Some of our studies were designed to investigate whether caloric restriction plays any role in decreasing oxy-radical generation upon mitochondrial complex II stimulation with succinate. Studies with animals of ages 6, 16, and 25 mo on either a calorically restricted diet or fed *ad libitum* showed significant age and succinate concen-

---

Fig. 10. *(opposite page)* Effect of age and caloric restriction changes in 5-NS signal amplitude (B_0) of spin-labeled lipid membranes following succinate-induced mitochondrial respiratory stimulation. There is an overall significant decrease in B_0 when 10 m*M* and 20 m*M* succinate-stimulated preparations are compared with control ($p < 0.0001$). The amplitude changes in 5-NS-labeled mixed population of synaptosomes and mitochondria isolated from rats cortices are indicative of pos-

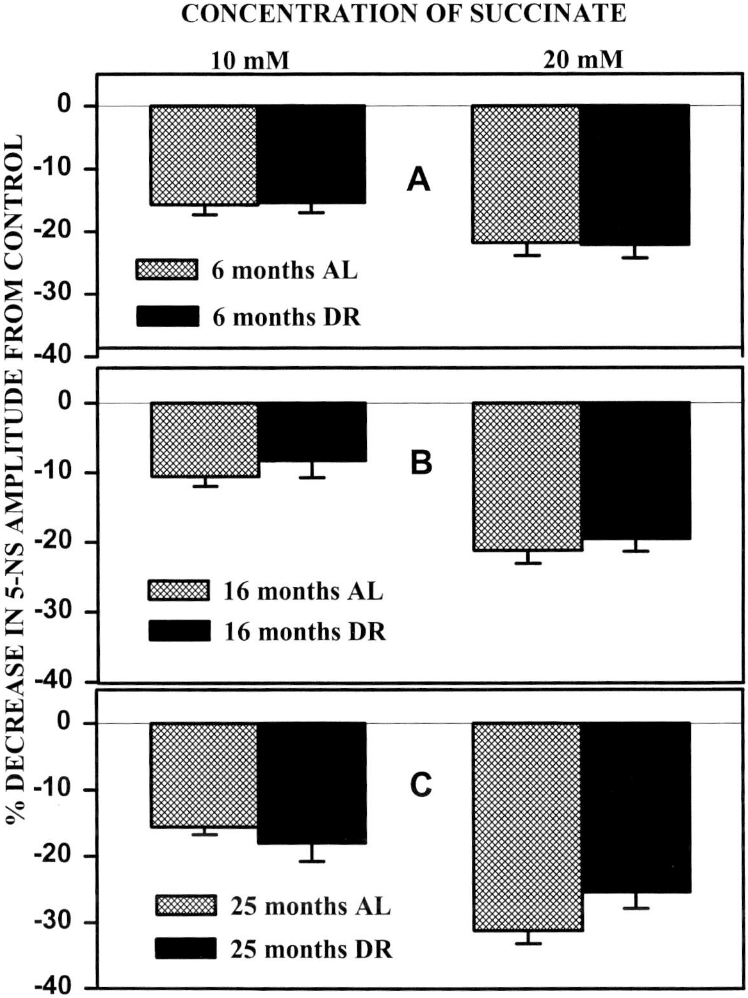

sible peroxidation-induced changes in the synaptosomal lipid membrane. A significant decrease in 5-NS signal amplitude was observed with increasing age ($p < 0.0001$) of animal following metabolic stimuli. No overall significant protective effect was observed with caloric restriction under similar conditions of metabolic stimulation. **(A)** A significant decrease in 5-NS amplitude that is consistent with oxy-radical generation was observed between control and 10 m*M* succinate ($p < 0.0001$) and 10 m*M* vs 20 m*M* ($p < 0.05$). **(B)** Significant decreases in B_0 in 10 m*M* ($p < 0.0001$) and 20 m*M* succinate treatments when compared to controls as well as when compared with each other ($p < 0.0005$). **(C)** Significant decrease in B_0 between control vs 10 m*M* succinate ($p < 0.0001$) and 20 m*M* ($p < 0.0001$). Although statistically nonsignificant ($p < 0.06$), there was a protective trend due to caloric restriction in terms of oxy-radical generation following stimulation with 20 m*M* succinate.

tration effects on the ability of the mitochondria to generate oxy-radicals upon succinate stimulation (Fig. 10). There was no significant effect of dietary restriction in altering the mitochondrial ability to generate oxy-radicals under identical metabolic stimulation conditions. However, in the 25-mo-old animals following succinate stimulation of mitochondrial respiration, a statistically nonsignificant trend for a decline in oxy-radical generation with caloric restriction was observed ($p < 0.06$, Fig. 10C).

EFFECTS OF AGE, DIET, AND SUCCINATE ON MEMBRANE FLUIDITY

Figures 11 and 12 depict the changes in membrane order with age, caloric restriction, and metabolic stimulation. The membrane order of the mixed population of synaptosomes and mitochondria, isolated from animals fed *ad libitum* (AL) and without any succinate stimulation, increased with age, consistent with earlier findings (Fig. 11) *(23,47)*. A highly significant lowering in the order parameter (increasing membrane fluidity) was observed with increasing succinate stimulation in animals of all age groups (Fig. 12). There is a significant decrease in order in AL and DR groups in the 6-mo-old group with 20 m*M* succinate stimulation (Fig. 12A). Figure 12B indicates the membrane lipids from the 16-mo-old animals treated with 20 m*M* succinate showed a greater increase in membrane fluidity in both dietary groups. A similar decrease in membrane fluidity following succinate stimulation is observed in the 25-mo-old animals (Fig. 12C). Stimulation 10 m*M* with caused a significant increase in membrane fluidity only in the 25 month AL group of animal.

The influence of diet on membrane order among 6-, 16-, and 25-mo-old animals was also investigated with respect to age, diet, and dose. Caloric restriction modulated membrane order only in the 25-mo-old animals when not metabolically stimulated with succinate (Fig. 12C). This result suggests that diet does play a role in decreasing the membrane rigidization in animals of the older age group. Following metabolic stimulation of the mitochondria with 20 m*M* succinate in animals fed either *ad libitum* or a calorically restricted diet, there was a significant decrease in membrane rigidity in animals of all age groups. Thus, the role of diet as a means of protection against membrane damage in the face of mitochondrial oxy-radical generation upon respiratory stimulation may be a consequence of the decreased endogenous substrate concentration seen in caloric restriction.

With increasing age of the animal, the oxy-radical generating capacity of the mitochondria increases, as evidenced by decreases in 5-NS signal amplitude and the order parameter for specific succinate concentrations. Thus, as

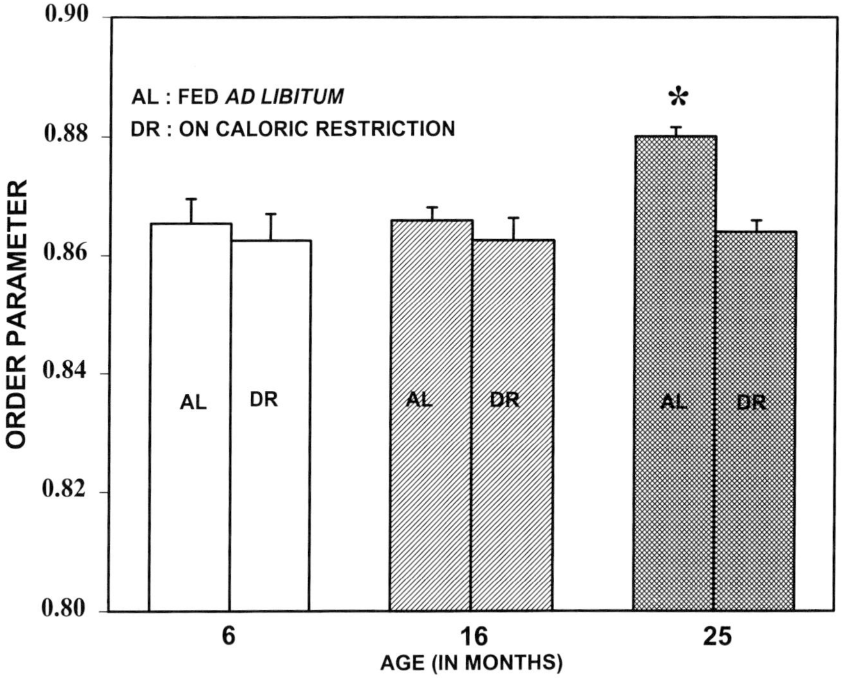

Fig. 11. With increasing age of the animal a significant increase in membrane order indicative of membrane rigidization was found ($p < 0.0007$). Significant increase in order of the mixed population of synaptosomal and mitochondrial membranes was found in 25-mo rats fed *ad libitum* when compared to age-matched rats on caloric restriction ($p < 0.0014$). No such protection from membrane rigidization was observed in the 6- or 16- mo-old rats.

the brain ages, it is likely that the mitochondria "leak" increased ROS upon stimulation. This could pertain to a lack of tighter coupling in the electron transport chain with increasing age. At 10 m*M* succinate stimulation, the observation that 6-mo-old rats showed similar decreases in comparison to 25-mo-animals may have to do with their higher basal metabolic rate resulting in a larger flux of oxy-radicals generated by the mitochondria. Differences in lipid composition with age may also contribute to the comparable decreases in B_0 of 5-NS. Mitochondrial respiratory stimulation with 20 m*M* succinate illustrates the increased leakage of oxy-radicals with age (Fig. 10). The lack of damage to membranes by oxy-radicals at 10 m*M* as compared to 20 m*M* stimulation may be due to sufficient levels of antioxidant protection in both systems to contend with the oxy-radical surge.

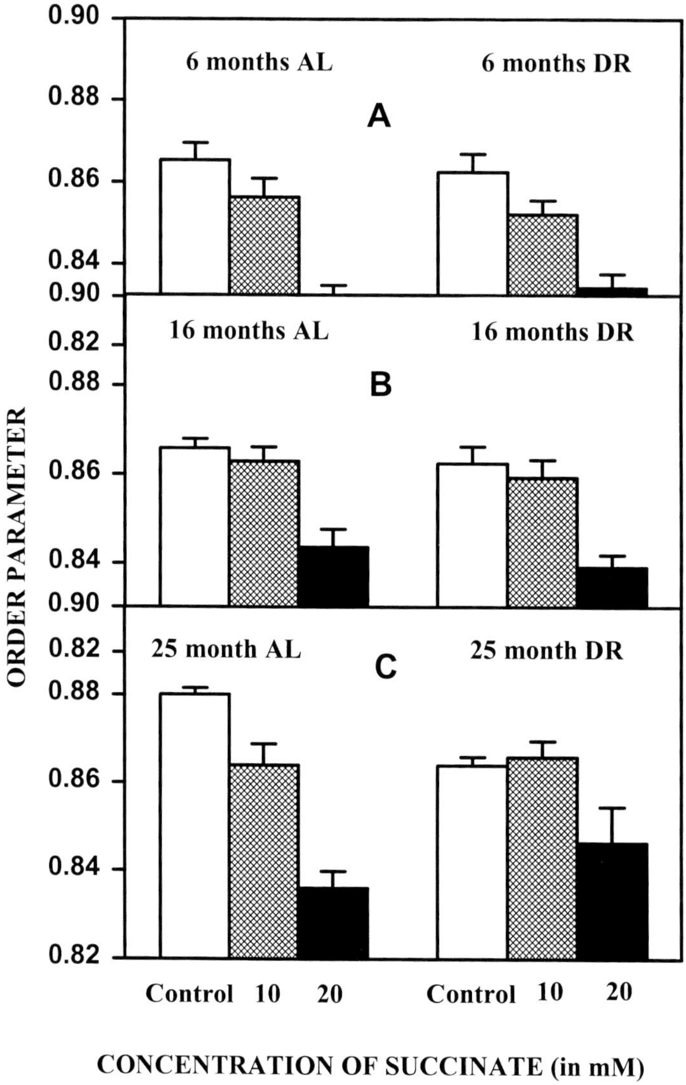

Fig. 12. Effect of age and caloric restriction and mitochondrial metabolic stimulation on the order parameter, an inverse indicator of lipid membrane fluidity. In all three age groups and the two dietary groups, a significant overall decrease in order was observed upon succinate stimulation of mitochondrial respiration (*$p < 0.0001$ indicates significance when compared with controls). **(A)** There is a significant decrease in order in AL and DR groups with 20 m*M* succinate stimulation ($p < 0.0007$). **(B)** Homogenates treated with 20 m*M* succinate showed a greater increase in membrane fluidity ($p < 0.001$) in both dietary groups. **(C)** Significant attenuation in membrane rigidity by caloric restriction is observed in 25-mo AL versus 25-mo

The effect of these free radicals on membrane fluidity during aging and its possible attenuation by dietary restriction has been of recent interest *(23,48)*. In our study, diet did not seem to play a protective role in the ability of the mitochondria to generate ROS in vitro. We have provided evidence in vitro that succinate-stimulated brain mitochondria, isolated from animals on a calorically restricted diet, generate equal amounts of oxy-radicals as compared to animals fed *ad libitum*. Caloric restriction did modulate rigidization of the membrane with age under control conditions as found in previous studies *(47)*, but did not influence the increase in membrane fluidity following mitochondrial stimulation with succinate. These findings support our hypothesis that caloric restriction by itself may not engender permanent structural or functional alterations in mitochondrial processes that occur with age. Upon subsequent metabolic stimulation, lipid membrane damage to the cell and organelle lipid membranes is induced, resulting in their dysfunction.

From our studies, we conclude that there is a dose-dependent decrease in 5-NS signal amplitude consistent with an increase in generation of oxy-radicals upon mitochondrial respiratory stimulation with succinate. This oxy-radical flux is able to cause lipid peroxidative damage to both the mitochondrial and synaptosomal membranes as evidenced by the changes in membrane fluidity. It should be qualified that these extreme metabolic conditions caused by high succinate concentrations may not reflect a situation present in vivo. The role of caloric restriction in attenuating the mitochondrial ability to generate radicals and prevent lipid peroxidation was found to be limited. Brain mitochondria isolated from calorically restricted animals, when subjected to similar metabolic stress conditions as mitochondria from animals fed *ad libitum* generate approximately the same flux of radicals and subsequent oxidative damage. A further evaluation will be required to determine whether the defects in the mitochondria are similar in animals from both sets of dietary conditions and the effect of oxy-radical leakage is masked until the stimulation of succinate dehydrogenase with high levels of substrate.

SUMMARY

This study investigated the effect of ROS, as generated by metabolic stimulation of mitochondria on brain membrane lipids and synaptosomal cytoskeletal proteins. We also evaluated the effect of age and dietary condi-

---

DR groups (**$p < 0.0014$). Stimulation with 20 m*M* succinate caused increased membrane fluidity in both dietary groups. 10 m*M* stimulation caused a significant increase in membrane fluidity only in the 25-mo-old AL group of animals ($p < 0.0007$).

tions on membrane lipid status with and without an activation of the mitochondrial electron transport chain using a metabolic substrate. Two EPR-active spin compounds, namely 5-nitroxyl stearate (5-NS), a lipophilic spin probe, and 2,2,6,6-tetramethyl-4-maleimidopiperidine-1-oxyl (MAL-6), a thiol-specific protein spin label, were utilized to assess membrane lipid damage and covalent modifications in the cytoskeletal proteins, respectively.

The brain is particularly vulnerable to oxidative damage because it contains relatively high concentrations of easily peroxidizable polyunsaturated fatty acids, and the brain is not highly enriched with protective antioxidant enzymes or small molecule antioxidants. Oxidative insult to normal neurons also results from catalytically active redox metal ions (i.e., iron and copper) and particular ROS-generating enzymes and peptides (e.g., nitric oxide synthase, xanthine oxidase, β-amyloid, etc.) present in the brain. Another factor that results in greater oxidative insult to brain relative to other tissues is that this organ consumes one fourth of the total O_2 intake, and, consequently, generates more oxy-radicals than most other organs in the body based on weight. Finally, owing to the postmitotic neuronal cellularity of the brain, further differentiation and/or cellular repletion does not occur. As a consequence, the brain's organelles are more likely to accumulate more oxidatively damaged biomolecules, as compared to cells that undergo mitosis, resulting in a loss of function *(16)*. Thus, the brain neurons and their organelles must survive for longer periods with oxidatively damaged dysfunctional organelles which occurs as a function of age and long-term metabolic stress *(16)*.

Our investigations using a mixed population of synaptosomes and mitochondria demonstrate an increase in lipid and protein oxidation, as assessed by specific EPR spin labels, following a metabolic stress, under state 4 conditions, induced by a high concentration of succinate. It is thus conceivable that an increased metabolic stress on the mitochondria causing an increased production of oxy-radicals can result in damage to biomolecular components, such as proteins and lipids, in a nerve cell. As elaborated, oxidant-induced mitochondrial damage resulting in energy depletion has been implicated in several neurodegenerative diseases. It is possible to envisage that respiration-derived oxy-radicals following a high metabolic challenge can lead to loss in the functional integrity of lipids and membranes of the mitochondria leading to such a damage.

For our study of the effects of mitochondrial respiration on membranes in aging and calorically restricted rats, we found that succinate stimulation caused an increased generation of oxy-radicals with increasing age of the animal. Further, the role of diet on protection of membrane lipids with age as assessed by measurement of membrane rigidization was significant. But under the conditions of an increased hypermetabolic stress, diet was unable to attenuate any of the damage caused to the membranes. This points to the possibility that membranes of the mitochondria and synaptosomes deteriorate with age and a calorically restricted diet aids only to the extent that the mitochondrial electron transport chain is not generating oxy-radicals as rapidly as in the case of the aged animals being fed *ad libitum*. The implications of this possibility include that the benefit of DR may be due to decreased oxy-radical damage to critical cellular components as a result of decreased mitochondrial activity and not necessarily to inherently altered membrane structure.

ACKNOWLEDGMENTS

This work was supported in part by grants from NIH.

REFERENCES

1. Perez-Campo R, Lopez-Torres M, Cadenas S, Rojas C, Barja G. The rate of free radical production as a determinant of the rate of aging: evidence from the comparative approach. J Comp Physiol B 1998;168:149–158.
2. Ku HH, Sohal RS. Comparison of mitochondrial pro-oxidant generation and antioxidant defenses between rat and pigeon: possible basis of variation in longevity and metabolic potential. Mech Ageing Dev 1993;72:67–76.
3. Ku HH, Brunk UT, Sohal RS. Relationship between mitochondrial superoxide and hydrogen peroxide production and longevity of mammalian species. Free Radical Biol Med 1993;15:621–627.
4. Harman D. The biological clock: the mitochondria? J Am Geriatr Soc 1972;20:145–147.
5. Browne SE, Beal MF. Oxidative damage and mitochondrial function in neurodegenerative diseases. Biochem Soc Trans 1994;22:1002–1006.
6. Beal MF. Energy, oxidative damage, Alzheimer's disease: clues to the underlying puzzle. Neurobiol Aging 1994;15:S171–174.
7. Sohal RS, Dubey A. Mitochondrial oxidative damage, hydrogen peroxide release, and aging. Free Radical Biol Med 1994;16:621–626.
8. Hiramatsu, M, Mori A. Exhaustive exercise affects fluidity and alpha-tocopherol levels in brain synaptosomal membranes of normal and vitamin E supplemented rats. Neurochem Res 1993;18:313–316.
9. Partridge RS, Monroe SM, Parks JK, et al. Spin trapping of azidyl and hydroxyl radicals in azide-inhibited rat brain submitochondrial particles. Arch Biochem Biophys 1994;310:210–217.
10. Gabbita SP, Butterfield DA, Hensley K, et al. Aging and caloric restriction affect mitochondrial respiration and lipid membrane status: an electron paramagnetic resonance investigation. Free Radical Biol Med 1997;23: 191–201.

11. Gabbita SP, Subramaniam R, Allouch F, et al. Effects of mitochondrial respiratory stimulation on membrane lipids and proteins: an electron paramagnetic resonance investigation. Biochim Biophys Acta 1998;1372:163–173.
12. Harman D. The aging process. Proc Natl Acad Sci USA 1981;78:7124–7128.
13. Harman D. Free radical theory of aging. Mutat Res 1992;275:257–266.
14. Halllwell B. Reactive oxygen species and the central nervous system. J Neurochem 1992;59:1609–1623.
15. Kehrer JP. Free radicals, mediators of tissue injury and disease. Crit Rev Toxicol 1993;23:21–48.
16. Stadtman ER. Protein oxidation and aging. Science1992;257:1220–1224.
17. Butterfield DA, Howard BJ, Yatin S, et al. Free radical oxidation of brain proteins in accelerated senescence and its modulation by *N-tert*-butyl-alpha-phenyinitrone (PBN). Proc Natl Acad Sci USA 1997;94:674–678.
18. Turturro A, Blank K, Murasko D, et al. Mechanisms of caloric restriction affecting aging and disease. Ann NY Acad Sci 1994;719:159–170
19. Sohal RS, Agarwal S, Candas M, et al. Effect of age and caloric restriction on DNA oxidative damage in different tissues of C57BU6 mice. Mech Ageing Dev 1994;76:215–224.
20. Sohal RS, Weindruch R. Oxidative stress, caloric restriction, and aging. Science 1996;273:59–62.
21. Sohal RS, Ku HH, Agarwal S, et al. Oxidative damage, mitochondrial oxidant generation and antioxidant defenses during aging and in response to food restriction in the mouse. Mech Ageing Dev 1994;74:121–133.
22. Wendruch R. The retardation of aging by caloric restriction: studies in rodents and primates. Toxicol Pathol 1996;24:642–745.
23. Yu BP, Suescun EA, Yang SY. Effect of age-related lipid peroxidation on membrane fluidity and phospholipase A_2: modulation by dietary restriction. Mech Ageing Dev1992;65:17–33.
24. Butterfield DA. Spin labeling in disease. In: Berliner LJ, Reuben J, eds. Biological Magnetic Resonance, Vol. 4. Plenum Press, New York, 1982, pp. 1–78.
25. Butterfield DA. Spectroscopic methods in degenerative neurological diseases. Crit Rev Neurobiol 1985;2:169–240.
26. Hubbell WL, McConnell HM. Molecular motion in spin-labeled phospholipids and membranes. J Am Chem Soc 1971;93:314–323.
27. Hensley K, Carney JM, Mattson MP, et al. A model for beta-amyloid aggregation and neurotoxicity based on free radical generation by the peptide: relevance to Alzheimer's disease. Proc Natl Acad Sci USA 1994;91: 3270–3274.
28. Hensley K, Carney JM, Hall N, et al. Electron paramagnetic resonance investigations of free radical-induced alterations in neocortical synaptosomal membrane protein infrastructure. Free Radical Biol Med 1993;17:321–331.
29. Butterfield DA, Carney JM, Umhauer S., et al. Changes in membrane cytoskeletal protein-protein interactions upon interactions of potential Alzheimer's disease therapeutic agents. In: Corain B, et al., eds. Advances in Clinical and Basic Research. John Wiley, New York, 1993.

30. Butterfield DA, Rangachari A. Membrane altering effects of velnacrine and *N*-methylacridinium: relevance to tacrine and Alzheimer's disease. Biochem Biophys Res Commun 1993;185:596–603.
31. Howard BJ, Yatin S, Hensley K, et al. Prevention of hyperoxia-induced alterations in synaptosomal membrane-associated proteins by *N-tert*-butyl-alpha-phenyinitrone (PBN) and 4-hydroxy-2,2,6,6-tetramethylpiperdidine-1-oxyl (Tempol). J Neurochem 1996;67:2045–2050.
32. Hall NC, Carney JM, Cheng MS, et al. Ischemia/reperfusion-induced changes in membrane proteins and lipids of gerbil cortical synaptosomes. Neuroscience 1995;64:81–89.
33. Butterfield DA, Hensley K, Harris M, et al. Beta-amyloid peptide free radical fragments initiate synaptosomal lipoperoxidation in a sequence-specific fashion: implications to Alzheimer's disease. Biochem Biophys Res Commun 1994;200:710–715.
34. Koppal T, Subramaniam R, Drake J, et al. Vitamin E protects against amyloid peptide (25–35)-induced changes in neocortical synaptosomal membrane lipid structure and composition. Brain Res 1998;786:270–273.
35. Belkin S, Mehlhorn RJ, Hideg K, et al. Reduction and destruction rates of nitroxide spin probes. Arch Biochem Biophys 1987;256:232–243.
36. Quintanilha AT, Packer L. Surface localization of sites of reduction of nitroxide spin-labeled molecules in mitochondria. Proc Natl Acad Sci USA 1977; 74:570–574.
37. Samuni A, Krishna MC, Mitchell JB, et al. Superoxide reaction with nitroxides. Free Radical Res Commun 1990;9:241–246.
38. Voest EE, van Faassen E, Marx JMM. An electron paramagnetic resonance study of the antioxidant properties of the nitroxide free radical TEMPO. Free Radical Biol Med 1993;15:589–595.
39. Voest EE., Van Faassen E, Neijt JP, et al. Doxorubicin-mediated free radical generation in intact human tumor cells enhances nitroxide electron paramagnetic resonance absorption intensity decay. Magnet Reson Med 1993; 30:283–288.
40. Chen K, Swartz HM. The products of the reduction of doxylstearates in cells are hydroxylamines as shown by oxidation by ^{15}N-perdeuterated Tempone. Biochim Biophys Acta 1989;992:131–133.
41. Subramaniam R, Roediger F, Jordan B, et al. The lipid peroxidation product, 4-hydroxy-2-trans-nonenal, alters the conformation of cortical synaptosomal membrane proteins. J Neurochem 1997;69:1161–1169.
42. Esterbauer H, Schaur, RJ, Zollner H. Chemistry and biology of 4-hydroxy nonenal, malondialdehyde and related aldehydes. Free Radical Biol Med 1991;11:81–128.
43. Chen JJ, Bertrand H, Yu BP. Inhibition of adenine nucleotide translocator by lipid peroxidation products. Free Radical Biol Chem 1995;19:583–590.
44. Hensley K, Carney JM, Subramaniam R et al. Brain regional correspondence between Alzheimer's disease histopathology and biomarkers of protein oxidation. J Neurochem 1995;65:2146–2156.

45. Smith CD, Carney JM, Starke-Reed PE, et al. Excess brain protein oxidation and enzyme dysfunction in normal aging and in Alzheimer's disease. Proc Natl Acad Sci USA 1991;88:10,540–10,543.
46. Laganiere S, Yu BP. Anti-lipid peroxidation action of food restriction. Biochim Biophys Acta 1987;779:89–137.
47. Choi JH, Yu BP. Brain synaptosomal aging: free radicals and membrane fluidity. Free Radical Biol Med 1995;18:133–139.
48. Aksenova MN, Aksenov MY, Carney JM, et al. Protein oxidation and enzyme activity decline in old brown norway rats are reduced by dietary restriction. Mech Ageing Develop 1998;100:157–168.

14

Malonate

Profile and Mechanisms of Striatal Toxicity

Alicia Meldrum, Keith J. Page, Barry J. Everitt, and Stephen B. Dunnett

INTRODUCTION

The experimental use of malonate as a toxin in the striatum began following speculation that neurodegeneration, specifically in Huntington's disease (HD), may be due to metabolic impairment. This hypothesis stems from observations of metabolic deficits in HD patients, including increased lactate levels and the presence of mitochondrial defects within the caudate nucleus *(1,2)*. Studies in vitro have substantiated the suggestion that metabolic compromise in neurons increases their vulnerability to excitotoxic insult *(3,4)*. These individual lines of investigation come together neatly to form a basis for the hypothesis that metabolic impairment may underlie the pathogenesis of HD *(5,6)*.

If HD is caused by a defect in energy metabolism then the inhibition of cellular metabolism with a metabolic toxin might produce features similar to those of HD pathology. Both 3-nitropropionic acid (3-NPA) and malonate inhibit the same enzyme and consequently have become candidate toxins for potential use in animal models of HD *(7,8)*.

EXPERIMENTAL PATHOLOGY OF MALONATE TOXICITY

The main pathological feature of HD is striatal atrophy. The neuronal loss is progressive and five pathological stages have been defined *(9)*. Grade 0 demonstrates no distinguishable abnormality at postmortem whereas grade IV brain can exhibit up to 95% loss of neurons. Within the striatum different neuronal populations are selectively affected by the disease process; medium spiny γ-aminobutyric acid-ergic (GABAergic) projection neurons are predominantly lost whereas there is a relative sparing of cholinergic and NADPH-diaphorase/nitric oxide synthase (NOS)-containing interneurons

From: *Mitochondrial Inhibitors and Neurodegenerative Disorders*
Edited by: P. R. Sanberg, H. Nishino, and C. V. Borlongan © Humana Press Inc., Totowa, NJ

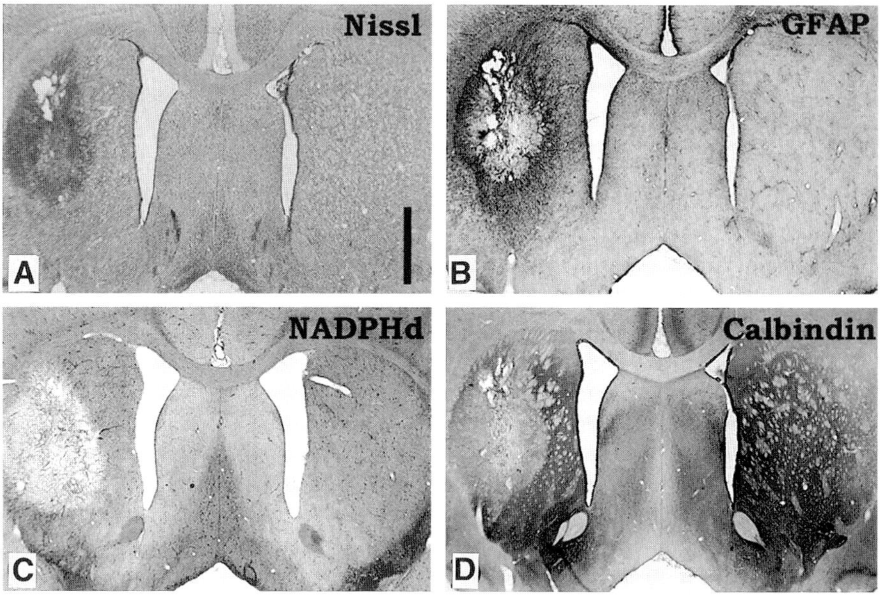

Fig. 1. Photomicrographs of unilateral malonate-induced lesions in the rat striatum. The lesions were produced by intrastriatal infusion of 1 µL of 1.0 *M* malonate. **(A)** Cresyl violet stain of cell loss and damage at the lesion site. **(B)** GFAP, a marker for reactive astrocytes. The most prominent astrocytosis is seen at the margins of the lesion. **(C)** NADPH-diaphorase, a marker for large aspiny interneurons. **(D)** Calbindin, one of several markers of medium spiny projection neurons. Scale bar = 1 mm.

(10,11). In early HD pathology is largely confined to the striatum; hence afferents projecting to the striatum are spared, reflected in the unchanged striatal levels of dopamine, serotonin, and norepinepherine *(12)*.

Intrastriatal infusion of optimal concentrations of malonate similarly produces neuronal loss confined to the striatum. Lesions are characterized by extensive gliosis around the lesion area and a loss of calbindin (a marker of medium spiny neurons) immunoreactivity within the lesion core. Lesions have been reported to demonstrate a sparing of both glia and axon bundles *(13)*; however, our observations indicate that reactive astrocytes are lost within the lesion core, demonstrated by the loss of glial fibrillary acidic protein (GFAP) immunoreactivity (Figs. 1 and 2). The toxicity of malonate in vivo would seem to be due to inhibition of the enzyme succinate dehydrogenase (SDH) insofar as coadministration of excess succinate successfully competes with malonate and significantly reduces lesion size *(14)*. This SDH

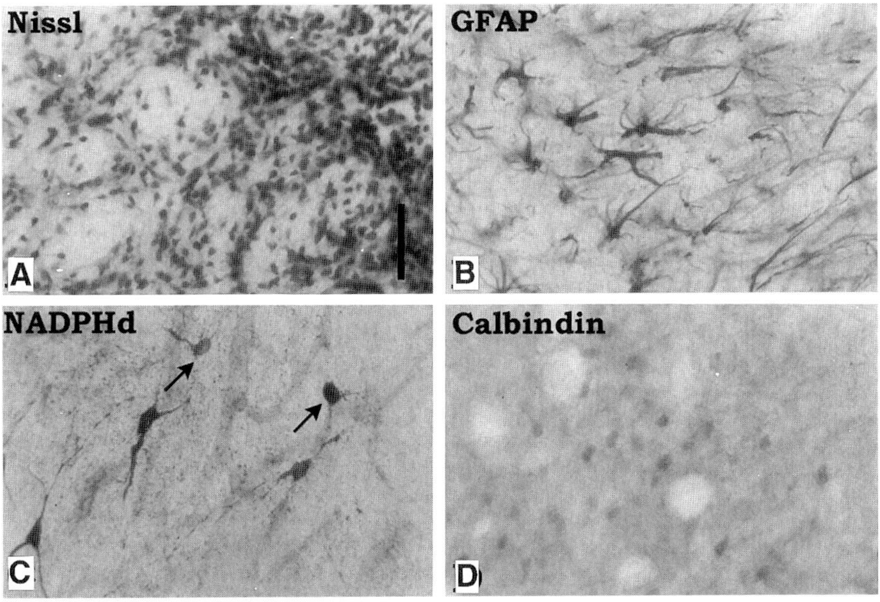

Fig. 2. Photomicrograph of lesion border. The lesion edge is clearly visible in the *top right corner* of Nissl-stained section (**A**); reactive astrocytes are present at the lesion border (**B**). NADPH diaphorase-containing interneurons do not survive in the lesion core and those surviving on the outskirts (*arrows*) do not extend processes into the area of neuronal loss (**C**). Calbindin-containing medium spiny neurons do not survive within the lesion (**D**). Scale bar = 200 μm.

inhibition leads to energy compromise demonstrated by a decrease in ATP concentrations and increases in lactate production *(8,15)*. Neurochemically, these lesions are characterized by significant decreases in markers for medium spiny neurons (GABA and substance P) and no change from control values in markers for medium aspiny neurons (somatostatin).

We have also demonstrated malonate toxicity to be dose dependent with a marked increase in lesion size at concentrations between 0.75 *M* and 1.0 *M* (Figs. 3 and 4). The extent of the lesion does not appear to be dependent on the volume of toxin infused or the rate of infusion.

Although metabolic toxicity would appear to replicate the main features of HD pathology *(7,8,16)* (and summarized in Table 1), a recent study has demonstrated that intrastriatal malonate does not spare dopaminergic afferents, indicated by a loss of tyrosine hydroxylase activity and decreased dopamine levels in the striatum *(17)*.

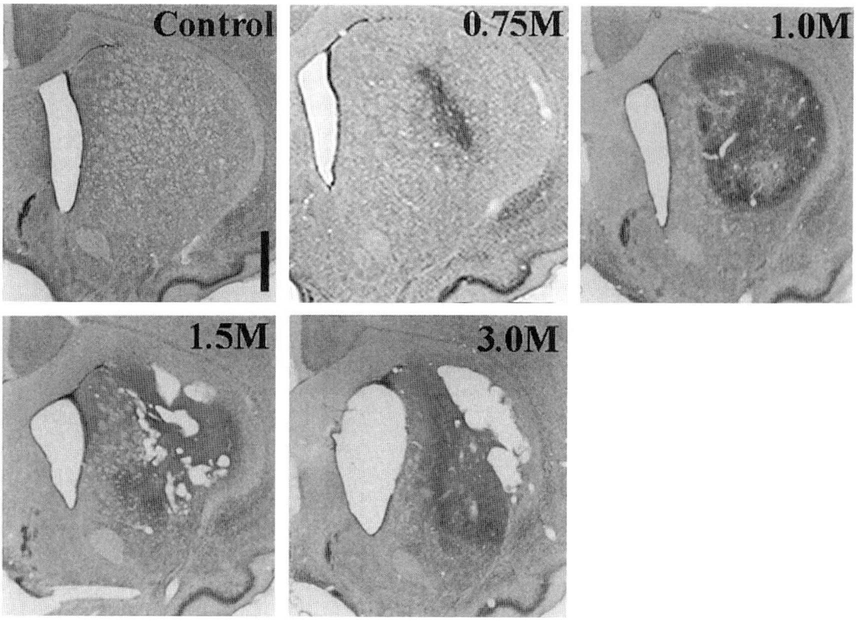

Fig. 3. Representative Nissl-stained sections at the level of the striatum to show the extent of lesions produced by injections of different concentrations of malonate. Scale bar = 1 mm.

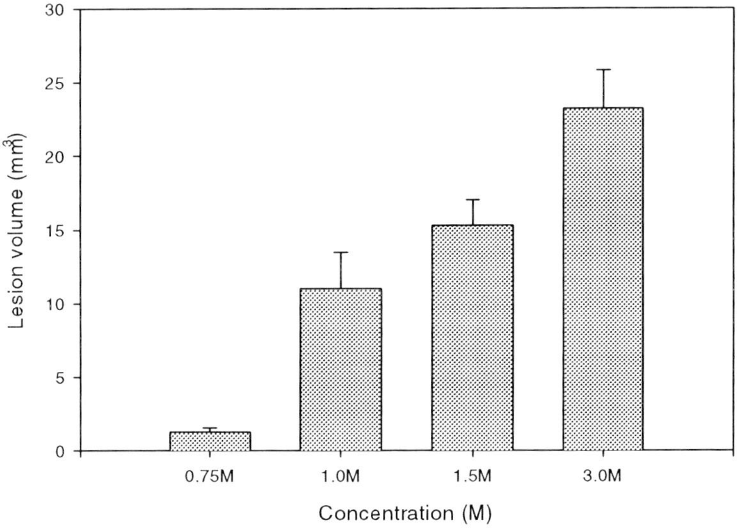

Fig. 4. Effects of malonate concentration on lesion volume as assessed 10 d postinfusion. Lesion volume was measured on Nissl-stained sections using the SeeScan image analysis system. Data are mean ± SE*M* values of eight animals

Table 1
Comparison of Huntington's Disease and Malonate-Induced Striatal Pathology

Huntington's disease	Malonate
Striatal vulnerability	Striatal vulnerability
Midlife onset	Middle-aged rats most vulnerable
Loss of medium spiny neurons	Loss of medium spiny neurons
Sparing of NADPH-diaphorase/ NOS neurons	Sparing of NADPH-diaphorase/ NOS neurons
Sparing of dopaminergic afferents	Loss of dopaminergic afferents

BIOCHEMICAL PATHWAYS OF MALONATE TOXICITY

The inhibition of the enzyme SDH by malonate stands as a classic example of competitive inhibition. SDH is an enzyme in the citric acid cycle responsible for the oxidation of succinate to fumarate. It is embedded in the inner mitochondrial membrane and is directly linked to the electron transport chain, forming part of the succinate-Q reductase complex (complex II). Thus, malonate inhibits both the production of energy-rich $FADH_2$ and the transfer of its electrons into the electron transport chain.

The mechanism by which malonate is thought to cause neuronal death is known as "weak," "indirect," or "secondary" excitotoxicity *(18)*. This theoretical mechanism is widely supported by both in vitro and in vivo studies *(19,20,13)*. Excitotoxicity has been proposed as a neurodegenerative mechanism of central nervous system (CNS) disorders *(21)* and can occur not only by means of the direct action of an agonist at a given glutamate receptor, but can also result from a defect in energy metabolism. Neurons that are metabolically compromised are unable to produce sufficient energy to carry out important ATP-dependent processes. A significant proportion of neuronal energy is used to power the sodium/potassium ATPase which maintains resting membrane potential. Failure of this pump results in a gradual membrane depolarization which leads to the loss of the voltage-dependent magnesium blockade of the *N*-methyl-D-aspartate (NMDA) receptor channel. With the removal of this blockade the NMDA receptor is readily activated by endogenous glutamate concentrations, resulting in the opening of the ion channel and consequent calcium influx (Fig. 5). Thus, a metabolically compromised neuron becomes more vulnerable to excitotoxicity. This mechanism is proposed to underlie malonate toxicity and is substantiated by observations that malonate toxicity is NMDA receptor-mediated *(13,15)* and that manipulation of membrane potential effects malonate toxicity such that membrane

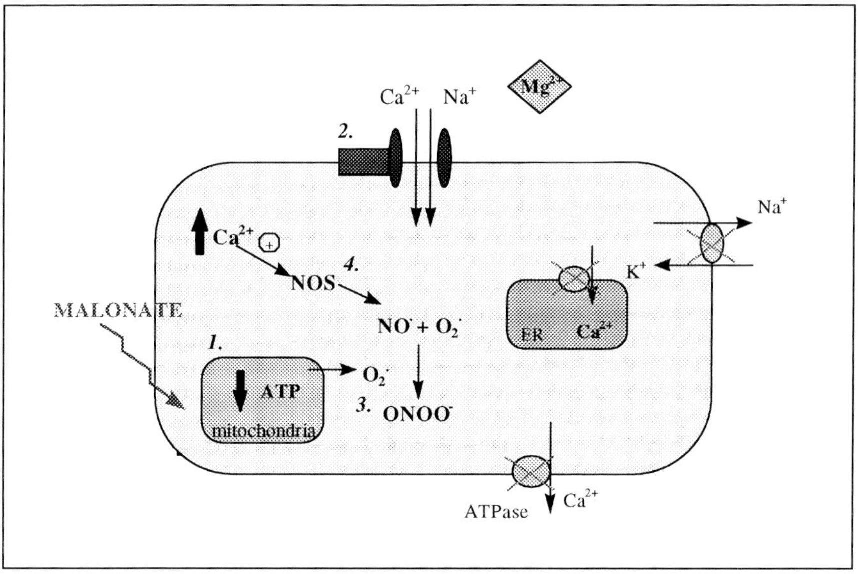

Fig. 5. Schematic diagram of intracellular mechanisms contributing to malonate-induced neuronal toxicity. The initial step in malonate-induced neuronal death involves mitochondrial inhibition leading to a decrease in energy production (**1**). One consequence of this reduced availability of energy is the loss of the voltage-dependent Mg^{2+} block of the NMDA receptor. This allows prolonged activation of the ion channel and increased influx of Na^+ and Ca^{2+} ions (**2**). The increase in intracellular Ca^{2+} is exacerbated in metabolically impaired neurons because they are unable to facilitate ATP-dependent extrusion or storage processes. A result of this Ca^{2+} increase is the activation of nitric oxide synthase (NOS) which increases production of the free radical $NO^{\cdot}$ (**4**). Metabolic inhibition also leads to the generation of $O_2^{\cdot}$ by mitochondria. $NO^{\cdot}$ and $O2^{\cdot}$ react together, generating peroxynitrite $ONOO^-$ which is thought to result in extensive oxidative damage to the neuron (**3**).

depolarization exacerbates neuronal death whereas hyperpolarization blocks malonate toxicity *(22)*.

MALONATE TOXICITY AND NEUROPROTECTIVE STRATEGIES

Malonate has been suggested to model HD disease pathogenesis and to replicate end-point HD striatal pathology. It therefore provides a model in which to investigate possible therapeutic strategies to block malonate toxicity. The excitotoxicity produced by malonate is secondary to mitochondrial impairment, but it is not fully understood how the inhibition of SDH by malonate results in ATP depletion or exactly how this metabolic compro-

Table 2
Summary of Ways in Which Malonate Toxicity Can Be Attenuated or Exacerbated

Compound	Mechanism of Action	Effect on Lesion Size	Reference
MK-801	Noncompetitive NMDA receptor antagonist	Decrease	*(8,13–15)*
Memantine	Noncompetitive NMDA receptor antagonist	Decrease	*(23)*
ARL-14896	Noncompetitive NMDA receptor antagonist	Decrease	*(14)*
LY274614	Competitive NMD receptor antagonist	Decrease Decrease (30%)	*(14)* *(15)*
7-Chlorokynurenate	Antagonist at glycine site of NMDA receptor	Decrease	*(14)*
NBQX	Non-NMDA receptor antagonist	No Change	*(14)*
Coenzyme Q_{10}	Enhances mitochondrial function	Decrease	*(33)*
Nicotinamide	Enhances mitochondrial function	Decrease	*(33)*
Thioctic/dihydrolipoic acid	Enhances mitochondrial function Free radical scavenging	Decrease	*(34)*
S-PBN	Free radical spin trap	Decrease No change ATP or lactate levels	(23)
NLA	Nonselective NOS inhibitor	Decrease (30%)	*(29)*
7-NI	Selective nNOS inhibitor	Decrease	*(30)*
S-Methylthiocitrulline	Selective nNOS inhibitor	Decrease (40%)	*(31)*
Lamotrigine	Gllutamate release inhibitor	Decrease (40–50%)	*(15)*
Quinolinic acid	Endogenous NMDA receptor agonist	Increase	*(37)*
NMDA	NMDA receptor agonist	Increase	*(36,38)*
AMPA	Non-NMDA receptor agonist	Increase	*(36)*
Glutamate	Glutamate receptor agonist	Increase	*(36)*

mise results in excitotoxic cell death. The mechanism of excitotoxicity itself, although widely accepted as central to the process of neurodegeneration, is itself not completely understood. Mitochondrial toxins are neurodegenerative via a cascade of events involving the electron transport chain, NMDA receptors, calcium influx, and free radical production. Thus, there is the potential for neuroprotective strategies acting at the level of mitochondrial inhibition by the toxin or by targeting the resultant excitotoxic processes (Table 2).

NMDA Receptor Antagonists

The NMDA receptor has been identified as a link between metabolic compromise and excitotoxicity; the proposed mechanism requires that this receptor plays a major role in malonate-induced toxicity. Pretreatment with 4 or 5 mg/kg of the noncompetitive NMDA receptor ion channel antagonist MK-801 30 min before intrastriatal infusion of 3 μmol malonate resulted in a 40–55% reduction in lesion volume, as assessed on Nissl and 2,3,5-triphenyltetrazolium (TTC) stained sections *(8,15,12)*. Treatment with 5 mg/kg both 30 min before and 210 min after infusions of 1 μmol or 2 μmol malonate resulted in a 80–90% attenuation *(13,14)*. Thus, notwithstanding the differences in the level of protection, MK-801 affords significant attenuation of malonate lesions, suggesting that malonate toxicity is mediated via NMDA receptors.

Other NMDA receptor antagonists have also demonstrated neuroprotective potential. ARL-15896, like MK-801, is a noncompetitive NMDA receptor antagonist which resulted in around 80% protection against 1.0 μmol malonate when administered intracerebrally or subcutaneously injection 30 min before and 210 min after malonate infusion *(24)*. LY274614, a competitive NMDA antagonist, and 7-chlorokynurenate, an antagonist at the glycine site of the NMDA receptor complex, were also neuroprotective against malonate lesions, demonstrating an approximate 70% reduction in lesion size *(14)*. These results indicate that antagonists at both the ion channel and the glycine site on the NMDA receptor attenuate malonate-induced striatal lesions.

Although excitotoxicity can also occur via other EAA receptors, the lack of protection by the competitive non-NMDA receptor antagonist 2,3-dihydroxoy-6-nitro-7-sulfamoylbenzo(*f*)-quinoxaline (NBQX) suggests little involvement of α-amino-3-hydroxy-5-methyl-4-isoxazole propionic acid (AMPA) and kainate receptors *(14)*. Collectively these studies substantiate the proposed mechanism of action of malonate and confirm the involvement of the NMDA receptor.

Free Radical Spin Traps

Free radical spin traps are compounds that have the ability to stabilize free radicals by forming stable adducts. Free radicals have been associated with NMDA receptor activation and excitotoxicity *(25)* and free radical spin traps have demonstrated neuroprotection against malonate toxicity *(26)*.

The free radical spin trap *n-tert*-butyl-α-(2-sulfophenyl)-nitrone (S-PBN) is neuroprotective against lesions produced by NMDA, AMPA, kainate, and malonate. The protective effects against malonate toxicity increased with the combination of S-PBN and MK-801, indicating an interaction between NMDA-mediated excitotoxic mechanisms and the generation of free radicals.

Intrastriatal infusion of malonate has been demonstrated to increase hydroxyl ($OH^{\cdot}$) radicals, indicating that indirect excitotoxicity involves the generation of free radicals *(26)*. S-PBN has no effect on the decreased ATP and increased lactate levels resulting from malonate toxicity *(26)*, indicating that S-PBN acts at excitotoxic processes secondary to metabolic impairment rather than protecting against the primary action of malonate as an SDH inhibitor.

Nitric Oxide Synthase Inhibitors

The generation of the free radical nitric oxide ($NO^{\cdot}$) has been implicated in the process of excitotoxic cell death following activation of NMDA receptors *(27)*. Nitric oxide synthase (NOS) inhibitors have been used to investigate the role of $NO^{\cdot}$ in neuronal death. The enzyme exists in various isoforms, neuronal (nNOS), endothelial (eNOS) and inducible NOS (iNOS) which may account for the inconsistency in the literature regarding the use of NOS inhibitors to show neuroprotection against excitotoxicity and ischaemia induced damage. The reaction of $NO^{\cdot}$ and superoxide ($O2^{\cdot}$) leads to the formation of the powerful oxidant peroxynitrite ($ONOO^{-}$). Peroxynitrite is a highly reactive mediator of both oxidative damage and the nitration of tyrosine to 3-nitrotyrosine *(28)*. Malonate-induced striatal lesions are partially blocked by the nonselective NOS inhibitor *N*-nitro-L-arginine (NLA), suggesting that $NO^{\cdot}$ has a role to play in indirect excitotoxicity *(29)*. Selective inhibitors of the neuronal form of NOS (nNOS), 7-nitroindazole (7-NI), and *S*-methylthiocitrulline provide complete attenuation of malonate toxicity *(30,31)* and studies using nNOS and eNOS knockout mice have substantiated these results in that nNOS mutant mice show attenuation of malonate striatal lesions, whereas eNOS mutant mice demonstrate significant increases in lesion size when compared with littermate controls *(32)*. Intrastriatal infusion of malonate induces increased $OH^{\cdot}$ and nitrotyrosine generation which is attenuated in nNOS mutant mice, suggesting that free radicals play a significant role in malonate-induced excitotoxicity. 7-NI, a selective inhibitor of nNOS in vivo, attenuated malonate-induced decreases in ATP and increases in lactate, indicating that $NO^{\cdot}$ is not only involved in excitotoxicity after activation of NMDA receptors but it is also involved in the action of malonate on energy depletion, suggesting a pivotal role of $NO^{\cdot}$ in malonate-induced toxicity.

Enhancers of Mitochondrial Function

Compounds that act as enhancers of mitochondrial function may alter malonate-induced toxicity by increasing the efficiency of oxidative phosphorylation, thereby overcoming the metabolic impairment caused by malonate.

Coenzyme Q_{10} is an electron donor and acceptor in the electron transport chain and nicotinamide is the precursor of NADH, the substrate for complex I. Both compounds have been demonstrated to block both the initial decrease in ATP and to prevent the resultant neuronal loss *(33)*. Thioctic acid and dihydrolipoic acid are endogenous cofactors that have been demonstrated to be neuroprotective against both NMDA- and malonate-induced striatal lesions *(34)*. Dihydrolipoic acid produced significant attenuation of the lesion volume with both NMDA and malonate, but dihydrolipoic acid and thioctic acid are readily interconvertible in vivo so the effect of each individual compound is difficult to ascertain. Subsequently thioctic acid has been shown to have no neuroprotective action against malonate-induced lesions when administered chronically over 3 wk prior to the lesion. The reason for this discrepancy is unclear but as thioctic acid is an endogenous cofactor, when it is in high concentrations other physiological regulatory mechanisms may come into play and reduce its neuroprotective efficacy *(35)*. The exact mechanism of action of thioctic acid and dihydrolipoic acid remains undefined but it is proposed to be due to either enhanced mitochondrial function, increased levels of free radical scavengers, or their own intrinsic ability to quench free radicals.

Malonate-induced neuronal death is suggested to result from excitotoxic processes secondary to mitochondrial inhibition, leading to the postulation that a combination of neuroprotective strategies acting at both the level of metabolic inhibition and at some point in the resultant excitotoxic cascade might provide an additive neuroprotective effect. Administration of a combination of MK-801 or lamotrigine, a glutamate release blocker, and coenzyme Q_{10} provided a level of protection greater than that seen with either compound alone, demonstrating the presence of interaction between NMDA receptor activation and mitochondrial function. A combination of nicotinamide and S-PBN also produces additive neuroprotective effects *(23)*.

Direct and Indirect Excitotoxic Interactions

One interesting aspect of the relationship between metabolism and excitotoxicity, relevant to the pathogenesis of HD, is that low-grade metabolic inhibition may increase the vulnerability of neurons to normally subtoxic concentrations of endogenous glutamate. Experimentally, subtoxic concentrations of malonate that produce a negligible neuronal loss when injected alone caused a significant increase in lesion volume when coinjected with low doses of NMDA, S-AMPA, and glutamate *(36)* or coinfused chronically with low-dose quinolinic acid (Quin) *(37)*. The chronic coinfusion of subthreshold QA and malonate resulted in lesions greater than either toxin infused alone and these lesions were blocked effectively by the coadministration of MK-801, supporting the hypothesis that malonate toxicity is NMDA mediated. The lesion induced by subthreshold concentrations of malonate and NMDA was completely blocked by MK-801. This attenuation is also evident with coinjection of malonate and glutamate although the lesion is reduced by only 40%, suggesting involvement of other EAA receptors in glutamate-mediated toxicity, although no additional protection was seen when both MK-801 and NBQX were coadministered. Consequently it was proposed that during mild metabolic compromise, glutamate toxicity is mediated only partly by ionotropic receptors *(36)*.

An experiment carried out in young (P7) rat brain also demonstrated synergism between malonate and NMDA *(38)*. Interestingly P7 rats receiving both NMDA, at a dose that results in a small lesion when infused alone, and 1.0 μmol malonate, exhibit a much larger extent of damage than NMDA alone, further demonstrating the capacity of metabolic impairment to render neurons more susceptible to NMDA-mediated toxicity. At this age the brain is relatively resistant to malonate toxicity; doses as high as 5 μmol did not produce significant neuronal loss. In our own studies (Fig. 4), adult rats with intrastriatal infusion of 1 μmol of malonate show a mean lesion volume of 11.5 ± 1.9 mm^3, which is comparable to data previously reported *(13)*. The authors suggest that neonatal rats may rely heavily on anaerobic metabolism and glycolysis, rather than oxidative phosphorylation to fulfil energy requirements or that P7 rat brain may have lower endogenous glutamate concentrations that are insufficient to activate NMDA receptors and trigger neuronal death.

Age-Dependent Characteristics of Malonate Toxicity in the Striatum

A striking feature of metabolic toxicity is its age dependence. Older animals have been demonstrated to be more vulnerable to metabolic toxicity than younger animals *(8,38–40)*. In a study by Beal et al. *(8)* malonate-induced toxicity was demonstrated to be greater in 4- and 12- month old rats than in 1-mo-old rats.

To investigate the age-dependence of malonate further, we have studied toxicity in groups of rats aged 6 wk, and also 3, 6, 9, 15, and 27 mo ($n = 5$ in each group). Intrastriatal injections of 1.0 *M* malonate were made into the right striatum at coordinates anterior +0.7 mm, lateral –2.9 mm with respect to bregma and –4.6 mm ventral from dura, with the nose bar set 2.3 mm below the interaural line. The toxin was infused via a stainless steel cannula connected to a Hamilton syringe mounted on a Harvard pump. Animals were perfusion fixed with 4% paraformaldehyde 10 d after toxin infusion. Brains were sectioned at 60 μm on a freezing microtome and stained for Nissl (Fig. 6). The lesion volume was assessed by stereology. The results confirm the age-dependence of malonate toxicity in that the 27-mo-group was more vulnerable than the 6-wk and 3-mo groups (Fig. 7). This correlates well with the midlife onset of HD. Evidence suggests that there is a decline in mitochondrial function with age *(41)* that underlies this increased vulnerability of older neurons to malonate toxicity. First, oxidative damage to mitochondrial DNA has been shown to increase with age in rat and human brain and muscle, *(42–44)* which may lead to an increase in mitochondrial DNA mutations and generalized mitochondrial dysfunction. Second, there is an age-associated loss of mitochondrial complex I and IV activity in primate cerebral cortex *(45)* and SDH activity in rat CA1 pyramidal cells *(46)*. Mitochondria also show age-related morphological changes *(46)*.

Many of these age-related changes in mitochondrial function are progressive, leading to the suggestion that older animals have a higher neuronal susceptibility to malonate toxicity, resulting from either increased levels of metabolic inhibition as a consequence of the interaction between the toxin itself and age-related mitochondrial decline or an increased vulnerability of the aged neuron to the intracellular events initiated by mitochondrial inhibition.

However, our results suggest that there is not a simple linear relationship between neuronal vulnerability and increasing age. There appears to be a threshold at around 6 mo at which point an increase in age-related mitochondrial decline does not increase vulnerability to 1.0 *M* malonate . Further studies into age-related changes in SDH activity may clarify these results.

The observation that 27-mo-old rats are no more vulnerable than 6-, 9-, or 15-mo-old rats, despite the fact that they would be expected to have reduced metabolic efficiency, may be explained by a decrease in the responsiveness of striatal neurons with age. Studies in the striatum of aged rats and cats demonstrate decreases in the generation of spontaneous action potentials and in the occurrence of spontaneous excitatory postsynaptic potentials *(47,48)*. This may be evidence for the loss of excitatory inputs to the striatum and age-related changes in the cortico-striatal pathway. Further, NMDA receptor function seems to be compromised during aging *(49)* and responses of striatal neurons to NMDA and glutamate are decreased in

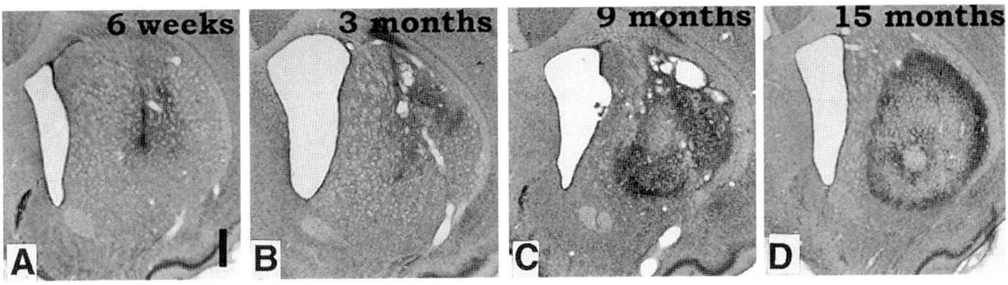

Fig. 6. Representative Nissl-stained sections showing the effects of age on striatal toxicity induced by the infusion of 1.0 *M* malonate. Scale bar = 1 mm.

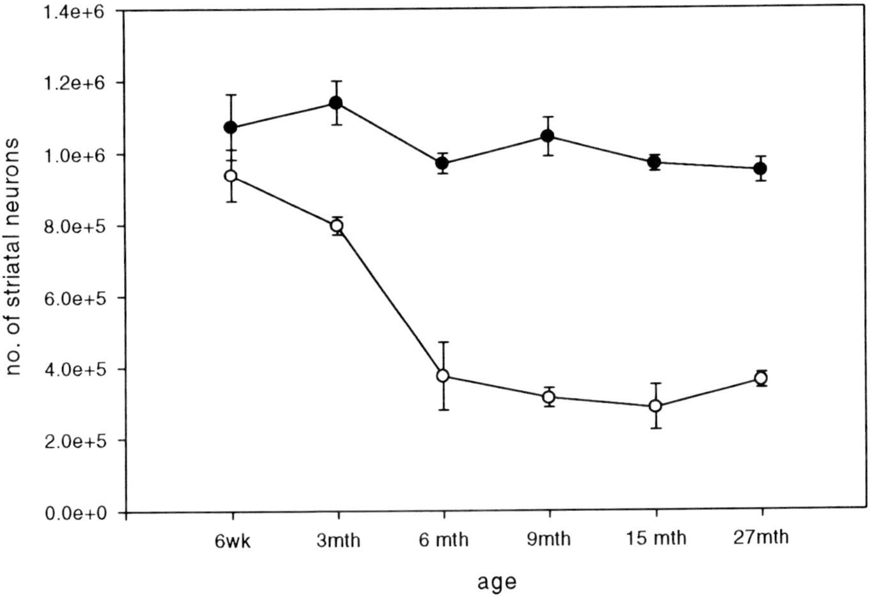

Fig. 7. Effects of age on malonate toxicity in the striatum. Data are mean ± SE*M* values of five animals. Note the increase in neuronal loss in age groups up to 6 mo.

aged animals *(50)*. EAA function is also altered during aging and there is evidence that EAA content in the brain decreases with age *(51,52)*. Together, these results suggest that the aged brain, although potentially more vulnerable to the initial metabolic inhibition, may not be as vulnerable as the younger brain to the secondary excitotoxicity which may be the final pathway leading to neuronal death.

SUMMARY

In neurodegenerative diseases, such as HD, where the neuronal loss is slow and progresses over many years, the use of subtoxic concentrations of metabolic toxins may approximately replicate the disease process. These toxins provide a more valuable model of neurodegenerative disease than those simply replicating the end-point. Pathology observed at postmortem, although a consequence of the disease process, may be almost entirely due to secondary mechanisms of neuronal death and give little insight into the initiation and progression of the disease.

The observation that metabolic inhibitors, including malonate and 3-NPA, induce in neurons a susceptibility to low concentrations of glutamate, taken together with observations of metabolic defects in patients with HD, Alzheimer's disease, and Parkinson's disease and the proposal that mitochondrial metabolic impairment can cause normal endogenous glutamate concentrations to become excitotoxic, provides us with the closest model yet of neurodegenerative processes.

REFERENCES

1. Jenkins BG, Koroshetz WJ, Beal MF, et al. Evidence for impairment of energy metabolism in vivo in Huntington's disease using localized H-1-NMR spectroscopy. Neurology 1993;432689–2695.
2. Gu M, Gash MT, Mann VM, et al. Mitochondrial defect in Huntington's disease on caudate nucleus. Ann Neurol 1996;39(3):385–389.
3. Novelli A, Reilly JA, Lysko PG, et al. Glutamate becomes neurotoxic via the NMDA receptor when intracellular energy levels are reduced. Brain Res 1988;451:205–212.
4. Henneberry RC, Novelli A, Cox JA, et al. Neurotoxicity at the NMDA receptor in energy compromised neurons. Ann NY Acad Sci 1989;568:225–233.
5. Beal MF. Does impairment of energy metabolism result in excitotoxic neuronal death in neurodegenerative illness? Ann Neurol 1992;31:119–130.
6. Beal MF, Hyman BT, Koroshetz W. Do defects in mitochondrial energy metabolism underlie the pathology of neurodegenerative diseases? TINS 1993;16:125–131.
7. Beal MF, Brouillet E, Jenkins BG, et al. Neurochemical and histologic characterisation of striatal excitotoxic lesions produced by the mitochondrial toxin 3-nitropropionic acid. J Neurosci 1993;13:4181–4192.
8. Beal MF, Brouillet E, Jenkins B, et al. Age-dependent striatal excitotoxic lesions produced by the endogenous mitochondrial inhibitor malonate. J Neurochem 1993;61:1147–1150.
9. Vonsattel JP, Myers RH, Stevens TJ, et al. Neuropathological classification of Huntington's disease. J Neuropathol Exp Neurol 1985;44:559–577.
10. Ferrante RJ, Kowall NW, Beal MF, et al. Selective sparing of a class of striatal neurones in Huntington's disease. Science 1985;230: 561–563.

11. Kowall NW, Ferrante RJ, Martin JB. Patterns of cell loss in Huntington's disease. TINS 1987;10:24–29.
12. Martin JB. Huntington's disease: new approaches to an old problem. Neurology 1984;34:1059–1072.
13. Greene JG, Porter RHP, Eller RV, et al. Inhibition of succinate dehydrogenase by malonic acid produces an excitotoxic lesion in rat striatum. J Neurochem 1993;61:1151–1154.
14. Greene JG, Greenamyre JT. Characterisation of the excitotoxic potential of the reversible succinate dehydrogenase inhibitor malonate. J Neurochem 1995;64: 430–436.
15. Henshaw R, Jenkins BG, Schulz JB, et al. Malonate produces striatal lesions by indirect NMDA receptor activation. Brain Res 1994;647:161–166.
16. Brouillet E, Hantrate P, Ferrante RJ, et al. Chronic mitochondrial energy impairment produces selective striatal degeneration and abnormal choreiform movements in primates. Proc Natl Acad Sci USA 1995;92:7105–7109.
17. Zeevalk GD, Manzino L, Hoppe J, et al. In vivo vulnerability of dopamine neurons to inhibition of energy metabolism. Eur J Pharmacol 1997;320:111–119.
18. Albin RL, Greenamyre JT. Alternative excitotoxic hypothesis. Neurology 1992;42:733–378.
19. Zeevalk GD, Nicklas WJ. Mechanisms underlying initiation of excitotoxicity associated with metabolic inhibition. J Pharmacol Exp Ther 1991;257:870–878.
20. Zeevalk GD, Nicklas WJ. Evidence that the loss of the voltage-dependent Mg2+ block at the NMDA receptor underlies receptor activation during inhibition of neuronal metabolism. J Neurochem 1992;59:1211–1220.
21. Schwarcz R. Excitotoxic models for neurodegenerative disorders. Life Sci 1984;35:19–32.
22. Greene JG, Greenamyre JT. Manipulation of membrane potential modulates malonate-induced striatal excitotoxicity in vivo. J Neurochem 1996;66:637–643.
23. Schulz JB, Matthews RT, Henshaw DR, et al. Neuroprotective strategies for treatment of lesions produced by mitochondrial toxins: implications for neurodegenerative diseases. Neuroscience 1996;71:1043–1048.
24. Greene JG, Porter RHP, Greenamyre JT. ARL–15896, a novel *N*-methyl-D-aspartate receptor ion channel antagonist: neuroprotection against mitochondrial metabolic toxicity and regional pharmacology. Exp Neurol 1996;137:66–72.
25. Dykens JA. Isolated cerebral and cerebellar mitochondria produce free radicals when exposed to elevated Ca^{2+} and Na^{+}: implications for neurodegeneration. J Neurochem 1994;63:584–591.
26. Schulz JB, Henshaw R, Siwek D, et al. Involvement of free radicals in excitotoxicity in vivo. J Neurochem 1995;64:2239–2247.
27. Gunasekar PG, Kanthasamy AG, Borowitz JL, et al. NMDA receptor activation produces concurrent generation of nitric oxide and reactive oxygen species-implication for cell-death. J Neurochem 1995;65:2016–2021.
28. Ischiropoulos H, Zhu L, Chen J, et al. Peroxynitrite-mediated tyrosine nitration catalysed by superoxide dismutase. Arch Biochem Biophys 1992;298:431–437.

29. Maragos WF, Silverstein FS. Inhibition of nitric oxide synthase activity attenuates striatal malonate lesions in rats. J Neurochem 1995;64:2362–2365.
30. Schulz JB, Matthews RT, Jenkins BG, et al. Blockade of neuronal nitric oxide synthase protects against excitotoxicity in vivo. J Neurosci 1995;15:8419–8429.
31. Matthews RT, Yang L, Beal MF. *S*-Methylthiocitrulline, a neuronal nitric oxide synthase inhibitor, protects against malonate and MPTP neurotoxicity. Exp Neurol 1997;143:282–286.
32. Schulz JB, Huang PL, Matthews RT, et al. Striatal malonate lesions are attenuated in neuronal nitric oxide synthase knockout mice. J Neurochem 1996; 67:430–433.
33. Beal MF, Henshaw DR, Jenkins BG, et al. Coenzyme Q10 and nicotinamide block striatal lesions produced by the mitochondrial toxin malonate. Ann Neurol 1994;36:882–888.
34. Greenamyre JT, Garcia-Osuna M, Greene JG. The endogenous cofactors, thioctic acid and dihydrolipoic acid, are neuroprotective against NMDA and malonic acid lesions of striatum. Neurosci Lett 1994;171: 17–20.
35. Angel R, Greenamyre JT. Chronic treatment with thioctic acid does not protect against malonate toxicity in vivo. Neurosci Lett 1995;196:125–127.
36. Greene JG, Greenamyre JT. Exacerbation of NMDA, AMPA and L-glutamate excitotoxicity by the succinate dehydrogenase inhibitor malonate. J Neurochem 1995;64:2332–2338.
37. Bazzet TJ, Falik RC, Becker JB, et al. Synergistic effects of chronic exposure to subthreshold concentrations of quinolinic acid and malonate in the rat striatum. Brain Res 1996;718:228–232.
38. Maragos WF, Silverstein FS. The mitochondrial inhibitor malonate enhances NMDA toxicity in the neonatal rat striatum. Dev Brain Res 1995;88:117–121.
39. Bossi SR, Simpson JR, Isacson O. Age-dependence of striatal neuronal death caused by mitochondrial dysfunction. NeuroReport 1993;4:73–76.
40. Brouillet E, Jenkins BG, Hyman BT, et al. Age-dependent vulnerability of the striatum to the mitochondrial toxin 3-nitropropionic acid. J Neurochem 1993;60:356–359.
41. Shigenaga MK, Hagen TM, Ames BN. Oxidative damage and mitochondrial decay in aging. Proc Natl Acad Sci USA 1994;91:10771–10778.
42. Ames DN, Shigenaga MK, Hagen TM. Oxidants, antioxidants and the degenerative diseases of aging. Proc Natl Acad Sci USA 1993;90:7915–7922.
43. Mecocci P, MacGarvey U, Kaufman AE, et al. Oxidative damage to mitochondrial-DNA shows marked age-dependent increases in human brain. Ann Neurol 1993;34:609–616.
44. Cortopassi GA, Shibata D, Soong N-W, et al. A pattern of accumulation of a somatic deletion of mitochondrial DNA in aging human tissues. Proc Natl Acad Sci USA 1992;89:7370–7374.
45. Bowling AC, Mutisya EM, Walker LC, et al. Age-dependent impairment of mitochondrial function in primate brain. J Neurochem 1993;60:1964–1967.
46. Bertoni-Freddari C, Fattoretti P, Caselli U, et al. Age-dependent decrease in the activity of succinate dehydrogenase in rat CA1 pyramidal cells: a quantitative cytochemical study. Mech Aging Dev 1996;90:53–62.

47. Levine MS, Lloyd RL, Hull CD, et al. Neurophysiological alterations in caudate nucleus in aged cats. Brain Res 1987;401:213–230.
48. Cepeda C, Walsh JP, Hull CD, et al. Intracellular neurophysiological analysis reveals alterations in excitation in striatal neurons in aged rats. Brain Res 1989: 494:251–226.
49. Ingram DK, Garofalo P, Spangler EL, et al. Reduced density of NMDA receptors and increased sensitivity to dizocilpine-induced learning impairment in aged rats. Brain Res 1992;580:237–280.
50. Cepeda C, Li Z, Levine MS. Aging reduces neostriatal responsiveness to *N*-methyl-D-aspartate and dopamine: an in vitro electrophysiological study. Neuroscience 1996;73:733–750.
51. Dawson R, Wallace DR, Meldrum MJ. Endogenous glutamate release from frontal cortex of adult and aged rats. Neurobiol Aging 1989;10:665–668.
52. Banay-Schwartz M, Lajtha A, Palkovits M. Changes in aging in the levels of amino acids in rat CNS structural elements. 1. Glutamate and related amino acids. Neurochem Res 1989;14:555–562.

15

Malonic Acid and the Chronic Administration Model of Excitotoxicity

Terence J. Bazzett, Roger L. Albin, and Jill B. Becker

INTRODUCTION

It is now recognized that neural death associated with acute insults such as trauma, hypoglycemia, seizures, global hypoxia, and stroke is due in part to calcium-mediated excitotoxic injury (for review *see* ref. *[1]*). More specifically, acute insult stimulates excessive release of excitatory amino acid neurotransmitters (EAAs) that in turn cause a rapid influx of calcium ions in affected neurons *(2)*. Intracellular calcium is essential for normal neural function; however, calcium overload may disrupt cell metabolism, cell excitability, gene expression, and other vital functions *(3)*. Although the process by which calcium-mediated excitotoxic injury occurs may vary *(4)*, the end result of compromised cell function inevitably contributes to cell death.

One focus of research in our laboratories is modeling neuronal death associated with chronic neurodegenerative disease. Within this area of research there is increasing speculation that, as with acute insult, excitotoxic damage may contribute to cell death associated with chronic neurodegenerative disorders. In particular, research has focused on the role of *N*-methyl-D-aspartate (NMDA) receptor sites. NMDA receptor sites form integral membrane calcium channels that are one source of calcium influx. Membrane depolarization and agonist stimulation of these receptors releases a magnesium ion channel block, allowing calcium to enter the cell *(5)*. Several laboratories have shown that stimulation of these receptors and subsequent calcium deregulation can result in excitotoxic cell death *(4,6–8)*. It follows that the physiological basis of neurodegenerative disease could be production of abnormally high concentrations of endogenous NMDA receptor agonists resulting in chronic calcium influx and ultimately cell death. This *excitotoxic hypothesis* of neurodegeneration is supported by in vivo

From: *Mitochondrial Inhibitors and Neurodegenerative Disorders*
Edited by: P. R. Sanberg, H. Nishino, and C. V. Borlongan © Humana Press Inc., Totowa, NJ

models showing that repeated *(9)* and chronic *(10)* exposure to EAAs, acting as NMDA receptor agonists, produces progressive cell death.

Of particular relevance to our research are excitotoxic models of Huntington's disease (HD). HD is an autosomal dominantly inherited genetic disease that results in profound degeneration of the striatum. This degeneration is characterized histopathologically by a progressive loss of medium spiny projection neurons with relative sparing of dopaminergic afferents and relative sparing of aspiny interneurons that cocontain somatostatin, neuropeptide Y, NADPH-diaphorase, nitric oxide synthase (SS/NPY/NADPH-NOS) *(11)*. Early excitotoxic models of HD striatal degeneration utilized acute intrastriatal injection of the non-NMDA receptor agonist kainic acid *(12)* and the NMDA/metabotropic glutamate receptor agonist ibotenic acid *(13)*. Although capable of producing excitotoxic striatal lesions, these compounds failed to produce an HD-like pattern of selective cell loss. Later, Beal and colleagues reported acute intrastriatal administration of the endogenous NMDA receptor agonist quinolinic acid (Quin) produced lesions with both neurochemical characteristics *(14)* and a selective pattern of cell loss reminiscent of HD *(15)*.

Initially, the finding that high concentrations of an endogenous NMDA receptor agonist could produce HD-like striatal degeneration suggested that the genetic aberration associated with HD might be causing overproduction of Quin. However, evaluation of HD patients has revealed little conclusive evidence of significantly elevated endogenous concentrations of Quin *(16)* or decreased Quin catabolism *(17)*. Likewise, assessment of a broader range of neurodegenerative disorders has failed to produce conclusive data linking elevated concentrations of any endogenous NMDA receptor agonist to Huntington's, Alzheimer's, or Parkinson's disease *(18,19)*. The inability to directly implicate elevated levels of an endogenous NMDA receptor agonist as the basis for these disorders raises questions about the validity of the excitotoxic hypothesis of chronic neurodegeneration. In addition, the pattern of selective cell loss seen in HD and after exposure to Quin suggests unique features of a striatal neuron subpopulation that result in increased vulnerability (or resistance) to the source of neurotoxicity. Taken together, these findings are difficult to reconcile within the parameters of the original excitotoxic hypothesis of chronic neurodegeneration.

Following acute brain trauma, increased glutamate release results directly in calcium influx, tonic cellular depolarization, and subsequent excitotoxic injury *(2)*. The absence of data to support such a direct theory of cell death in chronic neurodegenerative diseases prompted a revision of the original excitotoxic hypothesis. This revised theory of "weak" or "indirect" excitotoxic neurodegeneration suggests that in neurodegenerative diseases, normal concentrations of endogenous NMDA receptor agonists are capable of producing excitotoxicity in cells rendered susceptible by compromised metabolic activity *(20–22)*. For example, mitochondria are essential for maintaining cell membrane resting potential by providing metabolic energy for Na^+, K^+-ATPase. One possible result of impaired mitochondrial function is depolarization of the cell membrane and a subsequent reduction of the NMDA receptor magnesium blockade *(20)*. Deregulation of the NMDA receptor/calcium channel complex would then further depolarize the cell, presumably to an excitotoxic level. Further support for a role of mitochondria, and the indirect excitotoxic hypothesis, comes from the finding that in addition to their primary role in maintaining cellular energy stores *(23)*, mitochondria also buffer excess intracellular calcium *(24–26)*.

Several mitochondrial inhibitors have been found to produce excitotoxic lesions. Intrastriatal injection of aminooxyacetic acid inhibits aspartate transaminase, an enzyme essential for mitochondrial malate–aspartate membrane shunt function, producing excitotoxic lesions *(27)*. 3-Nitropropionic acid (3-NPA) produces excitotoxic lesions by irreversibly inhibiting succinate dehydrogenase, a key enzyme required for oxidative phosphorylation in mitochondria *(28)*. Similarly, intrastriatal injection of malonic acid (MA), a reversible inhibitor of succinate dehydrogenase, also produces excitotoxic lesions *(29)*. Further, blocking NMDA receptors sites can attenuate lesions produced by each of these mitochondrial inhibitors, suggesting a primary role for these receptor sites in the excitotoxic process *(27–29)*.

Perhaps the strongest evidence in support of the indirect excitotoxic theory comes from in vivo studies showing that metabolic inhibition can render cells susceptible to toxicity from concentrations of EAAs that do not produce cell death under normal conditions. For example, MA is reported to potentiate the neurotoxic effects of acute injection of the EAAs NMDA, α-amino-3-hydroxy-5-methyl-4-isoxazole propionic acid (AMPA), and L-Glutamate *(30)*.

To determine further the utility of an in vivo indirect excitotoxic model of neurodegeneration, our laboratories devised a series of experiments in which rats were exposed to chronic intrastriatal administration of Quin, MA, and a combination of these compounds. In particular, we hoped to produce and assess an HD-like pattern of striatal cell loss. In addition to testing the relative toxicity and toxic specificity of these compounds, we also wished to establish the reliability of a novel in vivo intracranial drug delivery system. This system was developed utilizing a chronic in vivo microdialysis probe through which a constant flow of solution is provided by a subcutaneously implanted osmotic minipump. This dialysis administration system has several potential advantages over traditional chronic drug infusion. For example, traditional drug infusion methods, via injection cannulae, provide point source delivery resulting in a small area of high drug concentration with regions of lower concentrations radiating from that point *(31)*. The dialysis system, on the other hand, provides drug diffusion along the entire length of a 4 mm dialysis fiber. Although a diffusion gradient radiating from the fiber is still present, consistent drug concentration exposure along the length of the fiber results in a relatively large region of the striatum in which to assess drug effects. In addition, because drug exposure results from diffusion through the fiber membrane, this system eliminates potential effects of osmotic changes associated with conventional chronic infusion systems *(32,33)*.

CHRONIC QUIN

Previous research suggests that acute injection of Quin is capable of producing a selective pattern of neurodegeneration similar to that seen in HD *(10,15,34)*. In particular, this pattern is marked by a significant loss of Nissl-stained cells with a relative sparing of SS/NPY/NADPH-NOS cells *(11)*. Our initial research using chronic intrastriatal dialytic administration of Quin resulted in a dose dependent neurotoxic effect *(35)*. Administration of approx 220 µL of a 4 m*M* concentration of Quin over 20 d resulted in no significant change in Nissl-stained cells compared to vehicle administration. In addition, 4 m*M* Quin did not significantly reduce cytochrome oxidase (CO) staining in the region surrounding the probe. At a concentration of 15 m*M*, Quin produced a significant lesion as measured by reduction in the number of Nissl-stained cells, and a significant reduction in CO staining in the region surrounding the probe. In addition, lesions produced by 15 m*M* Quin exhibited an HD-like pattern of relative sparing of SS/NPY/NADPH-NOS cells. At a concentration of 40 m*M*, Quin produced large lesions marked by loss of Nissl-stained cells and reduced CO staining. However, this dose also resulted in extensive striatal atrophy and a lack of quantifiable selectivity in cell destruction, reducing its usefulness as an animal model for HD neurodegeneration.

Perhaps more interesting than the production of selective HD-like lesions at the 15 m*M* concentration were results from immunohistochemical analysis of the 4 m*M* concentration. Although no significant decrease in Nissl-stained cells was apparent at this concentration, there was a significant decrease in calbindin (CALB) immunoreactive perikarya *(36)*. This finding is of particular interest when considering that intracellular calcium may be buffered by cellular proteins, including CALB *(37)*. In a subsequent experiment, striatal tissue from a group of rats killed immediately following chronic exposure to 4 m*M* Quin were compared to a group killed 4 wk after completion of chronic exposure. Results from this study showed a significant decrease in CALB immunoreactivity immediately following Quin exposure. However, CALB immunoreactivity was near normal levels in rats allowed the 4-wk recovery period following Quin exposure *(36)*. One explanation for this transient reduction is the possibility that CALB was redistributed from the cell body to dendritic processes during a time of NMDA activation. CALB exhibits characteristics of a fast mobile calcium buffering protein, capable of diffusing from surrounding cytoplasm to local sites of calcium influx. Dendritic processes contain a high concentration of NMDA receptors *(4,8,38)* and are capable of independent regulation of changes in calcium *(39,40)*, possibly through CALB recruitment. Ferrante and colleagues *(41)* have noted redistribution of CALB to distal processes in HD striatal tissue. Such transient changes, although not necessarily associated with cell death, do appear to indicate a compromised state of neuronal function.

CHRONIC MA

It is currently believed that NMDA induced excitotoxicity, such as that produced by Quin, results from prolonged depolarization and excess intracellular calcium *(42–44)*. It was thus reasoned that impairment of mitochondrial activity, causing a loss of NMDA receptor magnesium blockade and the subsequent inability of neurons to sequester intracellular calcium, could produce a pattern of excitotoxicity similar to that seen after Quin administration. In the rat, acute intrastriatal administration of MA produces dose-dependent lesions with features similar to those produced by Quin *(29,45–47)*. In addition, acute neurotoxic effects of MA can be blocked by NMDA antagonists *(29,46,47)*. The finding of neuroprotection through NMDA receptor antagonism is consistent with the hypothesis that mitochondrial function is one determinant of cellular susceptibility to excitotoxic injury.

To further assess MA-induced neurodegeneration, we devised a series of experiments to test the dose–response effects of MA toxicity *(48)*. Using the intrastriatal dialytic delivery system, we found that low concentrations of MA (100 m*M* and 400 m*M*) were ineffective in producing striatal lesions. However, at higher concentrations (1 *M*) MA produced lesions characterized by selective sparing of SS/NPY/NADPH-NOS cells in a region of reduced Nissl-stained cells. At a concentration of 4 *M*, MA produced large striatal lesions that lacked features of selective cell loss seen at the 1 *M* concentration.

As with a subthreshold concentration of Quin, subthreshold concentrations of 100 and 400 m*M* MA produced a significant dose dependent and transient decrease in CALB immunoreactive perikarya in a region that showed no change in Nissl cell staining *(48)*. The similarities between Quin and MA in the selective subpopulation of neurons affected, and the transient immunoreactive changes of CALB, strongly suggest that chronic exposure to MA produces primary neurotoxic effects via NMDA receptor/calcium channel deregulation.

QUIN + MA

It has been reported that MA is capable of potentiating the neurotoxic effects of acute injection of the EAAs NMDA, AMPA and L-glutamate *(30)*. To test further the indirect excitotoxic theory of chronic neurodegeneration, we wished to determine if chronic exposure to subthreshold concentrations of Quin and MA would have synergistic effects resulting in significant striatal cell degeneration. This experiment was designed to strengthen the contention that MA increases neuronal vulnerability to the excitotoxic effects of NMDA receptor stimulation in a chronic administration model.

For this experiment, rats were chronically exposed to 4 m*M* Quin, 400 m*M* MA, or a combination of 4 m*M* Quin plus 400 m*M* MA *(49)*. Results from animals receiving 4 m*M* Quin or 400 m*M* MA alone replicated early findings of no significant decrease in Nissl-stained cells in regions surrounding the dialysis probe. However, there was a significant decrease in Nissl-stained cells surrounding the probe in animals exposed to the combination of these drugs. Furthermore, there was a relative sparing of SS/NPY/NADPH-NOS neurons in this same region. This latter feature is of particular importance because relative sparing of SS/NPY/NADPH-NOS neurons, a hallmark of HD striatal neurodegeneration, is a feature shared also by both Quin and MA lesions.

We next wished to examine the relative contribution of NMDA receptors to the neurotoxic process associated with combined subthreshold Quin/MA treatment. MK-801 is a potent competitive antagonist at NMDA receptor sites that was previously found to block neurotoxicity associated with acute Quin injection *(50)*. For our experiment, the subthreshold concentrations of 4 m*M* Quin and 400 m*M* MA were chronically coadministered with a concentration of 1 m*M* MK-801. Using this paradigm, MK-801 exhibited complete neuroprotection against Quin/MA treatment *(49)*.

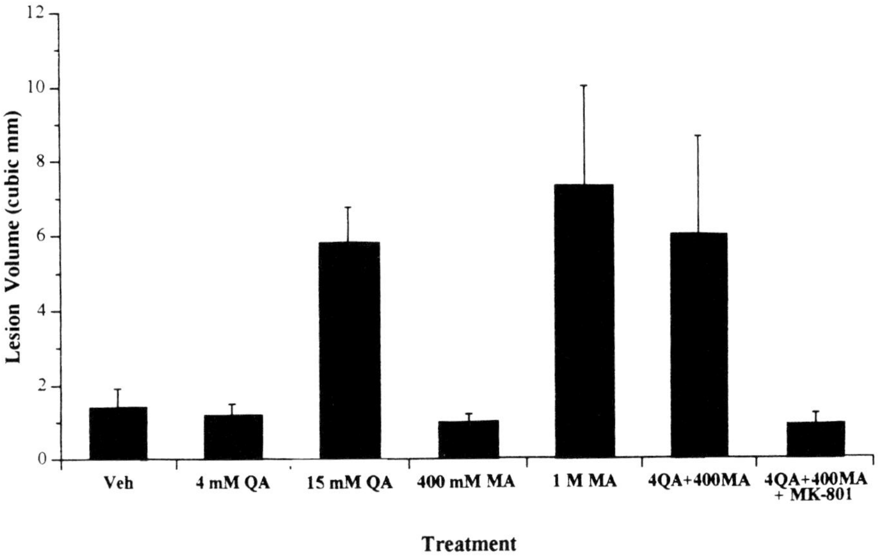

Fig. 1. Summary of lesion volume data from chronic dialytic administration studies *(35,48,49)*. A dose–response was revealed for both quinolinic acid (QA) and malonic acid (MA). The combination of subthreshold doses of QA (4 m*M*) and MA (400 m*M*) produced a significant lesion volume and selective cell loss similar to that produced by higher concentrations of QA and MA alone. The NMDA antagonist MK–801 (1 m*M*) provided neuroprotection against the combination of QA/MA, suggesting that NMDA receptor-mediated calcium influx is the basis for toxicity in this paradigm.

Selective sparing of SS/NPY/NADPH-NOS neurons in lesions produced by subthreshold concentrations of Quin and MA strongly suggests a synergistic effect, rather than development of a novel toxic effect of this combination. The finding of neuroprotection through blockade of NMDA receptors implicates these sites as a primary source of calcium deregulation associated with neurotoxicity in this model. Taken together these results appear to support the theoretical constructs of the indirect hypothesis of chronic neurodegeneration. The results of experiments described above are summarized in Fig. 1.

DISCUSSION

Although chronic administration of Quin and other EAAs seems to provide useful models for mimicking the histopathology and relative time course of neurodegenerative disorders, evidence that these disorders result from overproduction of endogenous EAAs is not compelling *(19)*. If excitotoxicity is the basis for specific neurodegenerative disorders, it is more likely that degeneration results from deficits in mitochondrial regulation of ionic influx or cation sequestration. Defective mitochondrial function has been implicated as a potential factor contributing to neuronal degeneration in numerous neurodegenerative diseases *(20–22)*. With regard to HD, decreased glucose metabolism is apparent in basal ganglia of both symptomatic *(51)* and at risk *(52)* patients. Elevated lactate concentrations, suggestive of decreased metabolism and abnormalities in mitochondrial oxidative phosphorylation, have also been noted in HD basal ganglia tissue *(53)*. Furthermore, postmortem HD caudate–putamen tissue shows a significant reduction in mitochondrial CO content compared to controls *(54)*. It is, however, still uncertain whether the CO reduction is a contributing factor to, or a consequence of, striatal degeneration.

It is difficult to speculate on the mechanisms of excitotoxicity that may be responsible for region specific degeneration. Initial topographical mapping of the recently identified HD gene *(55)* revealed widespread expression throughout the brain with little correlation to regions of increased degenerative susceptibility *(56–58)*. However, more recent mapping of huntingtin protein expression associated with this gene has shows a strikingly heterogeneous distribution within the striatum *(59)*. Of particular significance is the high correlation between huntingtin expression and the presence of CALB within striatal subregions that are particularly vulnerable to degeneration in HD.

The role of CALB in excitotoxicity and neurodegenerative disorders remains unclear. However, as with glucose metabolism, there is a negative correlation between age and the number of CALB-immunoreactive perikarya in humans *(60)* and in levels of CALB mRNA in both humans and rats *(61)*. Our findings of changes in CALB-immunoreactive perikarya in response to both Quin and MA suggest a role for this protein in regulating NMDA-induced excitotoxicity.

There is also evidence that excessive NOS-dependent nitric oxide release during high levels of NMDA receptor stimulation results in production of toxic hydroxyl radicals and nitrogen dioxide *(62,63)*. Striatal neurons spared in HD *(11)*, and after Quin *(10,15,34,35)* and MA *(29,45–47)* administration, contain NOS. It is possible that some intrinsic mechanism may protect these neurons from toxic effects of this particular free radical.

One hypothetical model of HD neurodegeneration is that an acceleration or exaggeration of age dependent reduction in basal ganglia mitochondrial activity may increase NMDA-mediated calcium influx and subsequent NOS release from one subpopulation of neurons. Over time, with preservation of SS/NPY/NADPH-NOS neurons, a relative increase in regional concentrations of NOS would be anticipated. In addition, normal EAA release from spared neurons combined with Quin production resulting from reactive gliosis *(64)* could produce a relative increase in EAA concentrations in this region. Finally, neurons undergoing both age dependent CALB *(60,61)*, and mitochondria *(65)* reduction would become particularly vulnerable to calcium influx. In agreement with the indirect excitotoxic hypothesis, this theoretical model proposes degeneration of a subpopulation of neurons resulting from a decrease in mitochondrial activity, resulting in membrane depolarization with concomitant NMDA receptor activation and/or impaired calcium homeostasis *(20–22,26)*. An additional component is the increased production of potentially toxic free radicals by a subpopulation of these cells. It thus appears that several factors may be associated with selective striatal degeneration in HD. If not itself the underlying cause, inhibition of mitochondrial function could potentially contribute to any of these additional factors.

Beal and colleagues *(45)* have noted that acute intrastriatal injection of MA produces lesions in 4-mo-old rats, but fails to produce neurotoxic effects in 1-mo-old rats. Systemic injection of the mitochondrial toxin 3-NPA similarly results in lesions that appear age dependent *(66)*. These findings suggest that MA and other mitochondrial inhibitors may provide a particularly useful model for neurodegenerative diseases of aging. On the other hand, such age-dependent differences in effects also require that particular attention be given this variable when developing experimental protocols for animal testing. Failure to replicate a particular age range when studying the effects of MA in animal models could potentially result in vastly different neurotoxic effects. In this regard, a researcher's decision to utilize a model of toxicity induced by mitochondrial inhibitors must be made with consideration to the potential for increased variability in results.

REFERENCES

1. Dubinsky JM. Examination of the role of calcium in neuronal death. Ann NY Acad Sci 1993;679:34–42.
2. Rothman SM, Olney JW. Excitotoxicity and the NMDA receptor—still lethal after eight years. Trends Neurosci 1995;18:57–58.
3. Gibbons SJ, Brorson JR, Bleakman D, et al. Calcium influx and neurodegeneration. Ann NY Acad Sci 1993;679:22–33.
4. Tymianski M, Charlton MP, Carlen PL, et al. Source specificity of early calcium neurotoxicity in cultured embryonic spinal neurons. J Neurosci 1993b;13:2085–2104.

5. Tsuzuki K, Iino M, Ozawa S. Change in calcium permeability caused by quinolinic acid in cultured hippocampal neurons. Neurosci Lett 1989; 105:269–274.
6. Hartley DM, Kurth MC, Bjerkness L, et al. Glutamate receptor-induced $^{45}Ca^{2+}$ accumulation in cortical cell culture correlates with subsequent neuronal degeneration. J Neurosci 1993;13:1993–2000.
7. Novelli A, Reilly JA, Lysko PG, et al. Glutamate becomes neurotoxic via the *N*-methyl-D-asparate receptor when intracellular energy levels are reduced. Brain Res 1988;451:205–212.
8. Tymianski M, Wallace MC, Spigelman I, et al. Cell-permeant Ca^{2+} chelators reduce early excitotoxic and ischemic neuronal injury in vitro and in vivo. Neuron 1993;11:221–235.
9. Jarrard LE, Meldrum BS. Selective excitotoxic pathology in the rat hippocampus. Neuropathol Appl Neurobiol 1993;19:381–389.
10. Beal MF, Ferrante RJ, Swartz KJ, et al. Chronic quinolinic acid lesions in rats closely resemble Huntington's disease. J Neurosci 1991;11:1649–1659.
11. Ferrante RJ, Kowall NW, Beal MF, et al. Selective sparing of a class of striatal neurons in Huntington's disease. Science 1985;230:561–564.
12. Coyle JT, Schwarcz R. Lesions of striatal neurons with kainic acid provide a model for Huntington's chorea. Nature 1976;263:244–246.
13. Schwarcz R, Hokfelt T, Fuxe K, et al. Ibotenic acid-induced neuronal degeneration: a morphological and neurochemical study. Exp Brain Res 1979; 37:199–216.
14. Beal MF, Kowall KW, Ellison DW, et al. Replication of the neurochemical characteristics of Huntington's disease by quinolinic acid. Nature 1986; 321:168–171.
15. Beal MF, Kowall NW, Swartz KJ, et al. Differential sparing of somatostatin-neuropeptide Y and cholinergic neurons following striatal excitotoxin lesions. Synapse 1989;3:38–47.
16. Schwarcz R, Tamminga CA, Kurlan R, et al. Cerebrospinal fluid levels of quinolinic acid in Huntington's disease and schizophrenia. Ann Neurol 1988;24:580–582.
17. Foster AC, Schwarcz R. Characterization of quinolinic acid phosphoribosyltransferase in human blood and observations in Huntington's disease. J Neurochem 1985;45:199–205.
18. Doble A. Excitatory amino acid receptors and neurodegeneration. Therapie 1995;50:319–237.
19. Young AB. Role of excitotoxins in heredito-degenerative neurologic diseases. Res Publ Assoc Res Nerv Ment Dis 1993;71:175–189.
20. Albin R L, Greenamyre JT. Alternative to excitotoxic hypothesis. Neurology 1992;42:733–738.
21. Beal MF. Does impairment of energy metabolism result in excitotoxic neuronal death in neurodegenerative illnesses? Ann Neurol 1992;31:119–130.
22. Beal MF. Aging, energy and oxidative stress in neurodegenerative disease. Ann Neurol 1995;38:357–366.

23. Henneberry RL, Novelli A, Cox JA, et al. Neurotoxicity at the *N*-methyl-D-aspartate receptor in energy-compromised neurons. An hypothesis for cell death in aging and disease. Ann NY Acad Sci 1989;568:225–233.
24. Abercrombie RF, Hart CE. Calcium and proton buffering and diffusion in isolated cytoplasm from *Myxicola* axons. Am J Physiol 1986;250:C391–C405.
25. Brinley FJ, Tiffert T, Scarpa A. Mitochondria and other calcium buffers of squid axon studied in situ. J Gen Physiol 1978;72:101–127.
26. Kiedrowski L, Costa E. Glutamate-induced destabilization of intracellular calcium concentration homeostasis in cultured cerebellar granule cells: role of mitochondria in calcium buffering. Mol Pharmacol 1995;47:140–147.
27. Beal MF, Schartz KJ, Hyman BT, et al. Aminooxyacetic acid results in excitotoxin lesions by a novel indirect mechanism. J Neurochem 1991; 57:1068–1073.
28. Beal MF, Brouillet E, Jenkins BG, et al. Neurochemical and histologic characterization of striatal excitotoxic lesions produced by the mitochondrial toxin 3-nitropriopionic acid. J Neurosci 1993;13:4181–4192.
29. Greene JG, Porter RHP, Eller RV, et al. Inhibition of succinate dehydrogenase by malonic acid produces an "excitotoxic" lesion in rat striatum. J Neurochem 1993;61:1151–1154.
30. Greene JG, Greenamyre JT. Exacerbation of NMDA, AMPA, and L-glutamate excitotoxicity by the succinate dehydrogenase inhibitor malonate. J Neurochem 1995;64:2332–2338.
31. Rice ME, Gerhardt GA, Hierl PM, et al. Diffusion coefficients of neurotransmitters and their metabolites in brain extracellular fluid space. Neuroscience 1985;15:891–902.
32. Bazzett TJ, Becker JB, Albin RL. A novel device for chronic intracranial drug delivery via microdialysis. J Neurosci Meth 1991;40:1–8.
33. Bazzett TJ, Becker JB, Albin RL. Construction and implantation of a chronic intracranial dialytic delivery system. Neurosci Protocols 1994;6: 1–10.
34. Figueredo-Cardenas G, Anderson KD, Chen Q, et al. Relative survival of striatal projection neurons and interneurons after intrastriatal injection of quinolinic acid in rats. Exp Neurol 1994;129:37–56.
35. Bazzett TJ, Becker JB, Kaatz, KW, et al. Chronic intrastriatal dialytic administration of quinolinic acid produces selective neural degeneration. Exp Neurol 1993;120:177–185.
36. Bazzett TJ, Becker JB, Falik RC, et al. Chronic intrastriatal quinolinic acid produces reversible changes in perikaryal calbindin and parvalbumin immunoreactivity. Neuroscience 1994;60:837–841.
37. Van Eldik LY, Zendegui JG, Marshak DR, et al. Calcium binding protein and the molecular basis of calcium action. Int Rev Cytol 1982;77:1–61.
38. Benke TA, Jones OT, Collingridge GL, et al. N-Methyl-D-aspartate receptors are clustered and immobilized on dendrites of living cortical neurons. Proc Natl Acad Sci USA 1993;90:7819–7823.
39. Guthrie PB, Segal M, Kater SB. Independent regulation of calcium revealed by imaging dendritic spines. Nature 1991;354:76–80.

40. Muller W, Connor JA. Dendritic spines as individual neuronal compartments for synaptic Ca^{2+} responses. Nature 1991;354: 73–76.
41. Ferrante RJ, Kowall NW, Richardson EP. Proliferative and degenerative changes in striatal spiny neurons in Huntington's disease: a combined study using the section-golgi method and calbindin D28k immunocytochemistry. J Neurosci 1991;11:3877–3887.
42. Choi DW. Ionic dependence of glutamate neurotoxicity. J Neurosci 1987; 7:369–379.
43. Rothman SM. Synaptic activity mediates death of hypoxic neurons. Science 1983;220:536–537.
44. Rothman SM. Synaptic release of excitatory amino acid neurotransmitter mediates anoxic neuronal death. J Neurosci 1984;7:1884–1891.
45. Beal MF, Brouilet E, Jenkins BG, et al. Age-dependent striatal excitotoxic lesions produced by the endogenous mitochondrial inhibitor malonate. J Neurochem 1993;61:1147–1150.
46. Greene JG, Greenamyre JT. Characterization of the excitotoxic potential of the reversible succinate dehydrogenase inhibitor malonate. J Neurochem 1995;64:430–436.
47. Henshaw R, Jenkins BG, Shultz JB, et al. Malonate produces striatal lesions by indirect NMDA receptor activation. Brain Res 1994;647:161–166.
48. Bazzett TJ, Falik RC, Becker JB, et al. Chronic administration of malonic acid produces selective neural degeneration and transient changes in calbindin immunoreactivity in rat striatum. Exp Neurol 1995;134:244–252. .
49. Bazzett TJ, Falik RC, Becker JB, et al. Synergistic effects of chronic exposure to subthreshold concentrations of quinolinic acid and malonate in the rat striatum. Brain Res 1996;718:228–232.
50. Beal MF, Kowall NW, Swartz KJ, et al. Systemic approaches to modifying quinolinic acid striatal lesions in rats. J Neurosci 1988;8:3901–3908.
51. Young AB, Penny JB, Sarosta-Rubinstein S, et al. Normal caudate glucose metabolism in persons at risk for Huntington's disease. Arch Neurol 1987;44:254–257.
52. Mazziotta JC, Phelps ME, Pahl JJ, et al. Reduced cerebral glucose metabolism in asymptomatic subjects at risk for Huntington's disease. N Engl J Med 1987;316:357–362.
53. Jenkins BG, Koroshetz WJ, Beal MF, et al. Evidence for impairment of energy metabolism in vivo in Huntington's disease using localized 1H NMR spectroscopy. Neurology 1993;43:2689–2695.
54. Brennan WA, Bird ED, Aprille JR. Regional mitochondrial respiratory activity in Huntington's disease brain. J Neurochem 1985;44:1948–1950.
55. The Huntington's Disease Collaborative Research Group. A novel gene containing a trinucleotide repeat that is expanded and unstable on Huntington's disease chromosomes. Cell 1993;72:971–983.
56. Landwehrmeyer GB, McNeil SM, Dure LS, et al. Huntington's disease gene: regional and cellular expression in brain of normal and affected individuals. Ann Neurol 1995;37:218–230.

57. Li SH, Schilling B, Young WS, et al. Huntington's disease gene (IT15) is widely expressed in human and rat tissues. Neuron 1993;1:985–993.
58. Strong TV, Tagle DA, Valdes JM, et al. Widespread expression of the human and rat Huntington's disease gene in brain and nonneuronal tissues. Nat Genet 1993;5:259–265.
59. Ferrante RJ, Gutenkunst CA, Persichette F, et al. Heterogeneous topographic and cellular distribution of huntingtin expression in the normal human neostriatum. J Neurosci 1997;17:3052–3063.
60. Nishiyama E, Ohwada J, Iwamoto N, et al. Selective loss of calbindin D28K-immunoreactive neurons in the cortical layer II in brains of Alzheimer's disease: a morphometric study. Neurosci Lett 1993;163:223–226.
61. Iacopino AM, Christakos S. Specific reduction of calcium-binding protein (28-kilodalton calbindin-D) gene expression in aging and neurodegenerative diseases. Proc Natl Acad Sci USA 1990;87:4078–4082.
62. Dawson VL, Dawson TM. Nitric oxide actions in neurochemistry. Neurochem Int 1996;29:97–110.
63. MacKenzie GM, Jenner P, Marsden CC. The effect of nitric oxide synthase inhibition on quinolinic acid toxicity in the rat striatum. Neuroscience 1995;1995:357–371.
64. Ceresoli G, Fuller MS, Schwarcz R. Excitotoxic lesions of the rat striatum: different responses of kynurenine pathway enzymes during ontogeny. Brain Res Dev Brain Res 1996;92:61–69.
65. Ragusa N, Turpeenoja L, Magri G, et al. Age-dependent modifications of mitochondrial proteins in cerebral cortex and striatum of rat brain. Neurochem Res 1989;14:415–418.
66. Brouilet E, Jenkins BG, Hyman BT, et al. Age-dependent vulnerability of the striatum to the mitochondrial toxin 3-nitropropionic acid. J Neurochem 1993;60:356–359.

16

Sodium Azide-Induced Neurotoxicity

Yun Wang and Cesario V. Borlongan

Neurodegeneration can be caused by dysfunction of the mitochondrial electron transport chain (ETC), and thus a mitochondrial etiology has been suggested for many neurodegenerative disorders, such as Parkinson's disease (PD), Huntington's disease (HD), and Alzheimer's disease (AD) *(1,2)*. Accordingly, mitochondrial toxins have been used in animal models to mimic neurodegeneration. For example, 1-methyl-4-phenyl-1,2,3,6-tetrahydropyridine (MPTP), via its active metabolite 1-methyl-4-phenylpyridinium (MPP^+), selectively inhibits NADH-coenzyme Q reductase (complex I) of the ETC, and induces Parkinsonism in humans, primates, and mice *(3–5)*. Aberrant free radical formation *(4)*, as well as impaired mitochondrial calcium metabolism *(6)*, have been observed to succeed ETC deficits. A cascade of events leading to cell death ensues following local infusion of MPP^+, and this involves increments in superoxide radicals which in combination with nitric oxide ($^{\cdot}NO$) can yield peroxynitrite anion ($ONOO^-$) and in turn can spontaneously decompose to produce the potentially destructive reactive hydroxyl free radicals ($^{\cdot}OH$) *(7,8)*.

Exposure of brain particles (submitochondrial fractions) to sodium azide, a complex IV (cytochrome oxidase) inhibitor, can yield $^{\cdot}OH$ and thus can produce ETC deficits similar to those seen in MPP^+ infusion *(9)*. Animals chronically treated with sodium azide exhibit a marked reduction in cytochrome oxidase activity and spatial learning deficit *(10)*. Because complex IV activity is altered in AD, chronic sodium azide administration may be a useful tool in investigating neuropathological damage and behavioral abnormalities associated with the disease.

The toxicity induced by sodium azide in animals has been used as a platform to investigate the process of neurodegeneration, as well as to develop treatment strategies. A recent study *(1)* using brain microdialysis revealed that in awake rats the elevated $^{\cdot}OH$ produced by MPP^+, but not by sodium azide, is inhibited by stereotaxic injections of either NO synthase inhibitor

From: *Mitochondrial Inhibitors and Neurodegenerative Disorders*
Edited by: P. R. Sanberg, H. Nishino, and C. V. Borlongan © Humana Press Inc., Totowa, NJ

nitro-L-arginine or the NMDA channel blocker MK-801. These results suggest that successful blockade of toxic ˙OH by treatment with oxygen free radical scavengers depends on which ETC complex is inhibited *(1)*. The presence of more free electrons, which can become oxygen free radicals, at the end (complex IV) of the ETC function may explain the inability of the free radical scavengers to block sodium azide-induced toxicity.

Despite its, neurotoxic feature, sodium azide is also an iron chelator and catalase inhibitor. Iron has been implicated in induction of lipid peroxidation, and levels of both iron and lipid peroxidation are elevated in PD *(11)*. Animals intranigrally infused with iron show elevation of iron staining in the nigra up to, at least, 1 mo post-infusion. Histological examination of brains from iron-infused animals revealed a shift of cellular iron staining characterized by early (1 h–1 d) iron positivity that is predominantly neuronal, followed by reactive glial and finally oligodendroglial high staining by 1 mo post-infusion *(11)*. This observation suggests that infused iron becomes a bound unreactive form over a short period of time, which coincides to acute (only 1 d) elevation in lipid peroxidation following intranigral iron infusion. Thus, iron-induced oxidative damage may contribute to the pathogenesis of PD. Indeed, iron metabolism in postmortem examination of PD is highly correlated with mortality *(12)*. The presence of oxidative stress markers (i.e., impairment of mitochondrial complexes) appears to be correlated with regions of marked pathological changes in AD *(13)*. In addition, there is an elevation of striatal iron during normal aging *(14)*. These data suggest that oxidative stress may be involved in disease process, as well as in aging, and iron chelators may prevent free radical-mediated neurodegeneration *(15)*. Of note, oxidatively sensitive apoptosis in cultured human cells induced by autologous monocytes is markedly prevented by sodium azide *(16)*. In addition, sodium azide has been found to markedly reduced a iron mobilization in a neuroblastoma cell line *(17)*. However, sodium azide by itself is highly neurotoxic and worsens rather than blocks free radical formation in the cell, and therefore limits its utility as therapeutic agent against neurodegeneration.

Catalase inhibitors have been implicated in inhibition of ethanol metabolism, and sodium azide is a known catalase inhibitor. Primary cultures of fetal hypothalamic neurons when treated for 5 h with ethanol (50 m*M*) enhances ethanol metabolism that is reversed by pretreating the cultured cells with sodium azide (5 m*M*) *(18)*. In addition, the primary metabolite of ethanol, acetaldehyde (AcHO), which is increased in homogenates of d 19 fetal rat brain after treatment with ethanol (50 m*M*), is also blocked by incubating the specimen with sodium azide *(19)*. Acetaldehyde, either exogenous or metabolically derived, binds to gastric mucosal proteins, so when the gastric mucosa are incubated with ethanol, AcHO production occurs that is significantly inhibited by sodium azide *(20)*. In vivo studies have similarly demonstrated that catalase inhibitors may protect against ethanol-induced toxicity. For example, 3 h after one oral dose of ethanol (4 g/kg) to a pregnant dam (gestation d 19), AcHO levels in fetal brain are significantly increased, but prevented by administration of another catalase inhibitor, 3-amino-1,2,4-triazole *(19)*. However, there is equally convincing evidence that sodium azide may modulate intracellular toxicity. Application of hydrogen peroxide (1.5 m*M*) in cultured chick embryo cardiac myocytes can induce cytotoxicity as revealed by increased release of lactate dehydrogenase (LDH), and inhibition of catalase with sodium azide increases LDH release *(21)*. Additional studies are warranted to provide the beneficial effects of sodium azide in catalase activity-mediated cell injury.

Studies using whole animals corroborate in vitro experiments showing that sodium azide is an inhibitor cytochrome oxidase. Brouillet and colleagues *(22)* demonstrated that intrastriatal and chronic systemic injections of sodium azide can induce oxidative stress as revealed by a significant increase in lactate and a significant decrease in ATP. A localized striatal damage accompanies systemic injections of sodium azide in rats *(22)* as well as in primates *(23)*. Interestingly, monkeys treated with chronic systemic sodium azide displayed progressive locomotor activity characterized by hyperkinesia for 2 wk following 8–10 wk post-treatment followed by hypokinesia *(23)*. The neuropathological damage and behavioral features observed in sodium azide-treated animals are reminiscent of HD.

As mentioned earlier, depending on which part of the ETC function is inhibited by a mitochondrial toxin, there is varying pattern of cellular vulnerability. In addition, a differential age dependent cellular vulnerability is noted with specific mitochondrial toxins. For example, sodium azide produces age-dependent striatal lesions *(22)*. Twelve-mo-old rats treated with 2 μmol of sodium azide intrastriatally exhibit the highest striatal lesion volume, and 4-mo-old treated rats display higher striatal lesion volume than 1-mo-old treated rats. This age-dependent striatal toxicity resembles that seen in systemic administration of another mitochondrial toxin 3-nitropropionic acid (3-NPA, complex II inhibitor *(23)*. Although both sodium azide and 3-NPA cause age-dependent striatal damage, the complex I inhibitor MPP^+ does not produce such an effect. The variability may be due to the degree of energy impairment elicited by these mitochondrial inhibitors *(22)*. Because sodium azide and 3-NPA are both irreversible inhibitors of cytochrome oxidase and succinate dehydrogenase, respectively, whereas MPP^+ is a reversible inhibitor, there is more widespread ETC inhibition associated with the former inhibitors *(24,25)*. Both sodium azide and 3-NPA destroy striatal interneurons, but these are spared by MPP^+ *(25)*. Furthermore, NMDA antagonists such as MK-801, as well as NO synthase inhibitors such as nitro-L-arginine, can block MPP^+ metabolic toxicity *(1)*, but these compounds are ineffective against sodium azide and 3-NPA. It is possible that synthase systems, which are not mediated by NMDA or NO, are altered due to more severe metabolic compromise produced by sodium azide and 3-NPA.

There exists some variability in brain regions affected by sodium azide, as noted also with 3-NPA, depending on the route of administration and dosing regimen *(26)*. Intraventricular or intraparenchymal injections and higher dosages (>400 mg/kg for intracerebroventricular and >3000 nmol for intrastriastal) of sodium azide can result to overall reduction of cytochrome oxidase activity in the brains of rats *(22)*. Even with systemic injections using fairly high doses of sodium azide can cause disruption of cytochrome oxidase throughout the brain in treated rats *(27)*. However, with subacute sytemic injections (i.p.) of 20 mg/kg for 5 d, the resulting central nervous system (CNS) damage is localized within the striatum *(22)*. In the case of widespread CNS damage arising from injections of high dosages of the drug, sodium azide may parallel the neuropathological damage of AD, whereas the localized striatal lesion caused by low dosage, systemic injection of sodium azide may reproduce the symptoms of HD. The knowledge of the route of administration and dosing regimen related to sodium azide, as well as other mitochondrial inhibitors, is important in developing animal models of neurodegenerative disorders *(26)*.

Recently, we investigated the dose dependency of sodium azide in neonatal rat pups and characterized the cytochrome oxidase activity using the triphenyltetrazolium chloride (TTC) staining. Brouillet and colleagues *(22)* have previously shown a dose-dependent effect using intrastriatal injections of 1500, 2000, and 3000 nmol of sodium azide in adult rats and found that γ-aminobutyric acid (GABA), substance P, somatostatin, and neuropeptide-Y immunoreactive neurons are depleted in the highest dosage (3000 nmol); only somatostatin is affected in 2000 nmol, while all markers are spared in 1500 nmol. In our study, we injected acutely (4 times in one d with a 2-h interval between injections) varying dosages (1, 2, 3, and 4 mg/kg, i.p.) of sodium azide in 2–3-d-old pups. Histological examination was conducted at 1 h after the last injection, and TTC staining revealed that all dosages used produced no visible alterations in metabolic activity throughout the brain. In contrast, 4-mo-old animals treated with the same dosages exhibited decreased overall TTC staining in all dosages, except in the lowest dosage. This observation indicates that the young animals may be resistant to sodium azide toxicity. This differential histological damage produced by sodium azide in young and adult rats parallels the age dependency of 3-NPA-induced neurotoxicity *(23,28)*. Interestingly, some animals that received the highest dosage started to die after the second injection, and the rest of the animals did not survive after the third injection.

Nevertheless, the brains from these animals did not show any observable TTC changes in overall staining. The mortality then could be due to some peripheral effects of the drug. To directly examine the effects of sodium azide in the CNS, brain slices (100 μm thick sections) were obtained from another set of age-matched pups and then incubated to 1, 2, 4, and 8 mg/L of sodium azide just prior to TTC staining. We noted that the two lower dosages had no visible effect on metabolic activity, but the two higher dosages clearly suppressed overall brain metabolic activity. It is possible that the blood–brain barrier in young animals may be less permeable to sodium azide than in older animals, but once the sodium azide had penetrated the CNS, as in the case of brain slices incubated with the drug, there is no age difference in the cellular vulnerability to sodium azide.

Previous studies have indicated that chronic application of sodium azide induces a toxic effect to the GABAergic and substance P-immunoreactive neurons, but not to the dopaminergic afferents, in striatum *(22)*. Using an in vivo voltametric technique, we were able to record the alterations in dopamine after acute sodium azide application. We found that sodium azide not only increases the dopamine release but also inhibits the clearance of dopamine (Figs. 1 and 2). These two effects may additively or synergistically potentiate the extracellular dopamine levels in the striatum. In concert with this finding, our preliminary experiments also have demonstrated that acute systemic administration of sodium azide increases locomotor behaviors in rodents (Fig. 3). The elevation in locomotor activity in sodium azide-treated animals may be due to increments in striatal dopamine levels. Whether these sodium azide-induced alterations in dopamine and locomotor activity are reversible remains to be determined. These observations open venues for designing treatment modalities to correct striatal dopamine-mediated abnormalities.

In summary, administration of sodium azide in vitro or in vivo may reflect many neuropathological as well as behavioral symptoms associated with neurodegenerative disorders. These models offer investigations into the process of cell injury via the ETC and provide possible treatment strategies based on the iron chelator and/or catalase inhibitor features and possible interaction with the dopaminergic pathway of sodium azide.

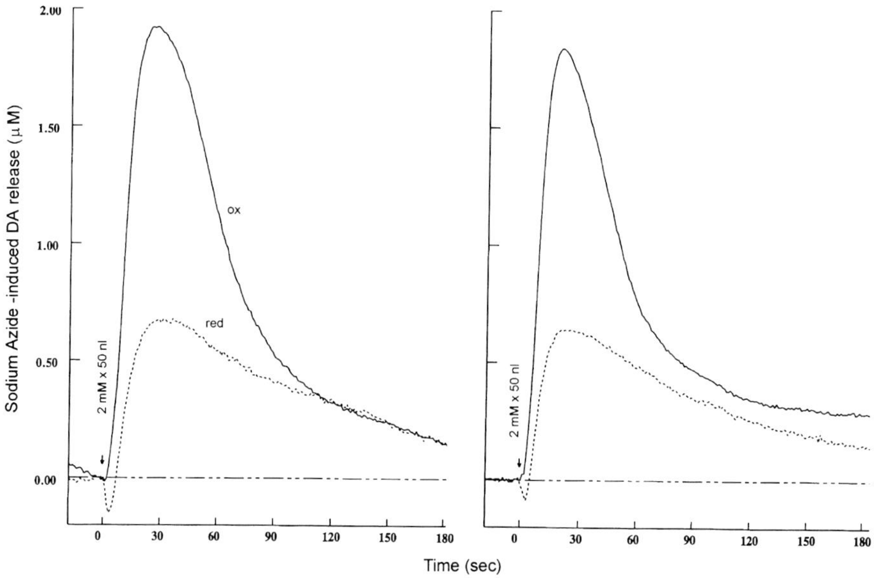

Fig. 1. Local application of sodium azide increases dopamine release in the striatum. Extracellular dopamine level was measured by in vivo voltammetry (IVEC10, Medical System). Sodium azide (2 m*M* × 50 nL in physiological saline) was directly applied, through pressure microinjection, to the dopamine terminals at striatum in a urethane-anesthetized Sprague–Dawley rat. Application of sodium azide in these two anterior striatal sites induces dopamine release. ox, oxidation current; red, reduction current.

REFERENCES

1. Smith TS, Bennett JP Jr. Mitochondrial toxins in models of neurodegenerative diseases. I: In vivo brain hydroxyl radical production during systemic MPTP. treatment or following microdialysis infusion of methylpyridinium or azide ions. Brain Res 1997;765:183–188.
2. Schulz JB, Matthews RT, Klockgether T, Dichgans J, Beal MF. The role of mitochondrial dysfunction and neuronal nitric oxide in animal models of neurodegenerative diseases. Mol Cell Biochem 1997;174:193–197.

Fig. 2. *(facing page)* Sodium azide inhibits dopamine clearance in striatum of a urethane–anesthetized rat. Dopamine clearance was measured indirectly by measuring the extracellular dopamine levels after its application. Dopamine or sodium azide was locally applied (*arrows*) to the striatum through a multibarrel pipet. **(A)** Application of dopamine induces a surge of extracellular dopamine. **(B)** Five min-

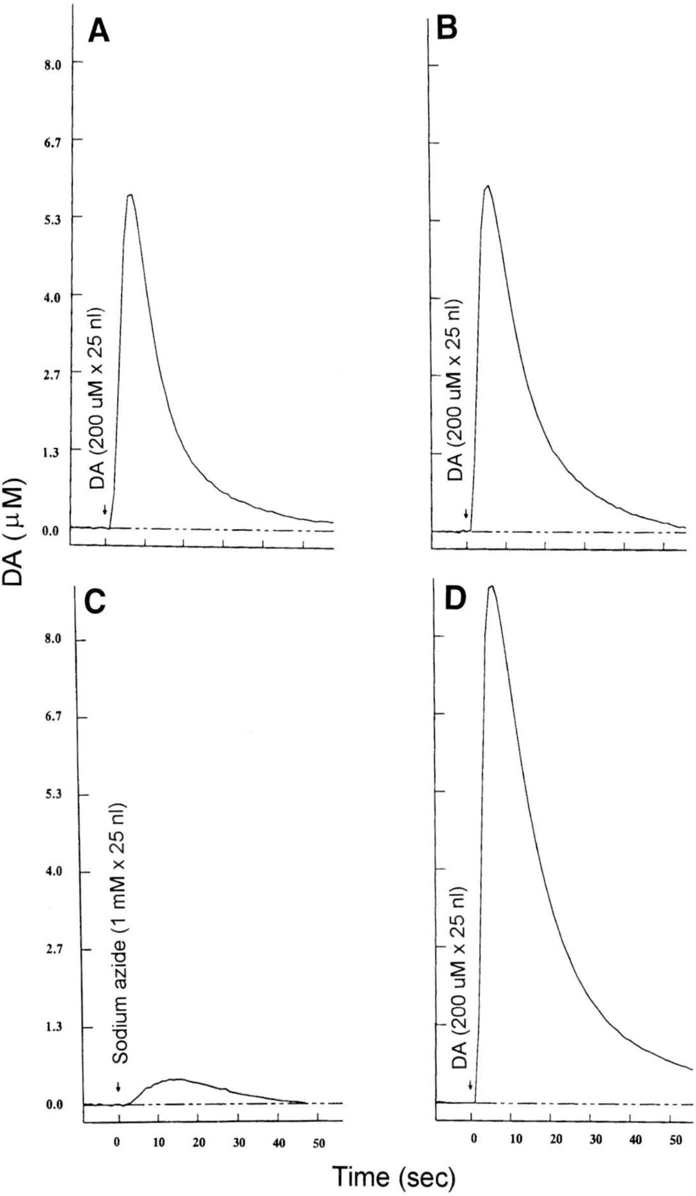

utes later, reapplication of the same dose of dopamine to this striatal site reproduces a similar overflow of dopamine. **(C)** Local application of sodium azide, at a lower concentration (1 m*M* × 25 nL), elicits a smaller increase of dopamine (as compared to that in Fig 1). **(D)** Five minutes later, reapplication of dopamine to this site induces a much higher dopamine overflow, suggesting that the clearance of dopamine is inhibited by locally applied sodium azide.

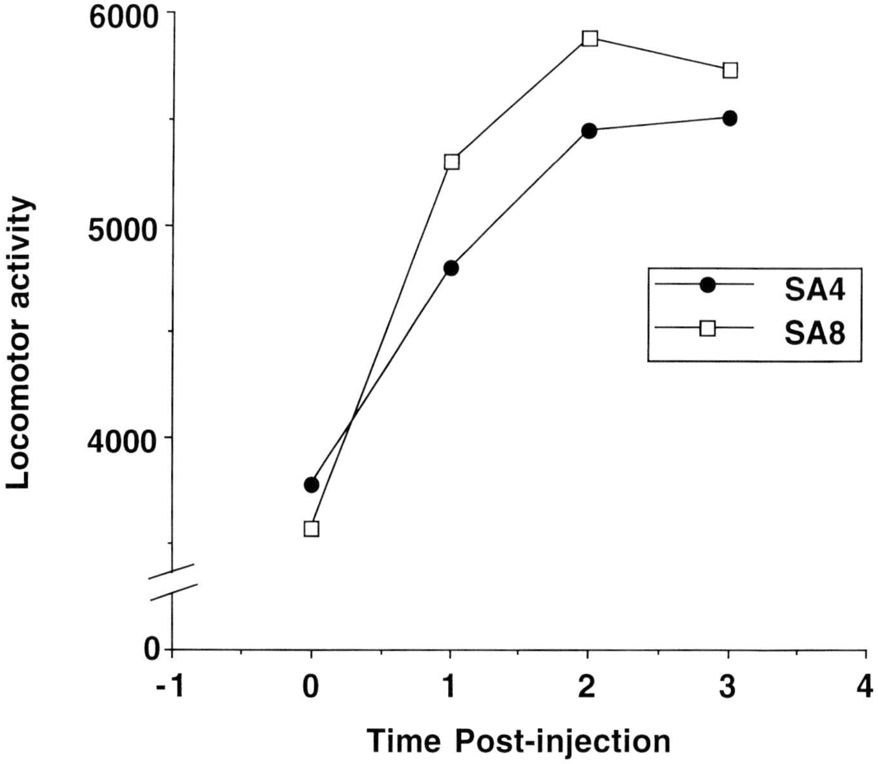

Fig. 3. Locomotor activity of sodium azide-treated rats. Animals received injections of 4 (SA4) or 8 (SA8) mg/kg of sodium azide (i.p., once every 2 h 4×). Locomotor activity (data on horizontal activity counts shown) was measured using the Digiscan monitor system (Omnitech, OH) after the last injection (time 0) and data were collected every hour over a 3-h period (times 1–3). Both dosages of sodium azide increase the locomotor activity of rats.

3. Gerlach M, Riederer P, Przuntek H, Youdim MB. MPTP. mechanisms of neurotoxicity and their implications for Parkinson's disease. Eur J Pharmacol 1991;208:273–286.
4. Swerdlow RH, Parks JK, Miller SW, Tuttle JB, Trimmer PA, Sheehan JP, Bennett JP Jr, Davis RE, Parker WD Jr. Origin and functional consequences of the complex I. defect in Parkinson's disease. Ann Neurol 1996;40:663–671.
5. Cassarino DS, Fall CP, Swerdlow RH, Smith TS, Halvorsen EM, Miller SW, Parks JP, Parker WD Jr, Bennett JP Jr. Elevated reactive oxygen species and antioxidant enzyme activities in animal and cellular models of Parkinson's disease. Biochim Biophys Acta 1997;1362:77–86.

6. Sheehan JP, Swerdlow RH, Parker WD, Miller SW, Davis RE, Tuttle JB. Altered calcium homeostasis in cells transformed by mitochondria from individuals with Parkinson's disease. J Neurochem 1997;68:1221–1233.
7. Beckman JS. Oxidative damage and tyrosine nitration from peroxynitrite. Chem Res Toxicol 1996;9:836–844.
8. Borlongan CV, Kanning K, Poulos SG, Freeman TB, Cahill DW, Sanberg PR. Free radical damage and oxidative stress in Huntington's disease. J Fla Med Assoc 1996;83:335–341.
9. Partridge RS, Monroe SM, Parks JK, Johnson K, Parker WD Jr, Eaton GR, Eaton SS. Spin trapping of azidyl and hydroxyl radicals in azide-inhibited rat brain submitochondrial particles. Arch Biochem Biophys 1994;310:210–217.
10. Bennett MC, Mlady GW, Fleshner M, Rose GM. Synergy between chronic corticosterone and sodium azide treatments in producing a spatial learning deficit and inhibiting cytochrome oxidase activity. Proc Natl Acad Sci USA 1996;93:1330–1334.
11. Sengstock GJ, Zawia NH, Olanow CW, Dunn AJ, Arendash GW. Intranigral iron infusion in the rat. Acute elevations in nigral lipid peroxidation and striatal dopaminergic markers with ensuing nigral degeneration. Biol Trace Elem Res 1997;58:177–195.
12. Marder K, Logroscino G, Tang MX, Graziano J, Cote L, Louis E, Alfaro B, Mejia H, Slavkovich V, Mayeux R. Systemic iron metabolism and mortality from Parkinson's disease. Neurology 1998;50:1138–1140.
13. Owen AD, Schapira AH, Jenner P, Marsden CD. Indices of oxidative stress in Parkinson's disease, Alzheimer's disease and dementia with Lewy bodies. J Neural Transm Suppl 1997;51:167–173.
14. Martin WR, Ye FQ, Allen PS. Increasing striatal iron content associated with normal aging. Mov Disord 1998;13:281–286.
15. Gassen M, Youdim MB. The potential role of iron chelators in the treatment of Parkinson's disease and related neurological disorders. Pharmacol Toxicol 1997;80:159–166.
16. Hansson M, Asea A, Ersson U, Hermodsson S, Hellstrand K. Induction of apoptosis in NK. cells by monocyte-derived reactive oxygen metabolites. J Immunol 1996;156:42–47.
17. Richardson DR. Mobilization of iron from neoplastic cells by some iron chelators is an energy-dependent process. Biochim Biophys Acta 1997;1320:45–57.
18. Reddy BV, Boyadjieva N, Sarkar DK. Effect of ethanol, propanol, butanol, and catalase enzyme blockers on beta-endorphin secretion from primary cultures of hypothalamic neurons: evidence for a mediatory role of acetaldehyde in ethanol stimulation of beta-endorphin release. Alcohol Clin Exp Res 1995;19:339–344.
19. Hamby-Mason R, Chen JJ, Schenker S, Perez A, Henderson GI. Catalase mediates acetaldehyde formation from ethanol in fetal and neonatal rat brain. Alcohol Clin Exp Res 1997;21:1063–1072.
20. Salmela KS, Sillanaukee P, Itala L, Vakevainen S, Salaspuro M, Roine RP. Binding of acetaldehyde to rat gastric mucosa during ethanol oxidation. J Lab Clin Med 1997;129:627–633.

21. Horwitz LD, Leff JA. Catalase and hydrogen peroxide cytotoxicity in cultured cardiac myocytes. J Mol Cell Cardiol 1995;27:909–915.
22. Brouillet E, Hyman BT, Jenkins BG, Henshaw DR, Schulz JB, Sodhi P, Rosen BR, Beal MF. Systemic or local administration of azide produces striatal lesions by an energy impairment-induced excitotoxic mechanism. Exp Neurol 1994;129:175–182.
23. Borlongan CV, Koutouzis TK, Randall TS, Freeman TB, Cahill DW, Sanberg PR. Systemic 3-nitropropionic acid: behavioral deficits and striatal damage in adult rats. Brain Res Bull 1995;36:549–556.
23. Metler FA. Choreoathetosis and striopallidonigral necrosis due to sodium azide. Exp Neurol 1972;34:291–308.
24. Brouillet E, Jenkins BG, Hyman BT, Ferrante RJ, Kowall NW, Srivastava R, Roy DS, Rosen BR, Beal MF. Age-dependent vulnerability of the striatum to the mitochondrial toxin 3-nitropropionic acid. J Neurochem 1993;61:1147–1150.
25. Henshaw R, Jenkins BG, Schulz JB, Ferrante RJ, Kowall NW, Rosen BR, Beal MF. Malonate produces striatal lesions by indirect NMDA. receptor activation. Brain Res 1994;647:161–166.
26. Borlongan CV, Polgar S, Freeman TB, Hauser RA, Cahill DW, Sanberg PR. Will fetal striatal transplants correct the akinetic end-stage of Huntington's disease? Neurodegeneration 1996;5:189–192.
27. Miyoshi K. Experimental striatal necrosis induced by sodium azide. A. contribution to the problem of selective vulnerability and histochemical studies of enzymatic activity. Acta Neuropathol (Berl) 1967;9:199–216.
28. Koutouzis TK, Borlongan CV, Freeman TB, Cahill DW, Sanberg PR. Intrastriatal 3-nitropropionic acid: a behavioral assessment. NeuroReport 1994;5:2241–2245

III
Treatment Interventions for Mitochondrial-Induced Neurotoxicity

17

Neuroprotective Strategies Against Cellular Hypoxia

Matthias W. Riepe

INTRODUCTION

Cellular hypoxia is a crucial event in the pathophysiologic cascade of several acute (e.g., cerebral ischemia, CO intoxication) and chronic (e.g., Parkinsons's disease, Huntington's disease, Alzheimer's disease) diseases of the central nervous system (CNS) *(1,2)*. Cellular hypoxia results from insufficient oxygen or substrate supply as in ischemia or hypoglycemia. In addition, it may be caused by impairment of mitochondrial energy metabolism with chemical inhibitors of oxidative phosphorylation and/or glycolysis such as cyanide, malonate, 3-nitropropionate, iodoacetate, and a manifold of other substances. Owing to their specificity, application of these substances provides a tool to understand the role of individual mitochondrial complexes or glycolytic enzymes in cellular hypoxia. Furthermore, several substances used in clinical practice have a partially inhibiting effect on mitochondrial energy metabolism, e.g., haloperidol on mitochondrial complex I *(3)* and acetylsalicylic acid on coupling of oxidation and respiration *(4)*.

For years, neuronal cell death under conditions of impaired energy metabolism was viewed as a result of increased susceptibility to excitatory overstimulation, which basically is glutamatergic overstimulation as glutamate is the most widespread endogenous excitatory neurotransmitter *(5,6)*. Since the first reports from experimental studies that glutamate antagonists are neuroprotective against cerebral ischemia and hypoxia *(7,8)*, a host of experimental studies investigated neuroprotection by application of glutamate antagonists in situations of acute energy failure. However, in clinical practice neuroprotection by glutamate antagonists did not keep the promise of the experimental studies *(9,10)*.

From: *Mitochondrial Inhibitors and Neurodegenerative Disorders*
Edited by: P. R. Sanberg, H. Nishino, and C. V. Borlongan © Humana Press Inc., Totowa, NJ

As for acute ischemia/hypoxia, evidence from animal models suggested that antagonists of glutamatergic neurotransmission might be beneficial for neurodegenerative diseases such as Huntington's disease, amyotrophic lateral sclerosis, and Parkinson's disease *(11–13)*. However, the relevance of these animal studies to the human pathophysiology is not clear and the actual pathophysiology is not very well understood. This applies even to the successful clinical trial with riluzole *(14)*, a drug interfering with glutamatergic neurotransmission.

In the current chapter I will present results of experimental studies on the pathophysiology of cellular hypoxia that can explain some of the shortcomings of the current focus on neuroprotective strategies aimed at antagonizing glutamatergic overexcitation. One of the main reasons is that experimental studies are performed mostly in control tissue. In contrast, actual patients suffer from an ongoing disease process (neurodegenerative diseases) or a disease with repeated transient episodes (vascular disease). Therefore a paradigm will be considered in which hypoxic tolerance is not merely tested in control tissue but in tissue that underwent mild hypoxic episodes previously. The endogenous ability to withstand hypoxia will be called primary hypoxic tolerance. It will be shown that hypoxic tolerance can be induced chemically also—induced hypoxic tolerance. In fact, induction of hypoxic tolerance by chemical preconditioning is a promising and practical neuroprotective strategy for diseases in which there is a high risk for cellular hypoxia.

BASIC PATHOPHYSIOLOGIC MECHANISMS DURING CELLULAR HYPOXIA

Resting membrane potential is of central importance for cellular physiology and pathophysiology. A change of resting membrane potential affects intra- and extracellular ionic composition and triggers intracellular cytotoxic and cytoprotective cascades. Partly, this is due to the reduced excitability of the neuronal cells. The further the resting membrane potential is hyperpolarized relative to the threshold for activation of fast sodium channels the fewer action potentials are generated and the less energy is needed for subseqent restoration of physiologic resting membrane potential. Neuronal cells benefit in particular because most of the energy expenditure in the CNS results from ion pumping needed for restoration and maintenance of physiological resting membrane potential around –60 mV to –80 mV. In addition, the effects of neurotransmitters may depend on resting membrane potential, e.g., the activation of glutamate receptor associated ion channels *(15)*.

Cellular hypoxia induces a biphasic change of membrane potential *(16,17)*. An initial hyperpolarization is followed by terminal depolarization.

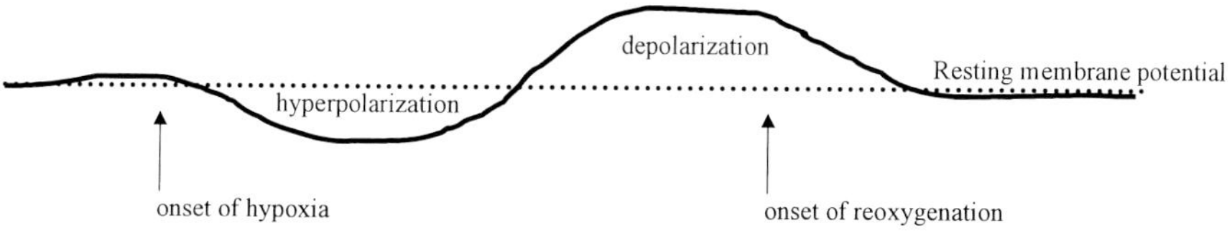

Prophylaxis	**Defense**	**Recovery**
Preconditioning	Adenosine	Free radical scavengers
NMDA-antagonists	Potassium channel agonists	Protease inhibitors
non-NMDA antagonists	Calcium channel agonists	
	Free radical scavengers	
	Protease inhibitors	
	non-NMDA antagonists	
	NMDA antagonists?	
	Neurotrophic factors?	

Fig. 1. Treatment strategies during different stages of hypoxia.

Most current pharmacological strategies aim at receptors activated at either of these different stages or at specific sequelae induced during these subsequent stages (Fig. 1).

Transient Hyperpolarization—Endogenous Neuroprotection

Impairment of cellular energy metabolism by cellular hypoxia or metabolic inhibition increases the transmembrane potassium conductance *(16,17)*. The flux of potassium ions across the cellular membrane hyperpolarizes the resting membrane potential. Early studies suggested predominant activation of calcium regulated potassium channels *(17)*. Using the tool of chemical inhibition of oxidative energy metabolism with a more gradual inhibition of energy metabolism it was demonstrated that the increased potassium conductance at onset of cellular hypoxia results from an opening of ATP-regulated potassium channels *(18,19)*.

K_{ATP}-channels serve as a metabolic sensor for neuronal cells and are activated when the intracellular ATP concentration decreases *(20)*. The opening of K_{ATP}-channels is specifically antagonized by sulfonylureas *(20)*. Activation of K_{ATP}-channels upon cellular hypoxia and the effects of agonists (e.g., diazoxide) and antagonists (e.g., glibenclamide) at K_{ATP}-channels depend on a variety of intracellular and extracellular modulators such as pH, intracellular ATP and ADP levels, and Mg^{2+} concentration *(21–23)*.

Terminal Depolarization B Induction of Pathophysiologic Cascades

Resting membrane potential depolarizes after prolonged inhibition of energy metabolism. Formerly, a cytotoxic mechanism was proposed that suggested potentiated activation of the *N*-methyl-D-aspartate (NMDA) subtype of glutamate receptors *(24,25)*. It had been shown in primary hippocampal cultures that the ion channel opened by NMDA, at which glutamate is the endogenous agonist, is blocked by physiologic concentrations of Mg^{2+} in a voltage-dependent fashion. Depolarization removes the Mg^{2+} ions and allows potentiated influx of both Na^+ and Ca^{2+} ions in control cultures *(15)*. Subsequently, this interpretation of the experimental results has been challenged. The more the studies looked directly at activation of NMDA receptors and not only at effects of applying antagonists to NMDA receptors, it turned out that under conditions of inhibited cellular energy metabolism the activation of NMDA receptors is reduced.

In primary cultures, currents activated by NMDA decrease under conditions of reduced intracellular ATP *(26,27)*, increased extracellular potassium concentration *(28)*, or increased intracellular sodium concentration *(29)*, conditions to be found when cellular energy metabolism is inhibited. After prolonged impairment of mitochondrial oxidative phosphorylation by 3-nitropropionate, the slope of the ongoing depolarization between –50 and –30 mV is only little affected by various combinations of glutamate antagonists, indicating that relief of the magnesium block which is important under physiologic conditions is of minor importance after prolonged inhibition of energy metabolism *(18)*. Finally, the effects of glutamate and NMDA were directly studied during chemical inhibition of energy metabolism. A time-dependent decrease of the depolarization induced by iontophoresis of NMDA and glutamate during inhibition of oxidative phosphorylation was observed *(30)*. This suggests that events other than glutamate-mediated depolarization are important in the final stages of metabolic inhibition and that the toxic events induced by endogenous excitatory agents go beyond a mere depolarization. It also suggests that either the Mg^{2+} blockade is altered during depolarization after prolonged blockade of oxidative phosphorylation, or that some other event accompanies metabolic inhibition that causes an impairment of the function of the NMDA-receptor associated ion channel. Final depolarization after chemical inhibition of energy metabolism seems largely determined by failure of activity of Na^+, K^+-ATPase and other ion pumps. After prolonged inhibition of oxidative phosphorylation neurons are not capable of meeting the demand for ATP-dependent ion exchange. Failure of Na^+, K^+-ATPase also provides an explanation for the increase in intracellular Ca^{2+} preceding cell death due to excitotoxic agents, because decreased sodium pump activity impairs Na^+/Ca^{2+} exchange *(31–33)*. Even though glutamate-induced excitation decreases with failure of energy metabolism *(30)*, a net depolarization of glutamate agonists results from an even more reduced capacity for repolarization due to failure of cellular energy stores. Neuronal depolarization is accelerated because the depolarizing effects of glutamate persist in glial cells at times when glutamate-induced depolarization in neurons is drastically reduced *(30)*.

Upon depolarization a complex cytotoxic cascade is triggered that includes increase of cellular calcium *(34)*, activation of proteases *(35,36)*, and production of free radicals *(37)*. The depolarization and the above mentioned sequelae are the adequate stimulus for activation of transcription factors such as AP-1 and nuclear factor κB which regulate the coordinated timely expression of stress-response genes that ultimately result in apoptosis or survival. Activation of transcription factors, on the other hand, modulates the function of voltage- and ligand-gated ion channels, e.g., voltage-dependent calcium channels and glutamate receptors *(38,39)*, which closes a loop between molecular modulators and ion channels.

Repolarization After Transient Hypoxia B Toxic Metabolites During Reoxygenation

Free radicals are constantly produced in eukaryotic cells. During hypoxic conditions due to inhibition of oxidative phosphorylation with electron transport inhibitors ("chemical hypoxia") and during recovery from hypoxic hypoxia, that is, cellular hypoxia due to reduction of pO_2, and chemical hypoxia the amount of free radicals increases manifold. Repeatedly, it has been shown that cell damage may result from reoxygenation mainly via increase of free radicals *(37,40–42)*. Thus free radicals seem to be a final common pathway of cell death during single, repeated or chronic hypoxia, ischemia, or excitatory challenge.

NEUROPROTECTION AGAINST TERMINAL PATHOPHYSIOLOGIC EVENTS

Classic neuroprotective strategies try to address the terminal pathophysiologic cascade, not the endgoneous protective phase. Most of the experimental studies investigate the effects of potential therapeutic drugs in control tissue. Previously healthy control animals are subjected to hypoxia, ischemia, chemical hypoxia, or excitatory agents. Studies in culture are even more ambiguous because the cultures they are routinely prepared from very young animals only. These paradigms may not be suitable for investigating neuroprotective treatments in humans because mostly patients are affected who suffer from an ongoing disease or have undergone previous episodes of cerebral cellular hypoxia (e.g., transient ischemic attacks). For investigation of the most promising protective strategies it seems therefore crucial that experimental paradigms are used in which the time frame for the pathophysiologic events is comparable to that of the disease to be studied and that has a history of repeated cellular hypoxia that is comparable to that of the disease to be investigated.

Glutamate Antagonists

Glutamatergic overexcitation is neurotoxic in situations of ischemia and hypoxia *(8)*. Glutamate antagonists have provided a potent experimental strategy for neuroprotection in situations of inhibited energy metabolism *(7,8)*. The success of this strategy in a variety of experimental models (note: all performed with control tissue) with a variety of substances binding to different locations at different glutamate receptor subtypes stimulated clinical studies. However, the results are ambiguous *(10)* and a large clinical trial had to be terminated early *(9)*.

Even though the sensitivity of neuronal glutamate receptors is reduced upon cellular hypoxia, the increased extracellular glutamate concentration observed in ischemia and other conditions of reduced energy metabolism may contribute to the pathophysiologic cascade culminating in cell death. Even the small amounts of sodium ions that enter the cell through glutamate channels under these conditions cannot be extruded when intracellular ATP is decreased *(30)*. The primary cause of depolarization in this situation, however, is failure of sodium transport and subsequent decay of ionic concentration gradients *(30)*. Consequently, glutamate antagonists can be expected to be more efficient in preventing cell death the more preserved the energy stores are or the better the energy regenerating systems work. This may be the reason why in most of the studies glutamate antagonists are applied before onset of ischemia or chemical hypoxia. Thus, glutamate antagonists may have a potential for prophylactic neuroprotection more likely than for post-event therapy.

Glutamate Release Inhibitors

Recently, several glutamate release inhibitors have been developed; e.g., lamotrigine and riluzole. Repeatedly they were investigated in situations with acute hypoxia. Glutamate release inhibitors protect against kainic acid induced lesions *(43)*, chemical hypoxia by 1-methyl-4-phenyl-1,2,3,6-tetrahydropyridine (MPTP) *(44)*, and focal cerebral ischemia *(45)*. More recently it has been shown that riluzole reduces progression of neurodegenerative diseases *(46,47)*. However, riluzole and other glutamate release inhibitors also affect fast sodium channels which according to the outline of the pathophysiology above may be a general and unspecific benefit due to a reduced excitatory drive.

Antioxidants and Free Radical Scavengers

The pharmacological strategy behind a variety of drugs is to scavenge free radicals generated during reoxygenation/reperfusion. A broad spectrum of substances has been investigated experimentally *(42,48)*. Recently, it was shown that it is important to take care that the antioxidants cross the blood–brain barrier *(49)*. The potential role of free radicals has also been shown in transgenic animals overexpressing MnSOD and upon dietary deficiency of Vitamin E. Ischemic damage was reduced in these animals *(50,51)*.

Growth Factors

Under experimental conditions in control tissue a variety of growth factors have been shown to be neuroprotective *(52)*. Nerve growth factor and basic fibroblast growth factor can protect cortical neurons against hypoglycemic and hypoxic damage via stabilization of calcium homeostasis *(53)*. Similar results have been reported for hepatocyte growth factor *(54)* and GDNF *(55)*. However, delivery to the CNS poses a challenge. While the experimental results are promising, CNTF did not show benefits in a large clinical trial with amyotrophic lateral sclerosis, a slowly progressing neurodegenerative disease *(14)*.

Calcium Channel Antagonists

It has been known for a long time that increase of intracellular calcium is a major hallmark preceding neuronal cell death *(56)*. Calcium channels are one of the sources through which intracellular calcium increases. It was shown years ago that antagonists at calcium channels can reduce neuronal damage *(57)*. After an intermediate focus on increase of intracellular calcium through glutamate receptor associated ion channels it was shown specifically that neuronal calcium channels are a potential target for anti-ischemic therapy *(58)*. Interestingly, calcium channel antagonists work best, when they are administered prior to onset of cellular hypoxia and are similar to glutamate antagonists in this respect.

Protease Inhibition

Activation of proteases is a comparatively late event in during cellular hypoxia and may determine late apoptotic cell death *(35)*. In some experimental paradigms it seems to be an obligatory event in the ischemic cell death cascade *(59,60)*. A promising feature of pharmacological strategies aiming at inhibition of proteases is that it may give an opportunity to start treatment even after onset of cellular hypoxia *(59)*.

Adenosine Agonists and Potassium Channel Activators

Adenosine is an endogenous neuromodulator that is released into the extracellular space upon cellular hypoxia *(61)*. Increase of adenosine hyperpolarizes neuronal cells through an increase of potassium channel conductance that includes activation of K_{ATP}-channels *(62)*. Activation of K_{ATP}-channels and other potassium channels repeatedly has been shown to mitigate lesions by chemical hypoxia, ischemia, or direct application of glutamatergic substances *(63–65)*.

CELLULAR HYPOXIA: NEUROPROTECTION THROUGH PRECONDITIONING

It has been proposed that inhibition of oxidative energy metabolism increases glutamate receptor mediated neurotoxicity *(24,66)*. Histopathologically, repeated inhibition of mitochondrial complex II with 3-nitropropionic acid (3-NPA) induces a distinct pattern that is reminiscent of Huntington's disease *(67)*. Histopathologic lesions correspond with behavioral abnormalities *(68)*. However, metabolic impairment also induces endogenous adaptations.

About 10 years ago it was reported for the first time that a short period of ischemia renders the heart muscle more tolerant against a subsequent prolonged ischemic episode *(69)*. Several years later it was shown that a similar phenomenon occurs in brain also *(70)*. A critical point for the effect of repeated hypoxic episodes is the time interval between subsequent episodes. Depending on time pattern of metabolic perturbations, a short ischemic or hypoxic episode may increase or decrease tolerance toward subsequent severe ischemia in heart and brain *(70–78)*—ischemic preconditioning. The time pattern is critical in heart as well as in brain. Ischemia during the time span used to induce early-onset tolerance in heart muscle causes an aggravation of morphologic lesions in brain *(76,79)*. On the contrary, the time window to induce tolerance against subsequent ischemia by a short preceding ischemic episode is limited to about 1 h in heart whereas it may last a few days in brain *(76,79)*.

Ischemic preconditioning is a practical experimental strategy in vascular territories without extensive collateralization. However, in view of therapeutic perspectives for humans a pharmacological strategy would be much more practical.

CHEMICAL HYPOXIA: A POTENT AND PRACTICAL NEUROPROTECTIVE STRATEGY

Chemical inhibition of energy metabolism allows selective inhibition of specific steps of oxidative phosphorylation and application of a controlled mode of inhibition either in vivo or in vitro. It also allows to observe cellular responses on a prolonged time scale.

3-NPA is a selective and long-lasting inhibitor of succinic dehydrogenase (SDH) *(80)*. It is known that 3-NPA induces a time- and dose-dependent decrease of SDH activity and ATP content in mouse cortical explant cultures *(81)*. Similar to hypoxic hypoxia application of 3-NPA causes a biphasic change of membrane potential. An initial hyperpolarization is followed by a (terminal) depolarization *(18)*. Toxicity of 3-NPA has a strong age-dependency *(67)*, setting a dosage window for potential therapeutic applications. A single dosage of up to 20 mg/kg in animals up to 400 g did not induce histological abnormalities in the striatum.

Increased tolerance by preconditioning can be shown functionally and histologically. This applies to necrotic as well as apoptotic cell death *(82,83)*. Similar to previous observations at an even higher dosage *(66)*, the treatment itself does not impair pathological or gross clinical parameters in vivo. Previously, it was demonstrated *(84)* that preconditionig ischemia protects against lethal ischemia when applied with a time interval of 3 d, but worsens the damage when applied with a time interval of 1 h. Upon chemical inhibition of oxidative phosphorylation with 3-NPA an early-onset and long-lasting neuroprotection is observed. The early onset of the protective effect is similar to the onset of ischemic preconditioning in heart muscle, the long duration is similar to ischemic preconditioning in brain *(82,85)*.

MECHANISMS OF PRECONDITIONING

Several mechanisms have been proposed to mediate increased tolerance against prolonged hypoxia or ischemia. For neuronal tissue, long-lasting tolerance has been related to a change in activity of ATPases, e.g., a reduced activity of Na^+/K^+-ATPase and increased activity of Mg^{2+}-ATPase *(86)*. Likewise, changes in the activity of respiratory enzymes *(87)* and expression of heat shock proteins *(88,89)* have been suggested. Early-onset ischemic preconditioning in heart muscle has been related to activation of ATP-regulated potassium channels *(90)*. Involvement of K_{ATP}-channels also has been shown in preconditioning of brain after a short ischemic episode *(84)* and chemical hypoxia *(91)*. Increased ability to preserve tissue energy levels after impairment of oxidative phosphorylation, preconditioning, has been observed in heart muscle *(71,92)*. Activation of ATP-regulated potassium channels seems to be involved in mediating preconditioning *(93)*. A variety of results, however, indicate that short periods of ischemia mediate long-lasting neuroprotection against a succeeding severe ischemia *(72,94,95)*. This suggests that *de novo* protein synthesis may be part of the spectrum of mechanisms involved in acquisition of the ischemic tolerance *(96)*. Currently, the following targets are most promising for pharmacologic intervention.

ATP-Regulated Potassium Channels

In vitro inhibition of energy metabolism by 3-NPA induces opening of ATP-regulated potassium channels in control tissue *(18)*. However, when upon application of 3-NPA resting membrane potential is measured in tissue that had undergone chemical hypoxia previously, this hyperpolarization is not observed *(91)* although ATP levels are not different. Depolarization upon in vitro application of glibenclamide, a selective antagonist at the K_{ATP}-channel, was reduced after 8 h of pretreatment. This indicates that no further potassium channels are being opened after a preceding hypoxic episode either because they are already fully activated or because the conditions of activation differ and/or because the density of K_{ATP}-channels changes *(97,98)*.

Pharmacologic induction of preconditioning with agonists at adenosine receptors has a protective effect when administered at a time interval of 15–30 min preceding lethal ischemia *(84)*. At least in part this seems mediated by hyperpolarization due to activation of K_{ATP}-channels. Induced hypoxic tolerance is partly reversed by application of sulfonylureas, antagonists at K_{ATP}-channels *(82)*. Reversal of induced hypoxic tolerance by sulfonylureas has already been demonstrated in humans *(102)*. In the future it will be needed to be investigated whether the benefits of sulfonylurea therapy in diabetics outweigh the potential risks caused by reversal of endogenous—or exogenous—induced hypoxic tolerance.

Energy Metabolism During Hypoxia

A relative preservation of cellular energy metabolism has been demonstrated with biochemical, histochemical, and fluorometric techniques. Increased ability to preserve tissue energy levels after impairment of oxidative phosphorylation, preconditioning, has been observed in heart *(71,92)* and brain *(91)*. Diminished decline of ATP results from preserved activity of respiratory enzymes *(91)*.

Chemical preconditioning by selective inhibition of mitochondrial complex II does not act only by a change in succinate related oxidation. As shown by a delayed increase of NADH upon application of cyanide, chemical preconditioning also influences NADH-related oxidation (as expected, NADH does not increase upon prolonged application of 3-NPA in hippocampal slices *[99]*) which is of major importance as mitochondrial complex I exerts control over mitochondrial energy metabolism in the brain *(100)*. Diminished increase of NADH upon induction of hypoxic tolerance could indicate facilitated adaptation to increased demand on preservation of energy production during disruption of oxidative metabolism in general. Alternatively, this could indicate decreased demand for maintenance of energy requiring processes, in particular for restoration of ionic homeostasis upon synaptic activity.

Glutamate Receptors

As outlined previously, glutamate antagonists and in particular NMDA antagonists were regarded as promising neuroprotective strategies. However, these results were obtained mainly in control tissue and therefore indicate only that glutamate antagonists can increase primary hypoxic tolerance. However, when NMDA antagonists are applied after successful induction of hypoxic tolerance this increase in induced hypoxic tolerance is abolished *(101)*. In contrast antagonists at non-NMDA receptors do not interfere with induced hypoxic tolerance *(101)*.

Free Radicals

Induction of hypoxic tolerance reduces posthypoxic free radicals, a known final pathway of neuronal cell death *(82)*. The reduction of posthypoxic free oxygen radicals is reversed by additional treatment of slices with glibenclamide prior to hypoxia and correlates with the reduction of the recovery of population spike amplitude in hippocampus, an established marker of neuronal integrity *(82)*.

CHEMICAL PRECONDITIONING—IN USE ALREADY?

Induction of hypoxic tolerance seems to be a physiologically relevant endogenous strategy of cytoprotection. It has been observed that in heart muscle infarct size is reduced in patients with preceding stenocardic symptoms, an indication of a transient ischemic episode *(104)*. Furthermore, ischemic preconditioning has been applied as a preventive cytoprotective strategy in the in situ human heart in coronary angioplasty, deliberately *(102)*. As in the animal studies, protection was reduced by glibenclamide *(102)*, an antagonist at the K_{ATP}-channel *(102)*.

Intermittent occlusion of cerebral arteries in humans, of course, is not practical. More promising as a prophylactic neuroprotective strategy is chemical preconditioning. In fact induction of hypoxic tolerance by chemical preconditioning may unknowingly be in use already also, as some pharmaceuticals with widespread therapeutic applications are inhibitors of mitochondrial energy metabolism. For many years it is known that acetylsalicylic acid (ASA) interferes with mitochondrial function *(4,105–107)*. A recent study showed that ASA increases neuronal hypoxic tolerance *(109)*. For the first time a time–effect curve for ASA was reported that mimics the clinical dosage of one administration daily. Induction of hypoxic tolerance by pretreatment with ASA in vivo is maximal with a time interval of 6 h, declines after 24 h, and is gone after 48 h. Pretreatment with ASA delays decline of energy metabolism upon severe hypoxia. Induction of hypoxic tolerance was observed with treatment in vitro, also, which proves that induction of hypoxic tolerance by ASA is independent from its action on platelets. It can be expected that chemical preconditioning can be induced repeatedly as has been shown for ischemic preconditioning *(108)*.

SUMMARY

In summary, neuroprotective strategies against cellular hypoxia need to be investigated in a paradigm relevant to clinical applications that is in tissue that suffers from an ongoing disease process or that repeatedly (at least once) has been subjected to transient pathologic conditions. In planning these experiments one carefully needs to account for time windows of protection and degeneration. Induction of hypoxic tolerance by chemical preconditioning is a potent and promising neuroprotective strategy with a proven rationale in humans. Further investigation might show drugs with broad dosage windows and greater efficiency than the drugs currently used without knowing in clinical practice.

REFERENCES

1. Beal MF. Does impairment of energy metabolism result in excitotoxic neuronal death in neurodegenerative illnesses? Ann Neurol 1992;31:119–130.
2. Ludolph AC, Riepe M, Ullrich K. Excitotoxicity, energy metabolism and neurodegeneration. J Inher Metab Dis 1993;16:1–8.
3. Burkhardt C, Kelly JP, Lim YH, et al. Neuroleptic medications inhibit complex I of the electron transport chain. Ann Neurol 1993;33:512–517.
4. Whitehouse MW, Haslam JM. Ability of some antirheumatic drugs to uncouple oxidative phosphorylation. Nature 1962;196:1323–1324.
5. Erecinska M, Nelson D. Amino acid neurotransmitters in the CNS. FEBS Lett 1987;213:61–66.

6. Mayer ML, Westbrook GL. The physiology of excitatory amino acids in the vertebrate central nervous system. Prog Neurobiol 1987;28:197–276.
7. Simon RP, Swan JH, Griffiths T, et al. Blockade of *N*-methyl-D-aspartate receptors may protect against ischemic damage in the brain. Science 1984; 226:850–852.
8. Rothman SM, Olney JW. Glutamate and the pathophysiology of hypoxic-ischemic brain damage. Ann Neurol 1986;19:105–111.
9. Davis SM, Albers GW, Diener HC, et al. Termination of acute stroke studies involving selfotel treatment. ASSIST Steering Committee Lancet 1997;349:32.
10. Muir KW, Lees KR. Clinical experience with excitatory amino acid antagonist drugs. Stroke 1995;26:503–513.
11. Beal MF, Brouillet E, Jenkins BG, et al. Neurochemical and histologic characterization of striatal excitotoxic lesions produced by the mitochondrial toxin 3-nitropropionic acid. J Neurosci 1993;13:4181–4192.
12. Couratier P, Hugon J, Sindou P, et al. Cell culture evidence for neuronal degeneration in amyotrophic lateral sclerosis bein linked to glutamate AMPA/ kainate receptors. Lancet 1993;341:265–268.
13. Turski L, Bressler K, Rettig K, et al. Protection of substantia nigra from MPP^+ neurotoxicity by *N*-methyl-D-aspartate antagonists. Nature 1991; 349:414–418.
14. ALS CNTF Treatment Study Group. A double-blind placebo-controlled clinical trial of subcutaneous recombinant human ciliary neurotrophic factor (rHCNTF) in amyotrophic lateral sclerosis. Neurology 1996;46:1244–1249.
15. Mayer ML, Westbrook GL, Guthrie PB. Voltage-dependent block by Mg^{2+} of NMDA responses in spinal cord neurones. Nature 1984;309:261–263.
16. Hansen AJ. Effects of anoxia on ion distribution in the brain. Physiol Rev 1985;65:101–148.
17. Fujiwara N, Higashi H, Shimoji K, et al. Effects of hypoxia on hippocampal neurons in vitro. J Physiol 1987;384:131–151.
18. Riepe M, Hori N, Ludolph AC, et al. Inhibition of energy metabolism by 3-nitropropionic acid activates ATP-sensitive potassium channels. Brain Res 1992;586:61–66.
19. Politi DM, Rogawski MA. Glyburide-sensitive K^+ channels in cultured rat hippocampal neurons: activation by cromakalim and energy depleting conditions. J Pharmacol Exp Ther 1991;40:308–315.
20. Ashcroft FM. Adenosine 5'-triphosphate-sensitive potassium channels. Annu Rev Neurosci 1988;11:97–118.
21. Findlay I. The effects of magnesium upon adenosine triphosphate-sensitive potassium channels in a rat insulin-secreting cell line. J Physiol 1987; 391:611–629.
22. Findlay I. Effects of pH upon the inhibition by sulphonylurea drugs of ATP-sensitive K^+ channels in cardiac muscle. J Pharmacol Exp Ther 1992; 262:71–79.
23. Aguilar-Bryan L, Clement JP, Gonzalez G, et al. Toward understanding the assembly and structure of KATP channels. Physiol Rev 1998;78:227–245.

24. Novelli A, Reilly JA, Lysko PG, et al. Glutamate becomes neurotoxic via the *N*-methyl-D-aspartate receptor when intracellular energy levels are reduced. Brain Res 1988;451:205–212.
25. Henneberry RC, Novelli A, Cox JA, et al. Neurotoxicity at the *N*-methyl-D-aspartate receptor in energy compromised neurons. An hypothesis for cell ceath in aging and disease. Ann NY Acad Sci 1989;568:225–233.
26. MacDonald JF, Mody I, Salter MW. Regulation of *N*-methyl-D-aspartate receptors revealed by intracellular dialysis of murine neurons in culture. J Physiol 1989;414 :17–34.
27. Mody I, Salter MW, MacDonald JF. Requirement of the NMDA receptor/channel for intracellular high-energy phosphates and the extent of intraneuronal calcium buffering in cultured mouse hippocampal neruons. Neurosci Lett 1988;93:73–78.
28. Ozawa S, Iino M, Tsusuki K. Suppression by extracellular K^+ of *N*-methyl-D-aspartate responses in cultured rat hippocampal neurons. J Neurophysiol 1990;64:1361–1367.
29. Yoneda Y, Ogita K. Inhibitory modulation by sodium ions of the *N*-methyl-D-aspartate recognition site in brain synpatic membranes. J Neurochem 1991; 57:2036–2046.
30. Riepe MW, Hori N, Ludolph AC, et al. Failure of neuronal ion exchange, not potentiated excitation, causes excitotoxicity after inhibition of oxidative phosphorylation. Neuroscience 1995;64:91–97.
31. Thompson SM, Prince DA. Activation of the electrogenic sodium pump in hippocampal CA1 neurons following glutamate-induced depolarization. J Neurophysiol 1986;56:507–522.
32. Lipton P, Lobner D. Mechanisms of intracellular calcium accumulation in the CA1 region of rat hippocampus during anoxia in vitro. Stroke 1990;21:III60–III64.
33. Barcenas-Ruiz L, Beuckelmann DJ, Wier WG. Sodium-calcium exchange in heart: membrane currents and changes in $Ca^{2+}i$. Science 1987;238: 1720–1722.
34. Kristian T, Siesjo BK. Calcium in ischemic cell death. Stroke 1998;29:705–718.
35. Chen J, Nagayama T, Jin K, et al. Induction of caspase-3-like protease may mediate delayed neuronal death in the hippocampus after transient cerebral ischemia. J Neurosci 1998;18:4914–4928.
36. Kampfl A, Posmantur RM, Zhao X, et al. Mechanisms of calpain proteolysis following traumatic brain injury: a review and update. J Neurotrauma 1997;14:121–134.
37. Siesjo BK, Agardh CD, Bengtsson F. Free radicals and brain damage. Cerebrovasc. Brain Metab. Rev. 1989;1:165–211.
38. Tong L, Toliver KT, Taglialatela G, et al. Signal transduction in neuronal death. J Neurochem 1998;71:447–459.
39. Furukawa K, Mattson MP. The transcription factor NF-kappaB mediates increases in calcium currents and decreases in NMDA- and AMPA/kainate-induced currents induced by tumor necrosis factor-alpha in hippocampal neurons. J Neurochem 1998;70:1876–1886.

40. Ambrosio G, Zweier JL, Duilio C, et al. Evidence that mitochondrial respiration is a source of potentially toxic oxygen free radicals in intact rabbit hearts subjected to ischemia and reflow. J Biol Chem 1993;268:18532–18541.
41. Dietrich WD. Morphological manifestations of reperfusion injury in brain. Ann NY Acad Sci 1994;723:15–24.
42. Chan PH. Role of oxidants in ischemic brain damage. Stroke 1996;27: 1124–1129.
43. McGeer EG, Zhu SG. Lamotrigine protects against kainate but not ibotenate lesions in rat striatum. Neurosci Lett 1990;112:348–351.
44. Jones-Humble SA, Morgan PF, Cooper BR. The novel anticonvulsant lamotrigine prevents dopamine depletion in C57 black mice in the MPTP animal model of Parkinson's disease. Life Sci 1994;54:245–252.
45. Kawaguchi K, Graham SH. Neuroprotective effects of the glutamate release inhibitor 619C89 in temporary middle cerebral artery occlusion. Brain Res 1997;749:131–134.
46. Gurney ME, Fleck TJ, Himes CS, et al. Riluzole preserves motor function in a transgenic model of familial amyotrophic lateral sclerosis. Neurology 1998;50:62–66.
47. Lacomblez L, Bensimon G, Leigh PN, et al. Dose-ranging study of riluzole in amyotrophic lateral sclerosis. Amyotrophic Lateral Sclerosis/Riluzole Study Group II. Lancet 1996;347:1425–1431.
48. Villa RF, Gorini A. Pharmacology of lazaroids and brain energy metabolism: a review. Pharmacol Rev 1997;49:99–136.
49. Schmid-Elsaesser R, Zausinger S, Hungerhuber E, et al. Superior neuroprotective efficacy of a novel antioxidant (U-101033E) with improved blood–brain barrier permeability in focal cerebral ischemia. Stroke 1997; 28:2018–2024.
50. Keller JN, Kindy MS, Holtsberg FW, et al. Mitochondrial manganese superoxide dismutase prevents neural apoptosis and reduces ischemic brain injury: suppression of peroxynitrite production, lipid peroxidation, and mitochondrial dysfunction. J Neurosci 1998;18:687–697.
51. van der Worp HB, Bar PR, Kappelle LJ, et al. Dietary vitamin E levels affect outcome of permanent focal cerebral ischemia in rats. Stroke 1998;29:1002–1005.
52. Hefti F. Pharmacology of neurotrophic factors. Annu Rev Pharmacol Toxicol 1997;37:239–267.
53. Cheng B, Mattson MP. NGF and bFGF protect rat hippocampal and human cortical neurons against hypoglycemic damage by stabilizing calcium homeostasis. Neuron 1991;7:1031–1041.
54. Miyazawa T, Matsumoto K, Ohmichi H, et al. Protection of hippocampal neurons from ischemia-induced delayed neuronal death by hepatocyte growth factor: a novel neurotrophic factor. J Cereb Blood Flow Metab 1998;18:345–348.
55. Kitagawa H, Hayashi T, Mitsumoto Y, et al. Reduction of ischemic brain injury by topical application of glial cell line-derived neurotrophic factor after permanent middle cerebral artery occlusion in rats. Stroke 1998; 29:1417–1422.

56. Siesjo BK. Cell damage in the brain: a speculative synthesis. Acta Psychiatr Scand Suppl 1984;313:57–91.
57. Mohamed AA, Gotoh O, Graham DI, et al. Effect of pretreatment with the calcium antagonist nimodipine on local cerebral blood flow and histopathology after middle cerebral artery occlusion. Ann Neurol 1985;18:705–711.
58. Pringle AK, Benham CD, Sim L, et al. Selective N-type calcium channel antagonist omega conotoxin MVIIA is neuroprotective against hypoxic neurodegeneration in organotypic hippocampal-slice cultures. Stroke 1996;27:2124–2130.
59. Markgraf CG, Velayo NL, Johnson MP, et al. Six-hour window of opportunity for calpain inhibition in focal cerebral ischemia in rats. Stroke 1998; 29:152–158.
60. Endres M, Namura S, Shimizu SM, et al. Attenuation of delayed neuronal death after mild focal ischemia in mice by inhibition of the caspase family. J Cereb Blood Flow Metab 1998;18:238–247.
61. Miller LP, Hsu C. Therapeutic potential for adenosine receptor activation in ischemic brain injury. J Neurotrauma 1992;9(Suppl 2):563–577.
62. Li H, Henry JL. Adenosine-induced hyperpolarization is depressed by glibenclamide in rat CA1 neurones. NeuroReport 1992;3:1113–1116.
63. Abele AE, Miller RJ. Potassium channel activators abolish excitotoxicity in cultured hippocampal pyramidal neurons. Neurosci Lett 1990;115:195–200.
64. Wind T, Prehn JH, Peruche B, et al. Activation of ATP-sensitive potassium channels decreases neuronal injury caused by chemical hypoxia. Brain Res 1997;751:295–299.
65. Heurteaux C, Bertaina V, Widmann C, et al. K^+ channel openers prevent global ischemia-induced expression of c-*fos*, c-*jun*, heat shock protein, and amyloid beta-protein precursor genes and neuronal death in rat hippocampus. Proc Natl Acad Sci USA 1993;90:9431–9435.
66. Simpson JR, Isacson O. Mitochondrial impairment reduces the threshold for in vivo NMDA-mediated neuronal death in the striatum. Exp Neurol 1993;121:57–64.
67. Brouillet E, Jenkins BG, Hyman BT, et al. Age-dependent vulnerability of the striatum to the mitochondrial toxin 3-nitropropionic acid. J Neurochem 1993;60:356–359.
68. Borlongan CV, Koutouzis TK, Freeman TB, et al. Behavioral pathology induced by repeated systemic injections of 3-nitropropionic acid mimics the motoric symptoms of Huntington's disease. Brain Res 1995;697:254–257.
69. Murry CE, Jennings RB, Reimer KA. Preconditioning with ischemia: a delay of lethal cell injury in ischemic myocardium. Circulation 1986;74:1124–1136.
70. Kitagawa K, Matsumoto M, Tagaya M, et al. 'Ischemic tolerance' phenomenon found in the brain. Brain Res 1990;528:21–24.
71. Murry CE, Richard VJ, Reimer KA, et al. Ischemic preconditioning slows energy metabolism and delays ultrastructural damage during a sustained ischemic episode. Circ Res 1990;66:913–931.

72. Aoki M, Abe K, Kawagoe J, et al. The preconditioned hippocampus accelerates HSP70 heat shock gene expression following transient ischemia in the gerbil. Neurosci Lett 1993;155:7–10.
73. Schurr A, Reid KH, Tseng MT, et al. Adaptation of adult brain tissue to anoxia and hypoxia in vitro. Brain Res 1986;374:244–248.
74. Lutz PL. Mechanisms for anoxic survival in the vertebrate brain. Annu Rev Physiol 1992;54:601–618.
75. Kato H, Kogure K, Nakano S. Neuronal damage following repeated brief ischemia in the gerbil. Brain Res 1988;479:366–370.
76. Kato H, Liu Y, Araki T, et al. Temporal profile of the effects of pretreatment with brief cerebral ischemia on the neuronal damage following secondary ischemic insult in the gerbil: cumulative damage and protective effects. Brain Res 1991;553:238–242.
77. Kato H, Araki T, Kogure K. Role of the excitotoxic mechanism in the development of neuronal damage following repeated brief cerebral ischemia in the gerbil: protective effects of MK-801 and pentobarbital. Brain Res 1990;516:175–179.
78. Tomida S, Nowak TSJ, Vass K, et al. Experimental model for repetitive ischemic attacks in the gerbil: the cumulative effect of repeated ischemic insults. J Cerebr Blood Flow Metab 1987;7:773–782.
79. Li Y, Kloner RAJ. Cardioprotective effects of ischemic preconditioning can be recaptured after they are lost. J Am Coll Cardiol 1994;23:470–474.
80. Alston TA, Mela N, Bright HJ. 3-nitropropionate, the toxic substance of indigofera, is a suicide inhibitor of succinate dehydrogenase. Proc Natl Acad Sci USA 1977;74:3767–3771.
81. Ludolph AC, Seelig M, Ludolph A, et al. 3-Nitropropionic acid decreases cellular energy levels and causes neuronal degeneration in cortical explants. Neurodegeneration 1992;1:155–161.
82. Riepe MW, Esclaire F, Kasischke K, et al. Increased hypoxic and ischemic tolerance by chemical inhibition of oxidative phosphorylation—'chemical preconditioning.' J Cerebr Blood Flow Metab 1997;17:257–264.
83. Chen J, Simon R. Ischemic tolerance in the brain. Neurology 1997;48:306–311.
84. Heurteaux C, Lauritzen I, Widmann C, Lazdunski M. Essential role of adenosine, adenosine A1 receptors, and ATP-sensitive K^+ channels in cerebral ischemic preconditioning. Proc Natl Acad Sci USA 1995;92: 4666–4670.
85. Riepe MW, Kasischke K, Gericke CA, et al. Increase of hypoxic tolerance in rat hippocampal slices following 3-nitropropionic acid is not mediated by endogenous nerve growth factor. Neurosci Lett 1996;211:9–12.
86. Benzi G, Gorini A, Arnaboldi R, Ghigini B, Villa R. Effect of intermittent mild hypoxia and drug treatment on synaptosomal nonmitochondrial ATPase activities. J Neurosci Res 1993;34:654–663.
87. Dagani F, Marzatico F, Curti D, et al. Effect of prolonged and intermittent hypoxia on some cerebral enzymatic activities related to energy transduction. J Cereb Blood Flow Metab 1984;4:615–624.

88. Kirino T, Tsujita Y, Tamura A. Induced tolerance to ischemia in gerbil hippocampal neurons. J Cereb Blood Flow Metab 1991;11:299–307.
89. Liu Y, Kato H, Nakata N, et al. Temporal profile of heat shock protein 70 synthesis in ischemic tolerance induced by preconditioning ischemia in rat hippocampus. Neuroscience 1993;56:921–927.
90. Gross GJ, Auchampach JA. Blockade of ATP-sensitive potassium channels prevents myocardial preconditioning in dogs. Circ Res 1992;70:223–233.
91. Riepe M, Niemi WN, Megow D, et al. Mitochondrial oxidation in rat hippocampus can be preconditioned by selective chemical inhibition of succinic dehydrogenase. Exp Neurol 1996;138:15–21.
92. Kida M, Fujiwara H, Ishida M, et al. Ischemic preconditioning preserves creatine phosphate and intracellular pH. Circulation 1991;84:2495–2503.
93. Van Winkle DM, Chien GL, Wolff RA, et al. Cardioprotection provided by adenosine receptor activation is abolished by blockade of the KATP channel. Am J Physiol 1994;266:H829–H839.
94. Kato H, Araki T, Murase K, et al. Induction of tolerance to ischemia: alterations in second-messenger systems in the gerbil hippocampus. Brain Res Bull 1992;29:559–565.
95. Liu Y, Kato H, Nakata N, et al.. Protection of rat hippocampus against ischemic neuronal damage by pretreatment with sublethal ischemia. Brain Res 1992;586:121–124.
96. Yoneda Y, Kuramoto N, Azuma Y, et al. Possible involvement of activator protein-1 DNA binding in mechanisms underlying ischemic tolerance in the CA1 subfield of gerbil hippocampus. Neuroscience 1998;86:79–97.
97. Deutsch N, Weiss JN. ATP-sensitive K^+ channel modification by metabolic inhibition in isolated guinea-pig ventricular myocytes. J Physiol 1993;465:163–179.
98. Mourre C, Smith M, Siesjo B, et al. Brain ischemia alters the density of binding sites for glibenclamide, a specific blocker of ATP-sensitive-channels. Brain Res 1990;526:147–152.
99. Riepe MW, Schmalzigaug K, Fink F, et al. NADH in the pyramidal cell layer of hippocampal regions CA1 and CA3 upon selective inhibition and uncoupling of oxidative phosphorylation. Brain Res 1996;710:21–27.
100. Davey GP, Peuchen S, Clark JB. Energy thresholds in brain mitochondria. Potential involvement in neurodegeneration. J Biol Chem 1998; 273:12,753–12,757.
101. Kasischke K, Ludolph AC, Riepe MW. NMDA-antagonists reverse increased hypoxic tolerance by preceding chemical hypoxia. Neurosci Lett 1996;214:175–178.
102. Tomai F, Crea F, Gaspardone A, et al. Ischemic preconditioning during coronary angioplasty is prevented by glibenclamide, a selective ATP-sensitive K^+ channel blocker. Circulation 1994;90:700–705.
103. Schmid-Antomarchi H, De Weille J, Fosset M, et al. The receptor for antidiabetic sulfonylureas controls the activity of the ATP-modulated K^+ channel in insulin secreting cells. J Biol Chem 1987;262:15840–15844.

104. Ottani F, Galvani M, Ferrini D, et al. Prodromal angina limits infarct size. A role for ischemic preconditioning. Circulation 1995;91:291–297.
105. Lutwak-Mann C. The effect of salicylate and cinchophen on enzymes and metabolic processes. Biochem J 1942;36:706–728.
106. Adams SS, Cobb R. A possible basis for the anti-inflammatory activity of salicylates and other non-hormonal anti-rheumatic drugs. Nature 1958; 181:773–774.
107. Martens ME, Lee CP. Reye's syndrome: salicylates and mitochondrial functions. Biochem Pharmacol 1984;33:2869–2876.
108. Chen T, Kato H, Liu XH, et al. Ischemic tolerance can be induced repeatedly in the gerbil hippocampal neurons. Neurosci Lett 1994;177:159–161.
109. Riepe MW, Kasischke K, Raupach A. Acetylsalicylate increases tolerance against hypoxic and chemical hypoxia. Stroke 1997;28:2006–2011.

18

Neuroprotective Effect of Perinatal Hypoxia Against 3-Nitropropionic Acid Neurotoxicity

Zbigniew K. Binienda and Andrew C. Scallet

INTRODUCTION

The decade of the 90s, although proclaimed by the US Congress as "the decade of the brain," might better be known as "the decade of the mitochondrion." A spontaneously emerging concept, "mitochondrial genetics" has become a distinct scientific field addressing various brain and muscle disorders associated with mutations of the mitochondrial DNA *(1)*. At the same time, an interaction between mitochondrial insufficiency and excitotoxicity in aging and neuronal diseases has been postulated *(2,3)*.

As mitochondrial function becomes impaired, whether due to toxicity or accumulation of DNA damage with age, cellular functions are diminished as well, due to energy depletion that may be followed by direct cell damage. Understanding the cellular events following alterations in mitochondrial energy metabolism is of prime importance for the treatment of a wide spectrum of clinical problems ranging from juvenile propionic encephalopathies through mid-age ischemic stroke to senile dementias. Mitochondrial energy disruption is also a frequently observed mechanism of therapeutic drug and neurotoxicant actions. Thus, treatment of mitochondrial energy disruption might help alleviate the neuropathies caused by anti-HIV therapeutics or 1-methyl-4-phenyl-1,2,3,6-tetrahydropyridine (MPTP) neurotoxicity.

Certain endogenous mechanisms for adaptation to cellular energy deficiency exist and may then act to reduce subsequent responses to bouts of energy disruption *(4)*. In this chapter, the potential neuroprotective effects of "hypoxic–ischemic preconditioning" and possible long-term protective effects of perinatal ischemia–hypoxia against neurotoxicity induced by the mitochondrial inhibitor, 3-nitropropionic acid (3-NPA), will be discussed.

From: *Mitochondrial Inhibitors and Neurodegenerative Disorders*
Edited by: P. R. Sanberg, H. Nishino, and C. V. Borlongan © Humana Press Inc., Totowa, NJ

MITOCHONDRIAL ENERGY DEFICIENCY

Evidence suggests a link between many neurodegenerative diseases and chronic defects in mitochondrial energy metabolism *(2)*. However, stressful conditions associated with mitochondrial energy deficiency may also trigger compensatory mechanisms and induce adaptation. Physiological adaptations to hypoxia alone ("acclimatization to hypoxia") have previously been demonstrated. These adjustments include somatic, respiratory adaptations such as increased pulmonary ventilation, modifications of the diffusing capacity for oxygen, and increases in blood hemoglobin or myoglobin concentrations *(5,6)*. In mammals native to high-altitude hypoxic environments, cellular adaptations are observed in addition to the somatic adaptations. The increase in the number of mitochondria and associated enhancement of enzymatic activity assure higher utilization of oxygen, enabling these species to survive the low, atmospheric pressures *(5)*. These compensatory responses may be accompanied by angiogenesis and proliferation of microvasculature in tissues including the heart and brain *(5,7)*. Such changes serve to "precondition" a resistance to cellular damage from subsequent episodes of energy disruption.

THE PATHOPHYSIOLOGY OF BRAIN DAMAGE INDUCED BY ENERGY DEFICIENCY

Cerebral hypoxia-ischemia results in a shift toward anaerobic glucose metabolism, an increase in lactate levels, a quick depletion of high-energy phosphates (ATP) in the brain, and acidosis. Subsequent failure of ion pumps leads to an increase in intracellular Na^+ followed by mitochondrial swelling, activation of K^+ channels, and depolarization of the nerve endings. Extracellular accumulation of glutamate and associated amino acids due to a reduction of their reuptake results in excitotoxicity, i.e., overstimulation of neuronal ionotropic (*N*-methyl-D-aspartate, α-amino-3-hydroxy-5-methyl-4-isoxazole propionic acid (AMPA)/quisqualate, kainate) and metabotropic glutamate receptors *(8)*. In turn, activated voltage and ligand-gated Ca^{2+} channels allow for an intracellular Ca^{2+} overload, marked by induction of enzymatic activity (proteases, endonucleases) that leads to cellular degradation and death *(9)*. Activation of phospholipases permits liberation and accumulation of free fatty acids (FFAs), including arachidonic acid, that may serve as substrates for generation of electron-rich oxygen radicals, e.g., superoxide $O_2^{\cdot-}$. Reactive oxygen species (ROS) initiate peroxidation of membrane lipids, oxidation of proteins, and breakdown of DNA and RNA *(10)*. The metabolism of arachidonate to eicosanoids may also prolong ischemia by maintaining hypoperfusion through vasoconstriction. Additional generation of nitric oxide (NO), through increased NO synthase activity and the reaction of NO with superoxide $O_2^{\cdot -}$ yields the peroxynitrite $ONOO^{\cdot -}$, another potent oxidant in the ROS group *(11)*. Evidence suggests that monoamine neurotransmitters, e.g., dopamine and serotonin, which increase during ischemia–hypoxia, are also involved in brain damage. The high sensitivity of the striatum to damage caused by energy deficiency may be related to the involvement of the dopaminergic nigrostriatal pathway and interactions of glutamate with dopamine *(12)*. In addition, degradation of dopamine and serotonin is associated with formation of hydroxyl radicals ($OH^{\cdot -}$), which may be additive to other free radicals in causing brain injury *(13)*. Hypoxic insult, which is generally less debilitating in the neonatal brain than in the adult brain, may induce compensatory changes in the factors discussed earlier, with the degree of induced change dependent on the specific factor.

COMPENSATORY MECHANISMS

Hypoxia–ischemia induced failure of energy production results in neuronal brain damage in the adult mammalian species. However, in the neonate, increased rates of glucose uptake and utilization of liver glycogen or decreased lactate production may each help to protect against damage from acute hypoxic insult *(14)*. Stress-related fetal catecholamine production and increase of blood flow to the brain help ensure not only protection from hypoxia, but also survival of the fetus after birth *(15)*.

Studies have also shown that postnatal day (PND) 3–11 rats, pretreated by exposure to subatmospheric air pressure, were able to survive subsequent anoxic challenge longer than controls *(16)*. This adaptive protection lasted up to 3 d. It seems that mitochondrial energy disruption in neonates may result in prolonged alterations in mitochondrial metabolism in the brain. For instance, when PND 12–13 rats were subjected to hypoxia–ischemia, immunoreactivity of γ-aminobutyric acid (GABA), mitochondrial cytochrome *c* oxidase, and ATP synthase activities in the cerebral cortex were enhanced for as long as 6.5 mo after the insult *(17)*.

LONG-TERM COMPENSATION

The concept that metabolic stress might have a long-term effect in the form of protective "preconditioning" is supported by evidence from a variety of experimental settings. It has been shown that perinatal hypoxia has neuroprotective effects with respect to insults to the central nervous system (CNS), at later stages of development *(18)*. Also, in adult rats, hypoxic insult resulted in subsequent protection against the neurotoxicity caused by kainate

(19) and seizures induced by the GABA antagonist bicuculline *(20)*. Intermittent hypoxia applied daily for several days resulted in a prolonged enhancement of succinate dehydrogenase (SDH) activity in the corpus striatum *(21)*.

Based on this background, we decided to assess the potential modulation of 3-NPA-induced neurotoxicity in adult rats exposed at birth to hypoxia–ischemia *(22)*. The plant and fungal toxin, 3-NPA, interrupts the mitochondrial electron transport chain via irreversible SDH binding and inhibition. A cellular ATP decrease is followed by membrane depolarization, excitotoxicity, and oxidative stress *(23–25)*. Sibling male Sprague–Dawley rats used in this study were delivered by cesarean section at 21 d of gestation, as either insulted ("I") or noninsulted ("NI") with hypoxia–ischemia. "I" rats were those in which perinatal hypoxia–ischemia was induced by submerging them as fetuses into warm saline for 15 min while they remained inside dissected uterine horns *(26)*. "NI" rats were delivered from the adjacent nonsubmerged horns. Rats at PND 70 were subjected to training in the behavioral tasks of an operant test battery. Behavioral testing was conducted 5 d a week for approx 50 min each day. At 12 mo of age, animals were injected intraperitoneally with 3-NPA immediately following each behavioral test session. The initial dose of 5 mg/kg/d was increased by 5 mg/kg/d each week, to a maximum of 30 mg/kg/d during the sixth week. At the end of the 3–NPA treatment, rats were anesthetized with pentobarbital sodium and perfused through the ascending aorta with 4% formaldehyde in phosphate buffer (0.1 *M*, pH 7.4). Coronal sections (50 μm) of various brain regions were stained using a modification of a silver staining procedure specific for degenerating axons, terminals, and neurons *(27)*.

Analyses of overt clinical signs as well as the behavioral and neurohistological endpoints suggested a long-term protective effect of perinatal hypoxic–ischemic insult against subsequent 3-NPA neurotoxicity. Only the "NI" rats exhibited the clinical symptoms of 3-NPA intoxication described by Hamilton and Gould *(28)*, namely incoordinated gait followed by a lateral or ventral recumbency. At high doses of 3-NPA, the performance of "NI" rats in the operant battery was worse than that of the "I" rats. Neurohistological examination of the brain revealed degenerating neurons, axons, and terminals throughout several parts of the diencephalon and cerebrum. In particular, degeneration was observed in the cerebral cortex, hippocampal subfield CA_1, thalamus, and caudate nucleus. Histological evaluation of four pairs of "I" rats and their "NI" siblings by an observer blind to their original treatment revealed that in three of the four pairs evaluated, the "NI" rats had greater damage than did the "I" rats (Fig. 1). In some cases, the "I" rats had no damage at all.

HYPOTHESES

The "ischemic–hypoxic preconditioning" or acquisition of resistance following an ischemic–hypoxic insult to subsequent ischemia–hypoxia was originally shown in the myocardium and then the in brain *(29,30)*. Thus far, several theories have been proposed to explain this phenomenon.

Selective Gene Expression

According to the selective gene expression theory, neuroprotection after hypoxia–ischemia would be induced by alteration in gene expression and protein synthesis *(30)*. C-*fos* and other immediate early genes, as well as heat shock proteins (HSPs), are expressed under conditions of metabolic stress in both adult and fetal brain *(31–33)*. Accelerated and enhanced *HSP70* gene expression was noted in the hippocampal CA_1 subfield in animals preconditioned with ischemia in response to subsequent ischemic insult and this effect was suggested as a possible mechanism underlying the development of ischemic tolerance *(34)*.

Permanent Neuronal GABA Elevation

GABA levels remain elevated in several brain regions particularly sensitive to energy deficiency (substantia nigra, hippocampus, and frontal cortex) for as long as 3 mo following ischemic insult. Increased GABA inhibitory function may be responsible for the diminished occurrence of bicuculline-induced seizures following ischemia–hypoxia *(20)* and may contribute to resistance against excitotoxic or metabolic stress.

Upregulation of Adenosine Receptors

Adenosine release during an episode of preconditioning cardiac ischemia, and the long-lasting up-regulation of adenosine A_1 type receptors in the myocardium were claimed to be protective against subsequent cardiac infarction *(35)*. Despite the short half-life of adenosine, stimulation of the adenosine A_1 receptor via intracoronary infusion of adenosine was effective for cardioprotection even after the infusion was stopped. Similarly, adenosine increases in the brain during energy disruption have been observed to occur concomitantly with a release of excitatory amino acids and GABA *(36)*.

Permanent K^+ Channel Activation

Activation of ATP-regulated potassium channels (K_{ATP}) combined with the upregulation of adenosine A_1 type receptors was suggested to contribute to the induction of brain tolerance to hypoxia–ischemia *(37)*. It is unknown whether the opening of K_{ATP} channels would be mediated by a decline of ATP concentration or, additionally, an increased adenosine level.

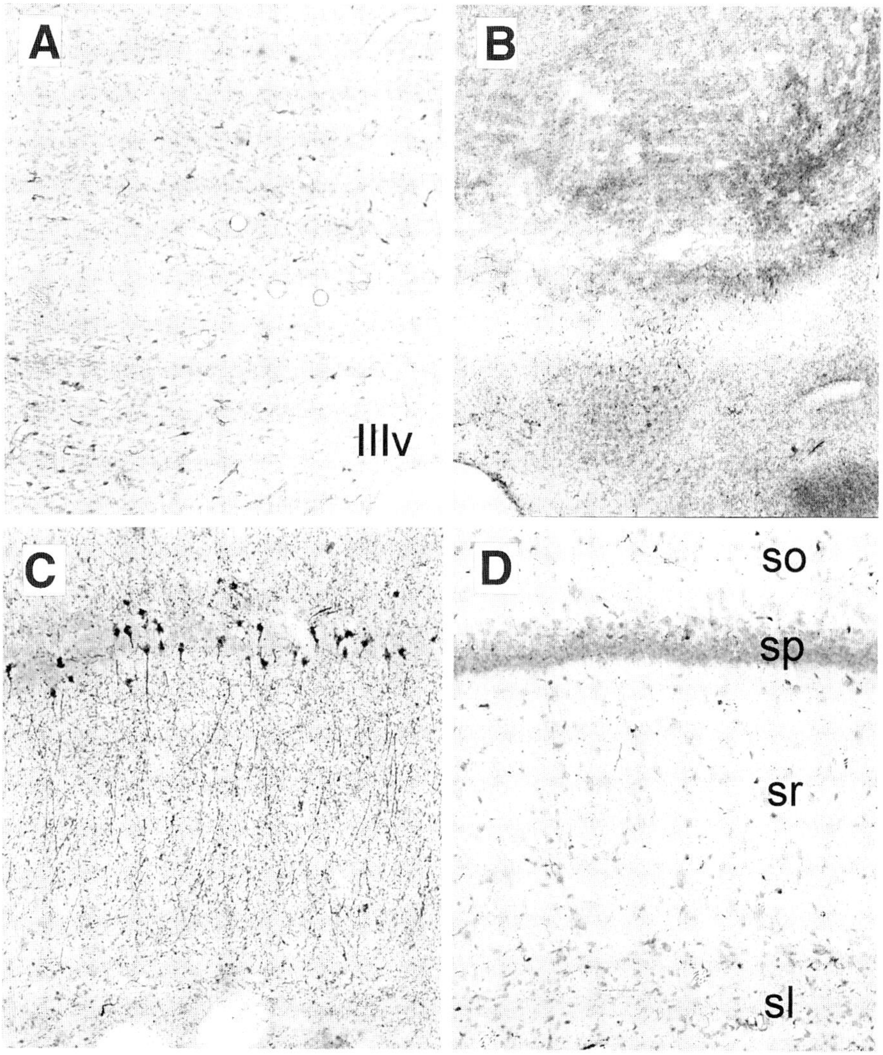

Fig. 1. **(A)** The ventral thalamus of an adult rat treated with multiple, increasing doses of 3-NPA (*see text for details*) appears completely normal when stained with a degeneration-selective reduced silver method of the Fink–Heimer type. This rat had been preconditioned (insulted, "I") by a perinatal episode of hypoxia–ischemia. The top of the third ventricle is visible in both **(A)** and **(B)** and is labeled as "IIIv." **(B)** By contrast, this nonpreconditioned (noninsulted, "NI") rat, which was not exposed to perinatal hypoxia–ischemia, sustained massive neurodegenerative damage in the thalamus (visible here as the circular necrotic region corresponding to the thalamic nucleus reuniens which is just dorsal to a small cluster of argyrophilic

Enhancement of Respiratory Enzyme Activity

The higher activities of neuronal and glial mitochondrial electron transport enzymes, such as cytochrome *c* oxidase or SDH, observed after hypoxic insult may also play a role in the mechanism of "preconditioning" *(17,21)*. The high mitochondrial energy production due to these enzymes may prevent rapid energy loss during hypoxic insult. This hypothesis is supported by an observed neuroprotective effect in chemical, malonate-induced hypoxia, following administration of the electron transport component, coenzyme Q_{10}. Coenzyme Q_{10}, known to increase the activity of the mitochondrial electron transport chain, also prevented energy (ATP) depletion in this study *(33)*.

CONCLUSION

Diminished neurotoxicity, observed after the chronic treatment with the mitochondrial inhibitor 3-NPA in adult rats exposed perinatally to hypoxia–ischemia, may result from a type of "preconditioning" that results in long-lasting neuroprotection. This enhanced neuroprotection may be especially potent because of the extremes of hypoxia that neonates can endure and still survive, as well as perhaps their increased plasticity at this developmental stage. A compensatory enhancement of various aspects of mitochondrial metabolism may be a major factor in the adaptation to metabolic stresses that occurs later in life. However, it appears that certain additional factors reviewed earlier, such as increased expression of neuroprotective genes, might also play a role in this phenomenon. Elucidation of the mechanisms underlying the acquired resistance to subsequent stresses of metabolic disruption has obvious implications for cardio and neuroprotection, as well as the clinical treatment of some neurodegenerative diseases.

ACKNOWLEDGMENTS

The authors thank Robert L. Rountree, Sherry A. Ferguson, David L. Frederick, Merle G. Paule, and William Slikker, Jr. for their collaboration in this study.

---

neurons near the top of the third ventricle). **(C)** Another 3-NPA-treated, but nonpreconditioned (noninsulted, "NI"), rat shows numerous necrotic neuronal cell bodies spread throughout stratum pyramidale (sp) of the CA1 subfield of the hippocampus. Note the degenerating dendrites of these dead neurons throughout stratum radiatum (sr). **(D)** A rat apparently completely protected by its preconditioning exposure to perinatal hypoxia–ischemia (insulted, "I") shows no damage whatsoever in its hippocampus when treated with 3-NPA as an adult. so, stratum oriens; sp, stratum pyramidale; sr, stratum radiatum; sl, stratum lacunosum.

REFERENCES

1. DiMauro S, Schon EA. Mitochondrial DNA and diseases of the nervous system: the spectrum. Neuroscience 1998;4:53–63.
2. Beal MF. Does impairment of energy metabolism result in excitotoxic neuronal death in neurodegenerative illnesses? Ann Neurol 1992;31:119–130.
3. Spencer PS, Allen CN, Kisby GE, et al. On the etiology and pathogenesis of chemically induced neurodegenerative disorders. Nerobiol Aging 1994; 15:265–267.
4. Landry J, Bernier D, Chretien P, et al. Synthesis and degradation of heat shock proteins during development and decay of thermotolerance. Cancer Res. 1982;42:2457–2461.
5. Guyton AC. Aviation, high altitude, and space physiology. In: Dreibelbis D, ed. Textbook of Medical Physiology, 7th edit. WB Saunders, Philadelphia, 1986, pp. 530–531.
6. Vaughan BE, Pace N. Changes in myoglobin content of the high altitude acclimatized rat. Am J Physiol 1956;185:549–556.
7. Harik SI, Hritz MA, LaManna JC. Hypoxia-induced brain angiogenesis in the adult rat. J Physiol 1995;485:525–530.
8. Rothman SM, Olney JW. Excitotoxicity and the NMDA receptor. Trends Neurosci 1987;10:299–302.
9. Choi D. Cerebral hypoxia: some new approaches and unanswered questions. J Neurosci 1990;10:2493–2501.
10. Halliwell B. Reactive oxygen species and the central nervous system. J Neurochem 1992;59:1609–1623.
11. Rubbo H, Radi R, Trujillo M, et al. Nitric oxide regulation of superoxide and peroxynitrite-dependent lipid peroxidation. J Biol Chem 1994;269:26,066–26,075.
12. Pastuszko A. Metabolic responses of the dopaminergic system during hypoxia in newborn brain. Biochim Med Metab Biol 1994;51:1–15.
13. Tipton KF. The prosthetic groups of pig brain mitochondrial monoamine oxidase. Biochim Biophys Acta 1968;159:451–459.
14. Vannucci RC, Mujece DJ. Effect of glucose on perinatal hypoxic-ischemic brain damage. Biol Neonate 1992;62:215–224.
15. Irestedt L, Lagercrantz H, Belfrage P. Cause and consequences of maternal and fetal sympathoadrenalactivation during parturition. Acta Obstet Gynecol Scand Suppl 1984;118:111–115.
16. Adolph EF. Physiological adaptation to hypoxia in newborn rats. Am J Physiol 1971;221:123–127.
17. Romijn HJ, Janezen AWJW, Van den Bogert C. Permanent increase of immunocytochemical reactivity for γ-aminobutyric acid (GABA), glutamic acid decarboxylase, mitochondrial enzymes, and glial fibrillary acidic protein in rat cerebral cortex damaged by early postnatal hypoxia–ischemia. Acta Neuropathol 1994;87:612–627.
18. Gidday JM, Fitzgibbons JC, Shah AR, et al. Neuroprotection from ischemic brain injury by hypoxic preconditioning in the neonatal rat. Neurosci Lett 1994;168:221–224.

19. Pohle W, Rauca C. Hypoxia protects against the neurotoxicity of kainic acid. Brain Res 1994;664:297–304.
20. Sieklucka M, Heim C, Block F, et al. Transient reduction of cerebral blood flow leads to longlasting increase in GABA content in vulnerable structures and decreased susceptibility to bicuculline induced seizures. J Neural Transm 1992;88:87–94.
21. Dagani F, Marzatico F, Curti D, et al. Effect of prolonged and enzymatic activities related to energy transduction. J Cereb Blood Flow Metab 1984;4:615–624.
22. Binienda Z, Frederick DL, Ferguson SA, et al. The effects of perinatal hypoxia on the behavioral neurochemical, and neurohistological toxicity of the metabolic inhibitor 3-nitropropionic acid. Met ab Brain Dis 1995;10:269–282.
23. Ludolph AC, Seelig M, Ludolph A, et al. 3-Nitropropionic acid decreases cellular energy levels and causes neuronal degeneration in cortical explants. Neurodegeneration 1992;1:155–161.
24. Beal MF, Ferante RJ, Henshaw R, et al. 3-Nitropropionic acid neurotoxicity is attenuated in copper/zinc superoxide dismutase transgenic mice. J Neurochem 1995;65:919–922.
25. Binienda Z, Kim CS. Increase in levels of total free fatty acids in rat brain regions following 3-nitropropionic acid administration. Neurosci Lett 1997;230:199–201.
26. Bjelke B, Andersson K, Ogren SO, et al. Asphyctic lesion: proliferation of tyrosine hydroxylase-immunoreactive nerve cell bodies in the rat substantia nigra and functional changes in dopamine neurotransmission. Brain Res 1991;543:1–9.
27. Scallet AC. Quantitative morphometry for neurotoxicity assessment. In: Chang LW, Slikker W Jr, eds. Neurotoxicology Approaches and Methods. Academic Press, Orlando, 1995, pp. 99–129.
28. Hamilton BF, Gould DH. Nature and distribution of brain lesions in rats intoxicated with 3-nitropropionic acid: a type of hypoxic (energy deficient) brain damage. Acta Neuropathol 1987;72:286–297.
29. Murry CE, Jennings RB, Reimer KA. Preconditioning with ischemia: a delay of lethal cell injury in ischemic myocardium. Circulation 1986;74:1124–1136.
30. Kitagawa K, Matsumoto M, Tagaya M, et al. 'Ischemic tolerance' phenomenon found in the brain. Brain Res 1990;528:21–24.
31. Sloviter RS, Lowenstein DH. Heat shock protein expression in vulnerable cells of the rat hippocampus as an indicator of excitation-induced neuronal stress. J Neurosci 1992;12:3004–3009.
32. Vendrell M, Curran T, Morgan JI. Glutamate, immediate-early genes, and cell death in the nervous system. Ann NY Acad Sci 1993;679:132–141.
33. Binienda Z, Scallet A. The effects of reduced perfusion and reperfusion on c-fos and hsp-72 protein immunohistochemistry in gestational day 21 rat brains. Int Dev Neurosci 1994;12:605–610.
34. Aoki M, Abe K, Kawagoe J, et al. The preconditioned hippocampus accelerates HSP70 heat shock gene expression following transient ischemia in the gerbil. Neurosci Let 1993;155:7–10.

35. Liu GS, Thornton J, Van Winkle DM, et al. Protection against infarction afforded by preconditioning is mediated by A_1 adenosine receptors in rabbit heart. Circulation 1991;84:350–356.
36. Fredholm BB, Johansson B, Van der Ploeg I, et al. Neuromodulatory roles of purines. Drug Dev Res 1993;28:349–353.
37. Heurteaux C, Lauritzen I, Widmann C, et al. Essential role of adenosine, adenosine A_1 receptors, and ATP-sensitive K^+ channels in cerebral ischemic preconditioning. Proc Natl Acad Sci USA 1995;92:4666–4670.
38. Schulz JB, Beal MF. Neuroprotective effects of free radical scavengers and energy repletion in animal models of neurodegenerative disease. Ann NY Acad Sci 1995;765:100–110.

19

Neural Transplantation and Huntington's Disease

What Can We Learn from the 3-Nitropropionic Acid Model?

Cesario V. Borlongan, Christine E. Stahl, Thomas B. Freeman, Robert A. Hauser, and Paul R. Sanberg

Neural transplantation as a treatment modality for patients suffering from neurodegenerative disorders (e.g., Parkinson's disease [PD]) has produced encouraging results. In recently conducted clinical trials, neural transplantation of human fetal ventral mesencephalic dopamine-secreting cells (the major type of cells that degenerates in PD) into the brains of PD patients has been demonstrated to ameliorate the clinical symptoms of the disease *(1,2)*. Concrete evidence detailing clinical improvement following fetal dopaminergic cell transplantation has been reported previously in PD patients *(3,4)*. For the first time, direct histopathological evidence became available from a transplanted PD patient, who died more than 16 mo posttransplantation of complications unrelated to the transplant procedure. Viable neural grafts were shown to integrate with the host tissue, and thus fetal tissue transplantation has been implicated as directly promoting symptomatic relief to the patient *(1,2)*. In an attempt to circumvent logistical and ethical problems with using human fetal grafts, porcine fetal cells have been directly transplanted in PD patients and positive preliminary results have been reported *(5)*. Other non-neural graft sources that are being examined at the preclinical level include Sertoli cells *(6)*, carotid body cells *(7)*, and kidney cells *(8)*, all of which have been suggested as dopamine- or neurotrophic factor-enriched cells.

Although neural transplantation holds great promise as a new therapeutic strategy to promote functional recovery in the human central nervous system (CNS), well-designed clinical trials are needed to justify the procedure in a large number of patients. Furthermore, a major effort is needed to pur-

From: *Mitochondrial Inhibitors and Neurodegenerative Disorders*
Edited by: P. R. Sanberg, H. Nishino, and C. V. Borlongan

sue animal experiments to understand in more detail the mechanisms of action of neural grafts (e.g., to clarify the level of integration of the grafts anatomically and functionally in the host brain). The pioneering work by Bjorklund and Stenevi *(9)* indicated that experimental Parkinsonism in adult rats could be reversed by implanted dopamine-rich tissue from rat fetuses. Subsequent animal research revealed that fetal dopamine-rich neural grafts can reinnervate the dopamine-denervated striatum, form synaptic connections with host neurons, release dopamine, and improve motor function, including the cardinal symptoms of PD, tremor, rigidity, and hypokinesia. It has become clear that in this PD animal model system, functional improvement depends primarily on the number of surviving dopamine grafted cells and the density and extent of the graft-derived reinnervation. Of note, there is a dismal survival rate of the transplanted cells in the PD patient of less than 10%. Although the observed survival and function of grafted dopamine neurons provide support toward performing clinical trials for PD, the symptomatic relief needs to be defined fully *(10)*. For example, what is the maximum symptomatic relief that can be obtained by increasing dopamine levels in a PD patient using cell transplantation therapy? What are the specific symptoms alleviated by dopamine grafts? Is disease progression permanently blocked by neural transplantation therapy? Careful investigations of these issues will offer parameters for enhancing the efficacy of the transplants as well as elucidate the safety/risk factors of the treatment strategy.

Notwithstanding, neural transplantation has emerged as an alternative clinical treatment for PD and has been proposed as a therapeutic approach for other neurodegenerative disorders. Recently, Peschanski and colleagues *(11)* outlined the rationale and accumulating evidence for proceeding with clinical trials of fetal neural transplantation as a treatment modality for Huntington's disease (HD). The positive results in neural transplantation for PD coupled with the demonstration of neurobehavioral effects of fetal neural transplantation in animal models of HD *(12–16)* have prompted several researchers to propose clinical trials for HD. Preliminary clinical investigations using fetal striatal cell transplantation for HD patients have been conducted in Mexico *(17)*, and most recently in the United States at Good Samaritan Hospital *(18)* and at the University of South Florida. While our experience with PD patients provided the foundation for extending neural transplantation therapy to other diseases, the implications of the scientific evidence from HD animal models should be closely evaluated prior to proceeding with the clinical trials. Do available laboratory data support neural transplantation as a treatment modality for HD? It is evident that the field of fetal neural transplantation has greatly relied upon laboratory investigations;

thus critical preclinical evaluation of each targeted disease should be a foremost prerequisite in extending fetal neural transplantation treatment to other neurodegenerative disorders.

ANIMAL MODELS OF HUNTINGTON'S DISEASE

HD is an inherited neurodegenerative disorder characterized by progressive locomotor (i.e., choreiform movements), psychological, and cognitive impairments *(19)*. Shoulson *(20)* described two distinct stages of motoric symptoms of HD, namely, the early stage characterized by hyperkinesia and chorea, and the late stage characterized by an akinetic, Parkinsonian-like syndrome. The early stage of HD may involve biochemical changes with subsequent compensatory growth of spiny neurons within the basal ganglia, specifically the caudate–putamen area, with degeneration or neuronal loss in this area strikingly evident in the later stage of the disease *(21–23)*. Chromosomal and genetic components of the disease have recently been identified and transgenic animal models are being developed. At present, however, animal models of HD have utilized excitotoxins. The dominant mechanistic hypothesis of the neuropathology seen in HD has been attributed primarily to elevated levels of endogenous excitotoxins *(24,25)*. Experimental paradigms of HD consisted of animal models of excitotoxins that include kainic acid (KA), ibotenic acid (IA), or quinolinic acid (QA). Rodent and primate studies utilizing excitotoxic lesions induced by KA, IA, or QA have reproduced many of the behavioral symptoms and pathologic changes observed in HD.

Most recently, the 3-nitropropionic acid (3-NPA), a fungal and plant toxin and suicide inhibitor of succinate dehydrogenase, has been proposed as an improved HD model. It targets primarily an impairment in cellular respiration, secondarily predisposes normal endogenous levels of neurotransmitters to become excitotoxic, and produces HD-like symptoms *(23)*. Consistent with the hypothesis of free radical damage and oxidative stress in HD, the 3-NPA model closely replicates the neuropathology and behavioral alterations associated with HD *(26)*. 3-NPA is known to produce striatal atrophy by irreversibly inhibiting the mitochondrial and citric acid cycle, thereby leading to depressed ATP and elevated lactate concentrations *(27,28)*. It has been reported that low-dose 3-NPA spares striatal afferent and NADPH-diaphorase neurons, but destroys striatal intrinsic neurons such as γ-aminobutyric acid (GABA), substance P, somatostatin, and neuropeptide Y-containing neurons *(23)*.

Chronic systemic injections of 3-NPA (10 mg/kg, i.p., once every 4 d for 28 consecutive days) in rats resulted in the evolution of deficits in spontaneous locomotor activity that resembled the progressive course of motor symptoms seen in HD *(29–31)*. Initially, 3-NPA treated animals exhibited significant hyperactivity, reaching a plateau after the third injection (d 12), then hypoactivity from the fourth injection (d 16) onwards. We further demonstrated that manipulating the time course of 3-NPA injections leads to sustained hyperactivity (early HD) or hypoactivity (advanced HD). This progressive behavioral pathology induced by 3-NPA is in contrast to that observed in excitotoxic animal models of HD which have reproduced only the hyperactive stage of HD *(24,30)*. Furthermore, a selective striatal lesion similar to that caused by excitotoxins has been demonstrated following intraparenchymal as well as systemic injections of 3-NPA *(23,24,30,32,33)*. Accordingly, the neurobehavioral pathology associated with the excitotoxins seems to correlate well with the early hyperactive stages of HD, while the later stages of 3-NPA toxicity appear to resemble the later hypoactive stages and, possibly, the juvenile onset (Westphal variant) of HD *(20,24,29,30)*. Because of the mechanistic and pathologic similarities between 3-NPA lesions and HD, 3-NPA has been proposed as an alternative HD model *(23,26–28)*. In addition, the 3-NPA model is a novel platform on which experimental treatments for promoting functional recovery can be verified across stages of the disease.

We raised some concerns on available data on transplantation studies that appear not to fully encompass the two-stage behavioral hallmarks of HD. In general, it is viewed that fetal striatal transplantation promotes "hypoactive effects" *(34)* because subsequent transplantation reverses the hyperactivity caused by excitotoxins *(35)*. A review of the laboratory evidence on neural transplantation for excitotoxin animal models of HD indicates gradual hypoactivity developing after some striatal transplantations. Polgar and colleagues *(36)* report that based on rodent studies, the transplants may not just "normalize" the locomotor abnormalities associated with the striatal lesion, but also precipitate a period of transient hypoactivity. It is possible that the growing transplant produces neurochemical changes in the host striatum that predisposes the transplanted animal to exhibit hypoactivity. Interestingly, the transplant-induced behavioral changes can be seen as early as 1 mo posttransplantation *(16,37,38)*. We proposed that this relatively rapid recovery of locomotor functions in the absence of clear indications of well-formed structural connections suggests a two-stage recovery following intrastriatal transplantation: a rapid, transient neurochemically mediated stage followed by slower, structural changes resulting in permanent recovery *(36,39)*. Unfortunately, this model is based on studies of neural transplantation in a hyperactive model of HD. Except for our recent report *(16)*, no study has directly investigated the effects of fetal striatal transplantation in a hypoactive model of HD. Most, if not all, fetal tissue transplantation in animal models of HD were carried out using the excitotoxin-induced hyperactive model of HD *(25,39)*. In this regard, the systemic 3-NPA model may provide direct investigations of the effect of fetal neural transplantation on the hypoactive stage of HD. We previously hypothesized that with the general view of graft-induced hypoactive effects, a transplanted hypoactive HD rat may exhibit increased hypoactivity. However, we observed that 3-NPA-treated hypoactive animals showed normalization of general locomotor activity following fetal striatal transplantation (*see below*), indicating that the alteration in behavior produced by neural transplantation is not a simple "hypoactive or hyperactive" effect.

Unlike those for PD, animal models for HD may not resemble clearly the disease. Because the hypoactive model of HD has become available only quite recently, most preclinical studies on the efficacy of fetal neural transplantation have used the hyperactive HD stage. Accordingly, if one has to follow the literature, proposed clinical trials should be carried out on HD patients in the early stages. Unfortunately, limited clinical trials have been conducted in late-stage HD patients, and because we have not fully examined the effect of transplantation at this stage, negative results from clinical data may be difficult to interpret. That is, absence of clinical improvement does not necessarily mean that the transplant is ineffective; rather, it could be due to wrong dosage of the cells or target site of the transplant. On the other hand, conducting early transplants for HD has its own drawbacks, in that the natural disease progression of HD (from dyskinesia to akinesia) may prevent delineation of true transplant-induced behavioral effects. More importantly, when motor endpoint is used as the primary index of successful neural transplantation, one must take into account the variability of movement activity in the transplant recipient, which fluctuates at different time points on a day-to-day basis. The inclusion of psychiatric index (i.e., CAPIT-HD; *[40]*) in the assessment of the transplanted patient may reveal a better view of transplant-induced effects.

An additional problem that complicates the timing of neural transplantation in HD is that second-order degeneration involving extrastriatal damage may follow the degeneration of the striatum. As such, late intrastriatal transplantation in HD may not be able to replenish the dying neurons associated with this second-order degeneration. Similarly, early transplantation in HD does not ensure that the grafted cells will promote a general retardation of the neurodegenerative process, instead the transplants may prevent the disease progression only in a limited fashion, that is, within the graftreinnervated striatal regions. In both late and early transplantation in HD, secondary cortical cell loss may not be totally reduced owing to possible graft size, reinnervation volume, or tissue heterogeneity problems *(41,42)*. Even if transplant-induced neuronal integration is accomplished, a near complete rescue from neuronal loss may be necessary, because dyskinesia or akinesia may result when a percentage of striatal neurons remain damaged *(39)*. For example, severe dyskinesia occurs when 25–60% of intrinsic striatal neurons (posterior putamen) are destroyed, but such dyskinesia disappears (accompanied by the onset of akinesia) when more than 70% striatal neurons degenerate *(43,44)*. Analyses of the volume of striatal atrophy and ensuing intrinsic as well as extrinsic neuronal loss may offer some clues on the histological degeneration and reconstruction before and after transplantation, and one can somehow predict the behavioral consequences of such CNS insult and repair status.

NEURAL TRANSPLANTATION PROTOCOL

The transplantation of rat fetal striatal cell suspension is conducted under aseptic conditions. Procedures for dissection and preparation of the rat fetal lateral eminence (16-d-old gestational age) have been described by Pakzaban and colleagues *(45)*. The dissected tissue pieces are first enzymatically (with trypsin) and then mechanically dissociated into a cell suspension. Approximately two striatal anlages per lesioned host striatum are transplanted, which correspond to six 8×10^5 viable rat cells. The viability and exact number of transplanted cells are analyzed using the trypan blue exclusion method *(29)*. The cell suspension is injected with a Hamilton syringe at the same coordinates as the neurotoxic lesion. Animals are first anesthetized with sodium pentobarbital (70 mg/kg, i.p.) and then mounted in the Kopf stereotaxic frame. Stereotaxic coordinates that targeted the striatum are chosen with reference to bregma: AP = +1.2 mm; ML = ±2.6 mm; and DV = –5.5 mm from dura. Each animal receives 3 μL of cells in medium infused over 5 min. An additional 5 min is allowed prior to retracting the needle. Sham-transplanted animals undergo the same transplant protocol, but receive the medium alone. After transplantation, the animals are placed in a heating pad until recovery. Body weights are monitored daily to ensure normal health conditions for the animals. At least 1 mo of a posttransplantation maturation period is allowed prior to behavioral testing.

NEURAL TRANSPLANTATION IN SYSTEMIC 3-NPA MODEL

To examine whether neural transplantation can correct the akinetic stage of HD, we used the chronic 3-NPA model, specifically the 3-NPA-induced hypoactive model of HD to demonstrate whether fetal tissue transplantation can ameliorate behavioral deficits associated with a more advanced stage of HD *(16)*. Twelve-week-old Sprague–Dawley rats were introduced to the 3-NPA dosing regimen (10 mg/kg, i.p., once every 4 d for 28 consecutive days). All animals were tested following the completion of the 28-d 3-NPA administration and at 3 mo posttransplantation using the Digiscan locomotor apparatus (Omnitech, Columbus, OH; Oasis program). The apparatus consisted of a box (40 × 40 × 35.5 cm) surrounded by two levels of infrared beams. Data were then collected through a Compudyne MS DOS computer. The following 14 locomotor parameters were evaluated: horizontal activity, total distance, movement time, rest time, speed, number of movements, average distance, vertical activity, vertical time, vertical movement, stereotypy, number of stereotypies, clockwise rotation, and anticlockwise rotation. The test session was conducted throughout the dark phase of a 12–12-h diurnal cycle (6 PM–6 AM). It has been demonstrated previously that rats being nocturnal animals normally display most abnormal locomotor activity changes during the night *(46)*; thus tests were performed during their awake period. Both the housing cage and the Digiscan box were made of Plexiglas and were supplied with the same bedding materials, thus ensuring that any locomotor changes were of treatment, and not environmental, effects on the animals. In addition, a 1-h habituation period (5 PM–6 PM) to the test apparatus was allowed prior to data collection. Measurements were taken every 60 min, and the common peak activity period was selected (12 midnight–1 AM) for evaluation of locomotor activity.

Both groups of animals did not significantly differ from each other in their locomotor behavior during either pretreatment or posttreatment of 3-NPA ($p > 0.05$). As previously reported *(29–31)*, all animals exhibited significant hypoactivity at posttreatment of 3-NPA in all locomotor parameters, except clockwise and anticlockwise rotations ($p > 0.05$). At 3 mo posttransplantation, 3-NPA-treated animals that received lateral ganglionic eminence (LGE) grafts had a significant increase in their locomotor activity compared to animals that received medium alone ($p > 0.05$) (Fig. 1). The increase in the locomotor activity of 3-NPA-treated animals that received the LGE grafts was also significantly higher than their post-3-NPA injection (pretransplant) activity ($p > 0.05$). In contrast, animals that received medium alone did not differ significantly from their post-3-NPA injection activity.

At the end of the behavioral testing, animals were deeply anesthetized then perfused with 1% heparinized saline (200–300 mL) followed by 4% paraformaldehyde in 0.1 *M* phosphate buffer (300–400 mL). Brains were extracted and stored in fixative overnight, then serially sectioned at 30 µm.

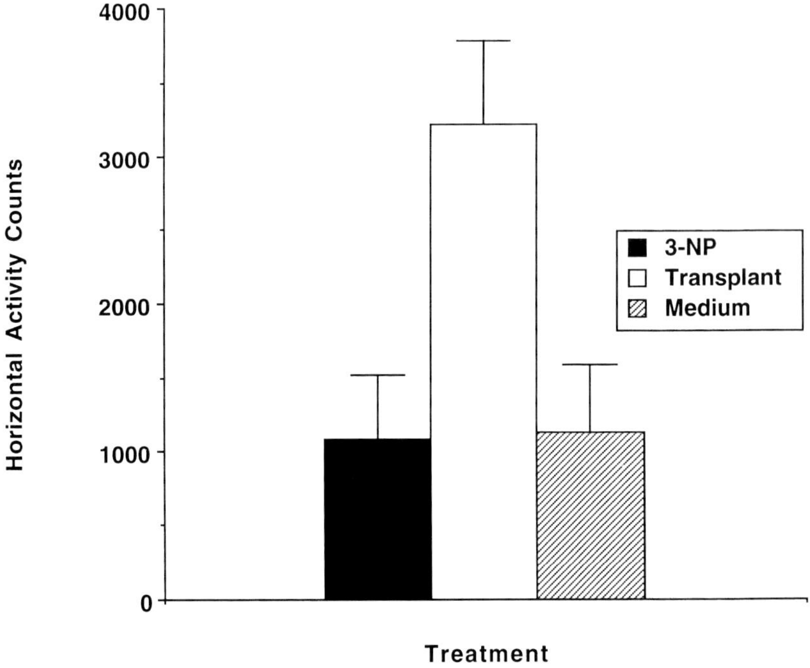

Fig. 1. Locomotor activity (horizontal activity data shown) produced by chronic 3-NPA administration in animals prior to transplantation (3-NP) and after transplantation (Transplant or Medium). Fetal striatal grafts significantly reversed the hypoactivity produced by 3-NPA at 3 mo posttransplantation.

Sections were serially stained with cresyl violet and acetylcholinesterase (AChE) *(29,30,38,45)*. Brain structures were then examined by light microscopy for qualitative assessment of graft survival. Surviving striatal grafts were identified in functionally recovered animals (*see below* for further discussion on graft survival). This observation supports the use of fetal striatal transplants to correct the akinesia stage deficits associated with the advanced stage of HD. To the best of our knowledge, this is the first study that has investigated the effects of fetal striatal transplantation in a hypoactive model of HD.

NEURAL TRANSPLANTATION IN THE INTRAPARENCHYMAL 3-NPA MODEL

Animals with unilateral neurotoxic lesions in the striatum exhibit a stereotypical biased rotational behavior in response to dopamine agonists, and we have injected the 3-NPA intraparenchymally to examine the direct effects of the toxin on the striatum *(47)*. Because the rotational test may be subject to sensitization effects of the drug, we proposed the drug-free elevated body swing test (EBST) as an alternative behavioral index of motor asymmetry in striatal lesioned animals *(48)*. This technique involves elevating the animal from the ground by holding its tail and simply recording the number of swings to either side made by the animal over 30 s or after 20 repeated trials of lifting the animal by the tail *(49)*.

To create a unilateral striatal lesion, Sprague–Dawley rats were first anesthetized with sodium pentobarbital (70 mg/kg, i.p.), then mounted in a Kopf stereotaxic frame. Stereotaxic coordinates were based from the bregma point with the tooth bar adjusted at the level of the interaural line. Animals were stereotaxically lesioned in the right striatum (AP = +1.2; ML = –2.8 mm; DV = –5.5) by injecting 500 nmol of 3-NPA in 2 μL of 0.9% saline with pH adjusted to 7.4 using 6M NaOH. Each animal received 2 μL of the neurotoxic solution which was injected over a 2-min period. An additional 5 min was allowed before the cannula was retracted. The body temperature of the animals was kept at normal limits throughout the surgical procedure and until recovery.

Animals that received 3-NPA lesion surgery exhibited biased swing activity at 7, 14, 21, and 28 d postlesion *(49)*. Over a period of 30 s, 3-NPA-lesioned animals displayed about nine mean ipsilateral (to the lesion) swings and only about one contralateral swing, starting at 7 d and extending throughout the postlesion test days. When using the 20-lift trial EBST, 3-NPA lesioned animals exhibited >70% ipsilateral swing activity *(49)*. Normal or sham-lesioned animals displayed about 50% ipsilateral swing activity *(48,49)*.

Neural transplantation was conducted after behavioral testing at 28 d postlesion and a 1 mo maturation period was allowed prior to introducing the animals again to the EBST. 3-NPA-lesioned animals showed a marked reduction in the mean ipsilateral swings (about five swings) and this almost equaled the number of contralateral swings (Fig. 2). In addition, they had about 50% ipsilateral swing activity at posttransplant test days using the 20-lift trial EBST indicating that the transplanted cells promoted a normalization of biased swing activity. The recovery from asymmetrical motor activity was noted as early as 1 mo and persisted up to 3 mo posttransplantation. In contrast, animals transplanted with medium alone continued to exhibit a biased swing activity. These results demonstrate the efficacy of neural tranplantation therapy for treating the 3-NPA-induced motor dysfunctions and also demonstrate that the EBST is sensitive in monitoring recovery of motor function in unilaterally lesioned animals following fetal striatal transplants.

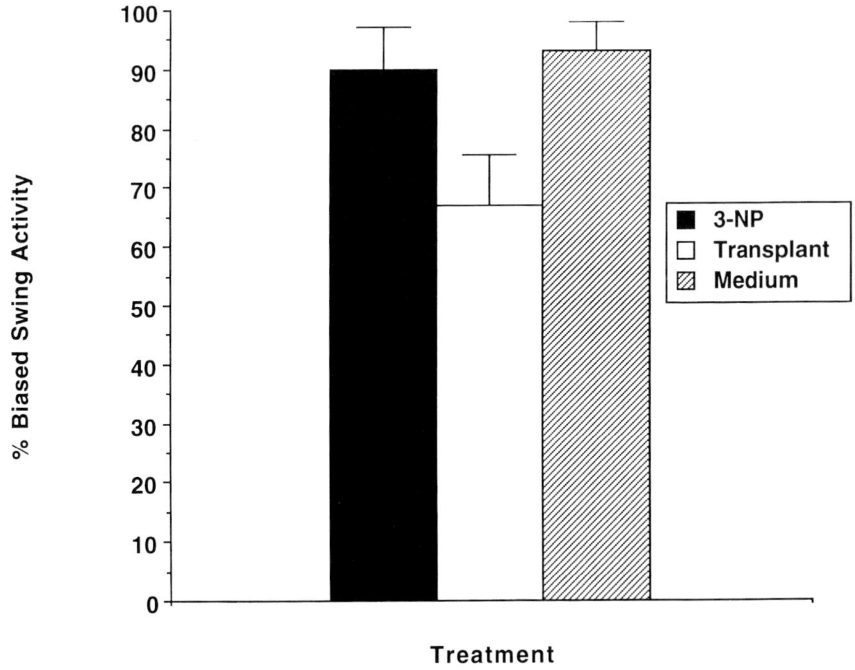

Fig. 2. Asymmetric motor activity as revealed by elevated body swing test produced by unilateral intrastriatal 3-NPA in animals prior to transplantation (3-NP) and after transplantation (Transplant or Medium). Fetal striatal grafts normalized the biased swing activity produced by 3-NPA as early as 1 mo posttransplantation.

NEURAL TRANSPLANTATION AND 3-NPA-INDUCED STRIATAL DAMAGE

We have characterized the survival of grafted fetal striatal cells at 3 mo posttransplantation *(16)* and noted that the observed functional recovery is dependent on the prolonged survival of these cells in the brain. Pundt and colleagues *(38)* observed a similar albeit behavioral recovery following fetal striatal transplants in an excitotoxic rat model of HD, and they reported consistent AChE and dopamine- and adenosine-3',5'-monophosphate-regulated phosphoprotein (DARPP-32) patches within the transplants. AChE and DARPP-32 immunocytochemical analyses have served as suitable methods for identifying striatal tissue within the transplants *(30,51)*, as well as transplant-induced host reconstruction *(15)*. In our study, surviving LGE grafts were identified in 75% of grafted animals at 3 mo posttransplantation, and the grafts were located next to the lateral ventricle and grossly resembled normal striatum. The transplanted LGE tissues appear to reconstitute neural

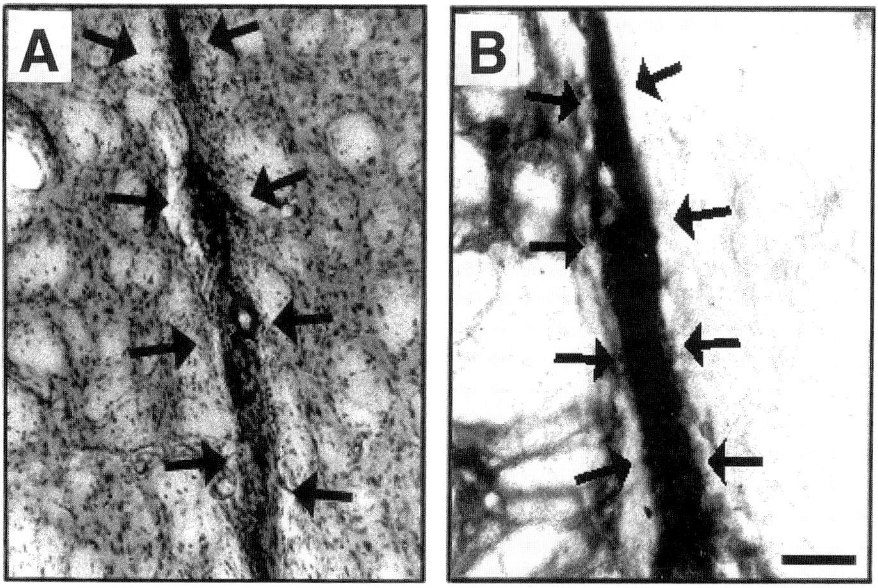

Fig. 3. Nissl and AChE stains of a 3-NPA-lesioned animal that received a transplant. Densely packed immature cells can be seen in **(A)**, while a bundle of AChE-positive neurons is noted in B. Scale bar equals 175 μm and 150 μm for **(A)** and **(B)**, respectively.

tissue and are mostly homogeneous with densely packed cells and minimal obvious inflammatory/immunological reactions. AChE stains revealed dark bands of positively stained cells within the grafts (Fig. 3), and in a few cases, the darkly stained bundles of AChE-positive graft extended toward the lesioned host striatum. Compared with those reported by Pundt and colleagues *(38)*, the outgrowth of AChE from our grafts did not form intense connections with the intact AChE patches of the host striatum. The negative cases that did not stain equivocally or not at all for AChE had very small grafts which may be due either to a low number of stereotaxically implanted striatal cells or to implantation of some cells in the lateral ventricle. In contrast, the striata from animals that did not receive the fetal grafts exhibited very light Nissl and AChE stains. In these brains, there was a marked striatal atrophy together with the loss of intrinsic neurons in the dorsolateral striatum.

Is AChE-positivity in grafts not a prerequisite for behavioral recovery in HD? We demonstrated that fetal striatal grafts can normalize general locomotor activity in the absence of complete AChE-positive patches within the lesioned striatum *(16)*. This would indicate that there is no positive correlation between functional recovery and the appearance of AChE-positive patches following transplantation. Previously, we also reported normalization of drug-induced rotational behavior in rats with unilateral excitotoxin-lesioned striatum in the absence of AChE-positive patches *(16,50)*. Furthermore, Brundin and colleagues *(52)* have recently noted that a more complex motor and cognitive deficit recovery may be related to increased density of AChE-rich patches in human fetal tissue striatal transplants. This view is shared by Dunnett and colleagues *(51)* and they reported that the observation of striatal-like AChE-rich patches in striatal transplants correlates highly with recovery of a complex behavioral task. Nevertheless, Brundin and colleagues *(52)* found only 10% AChE-positivity in their cell suspension transplants of human lateral ganglionic eminence into the rodent model, suggesting that a high number of AChE-positive grafted neurons may be difficult to obtain, but behavioral recovery may still be possible. Thus, AChE-positivity in the grafts may not be necessary for amelioration of deficits in simple tasks such as the locomotor activity *(16,50)*.

Our transplantation study using the 3-NPA model is the first investigation of the effects of such therapy in the hypoactive model of HD. Because most transplant studies using the excitotoxin-induced hyperactive model of HD have reported decreased locomotor activity following transplantation *(39)*, the general view on the effects of transplants has been that fetal striatal grafts will promote "hypoactivity" in transplant recipients. In HD animals, a decrease in generalized hyperactivity is noted consistently in most fetal striatal transplantation studies with or without striatal-like patches. This would further suggest that transplantation in a hypoactive model (the late stage HD) may produce detrimental instead of beneficial effects. However, our observation of functional recovery in 3-NPA hypoactive animals following transplantation indicates that the graft-induced normalization of locomotor behavior may not be a simple hyperactive or hypoactive effect but a complex mechanism that entails neurochemical as well as anatomical reorganization in the lesioned striatum of the host brain *(16)*. We propose that, at least in the early stages following transplantation, a graft-derived trophic effect may mediate the observed functional recovery because of the incomplete integration between graft and host striatum. The observation of early general motoric behavioral recovery in animals may be related to possible clinically significant improvement in the generalized choreiform movements associated with HD patients *(16)*. Of note, similar robust recovery has been observed in transplanted Parkinsonian animals *(32)*, and the release of trophic factors from the grafts has been implicated as a mechanism by which transplants may induce behavioral changes *(6)*. Preliminary reports of early significant recovery described in patients who have received fetal striatal transplantation may be related to nonspecific effects of the transplants (i.e., trophic factors) as compared to specific long-term circuitry changes and concomitant increases in striatal-like patches *(18)*. Although we examined only the hypoactive stage produced by 3-NPA, it is also possible to investigate the effects of neural transplantation during the progression of 3-NPA neurotoxicity; thus one will be able to gauge the optimal window of opportunity for conducting transplantation therapy in HD.

With preliminary clinical trials of neural transplantation for HD patients underway *(18,53)*, laboratory studies need to address critical issues directly related to HD, especially the disease progression. Previous transplant studies have concentrated on the early hyperkinetic stage of HD, and have reported transplant-induced recovery. Our present data on the 3-NPA-induced hypoactive model is the first report demonstrating that fetal striatal grafts can also promote functional recovery in the late akinetic stage of HD. Animal models need to resemble closely the pathology as well as the behavioral symptoms of the disease if treatment strategies are to be conclusively verified for clinical applications *(54)*. The dosing regimen, route of administration, and timing of therapeutic intervention are important factors in promoting 3-NPA as an improved HD model *(55,56)*, and these same factors need to be carefully examined to elucidate the true effects of the transplant *(57)*.

CONCLUSION

Preclinical data from our laboratory indicate that neural transplantation can correct the akinetic stage of HD; however, additional studies are warranted to correlate the observed functional recovery with graft survival and integration with the host tissue. We raised some concerns that although the clinical trials have begun in HD patients, more studies are needed to address issues on the optimal timing, the accurate location, and the sensitive behavioral assessment of the transplanted tissue in HD animal models. The rapid turnover of experimental therapies into the clinic should not compromise the possible harm they may entail to the patient. The search for appropriate animal models for investigations of yet undefined or not fully understood stages of human disorders as they relate to treatment intervention should be a major research endeavor prior to proceeding with clinical trials. The 3-NPA model of HD offers new venues on understanding the disease progression, as well as on optimizing treatment strategies.

REFERENCES

1. Kordower JH, Freeman TB, Snow BJ, Vingerhoets FJ, Mufson EJ Sanberg PR, Hauser RA, Smith DA, Nauert GM, Perl DP, Olanow CW. Neuropathological

evidence of graft survival and striatal reinnervation after the transplantation of fetal mesencephalic tissue in a patient with Parkinson's disease. N Engl J Med 1995;332:118–1124.
2. Freeman TB, Olanow WC, Hauser RA, Nauert GM, Smith DA, Borlongan CV, Sanberg PR, Holt DA, Kordower JH, Vingerhoets FJ. Bilateral fetal nigral transplantation into postcommissural putamen in Parkinson's disease. Ann Neurol 1995;38:379–388.
3. Lindvall O, Brundin P, Widner H, Rehncrona S, Gustavii B, Frackowiak R, Leenders KL, Sawle G, Rothwell JC, Marsden CD. Grafts of fetal dopamine neurons survive and improve motor function in Parkinson's disease. 1990;247:574–577.
4. Lindvall O, Rehncrona S, Brundin P, Gustavii B, Astedt B, Widner H, Lindholm T, Bjorklund A, Leenders KL, Rothwell JC. Human fetal dopamine neurons grafted into the striatum in two patients with severe Parkinson's disease. Arch Neurol 1989;46:615–631.
5. Isacson O, Deacon TW, Pakzaban P, Galpern WR, Dinsmore J, Burns LH. Transplanted xenogeneic neural cells in neurodegenerative disease models exhibit remarkable axonal target specificity and distinct growth patterns of glial and axonal fibres. Nat Med 1995;1:1189–1194.
6. Borlongan CV, Cameron DF, Saporta S, Sanberg PR. Intracerebral transplantation of testis-derived sertoli cells promotes functional recovery in female rats with 6-hydroxydopamine-induced hemiparkinsonism. Exp Neurol 1997; 148:388–392.
7. Espejo EF, Montoro RJ, Armengol JA, Lopez-Barneo J. Cellular and functional recovery of Parkinsonian rats after intrastriatal transplantation of carotid body cell aggregates. Neuron 1998;20:197–206.
8. Granholm AC, Henry S, Herbert MA, Eken S, Gerhardt GA, van Horne C. Kidney cografts enhance fiber outgrowth from ventral mesencephalic grafts to the 6-OHDA-lesioned striatum, improve behavioral recovery. Cell Transplant 1998;7:197–212.
9. Bjorklund A, Stenevi U. Reconstruction of the nigrostriatal dopamine pathway by intracerebral nigral transplants. Brain Res 1979;177:555–560.
10. Lindvall O. Neural transplantation. Cell Transplant 1995;4:393–400.
11. Peschanski M, Cesaro P, Hantraye P. Rationale for intrastriatl grafting of striatal neuroblasts in patients with Huntington's disease. Neuroscience 1995;68:273–285.
12. Sanberg PR, Coyle JT. Scientific approaches to Huntington's disease. CRC Crit Rev Neurobiol 1984;1:1–44.
13. Isacson O, Brundin P, Gage FH, Bjorklund A. Compensation of lesion-induced changes in cerebral metabolism and behavior by striatal neural implants in a rat model of Huntington's disease. In: Will BE, Schmitt P, Dalrymple-Alford JC, eds. Brain Plasticity, Learning and Memory, Vol. 28. Plenum Press, New York, 1985, pp. 781–785.
14. Isacson O, Brundin P, Gage FH, Bjorklund A. Neural grafting in the rat model of Huntington's disease: progressive neurochemical changes after neostriatal ibotenate lesions and striatal tissue grafting. Neuroscience 1985;16:799–817.

15. Wictorin K, Ouimet CC, Bjorklund A. Intrinsic organization and connectivity of intrastriatal transplants in rats as revealed by DARPP-32 immunohistochemistry: specificity of connections with host brain. Eur J Neurosci 1989; 1:690–701.
16. Borlongan CV, Koutouzis TK, Poulos SG, Saporta S, Sanberg PR. Bilateral fetal striatal grafts in the 3-nitropropionic acid-induced hypoactive model of Huntington's disease. Cell Transplant 1998;7:131–135.
17. Madrazo I, Franco-Bourland RE, Cuevas C. Fetal neural grafting for the treatment of Huntington's disease (HD)—report of the first case. Soc Neurosci Abstr 1991;17:902.
18. Philpott LM, Kopyov O, Lee AJ, Jacques S, Duma CM, Caine S, Yang M, Eagle KS. Neuropsychological functioning following fetal striatal transplantation in Huntington's chorea: three case presentations. Cell Transplant 1997;6:203–212.
19. Bruyn GW. Huntington's chorea: historical, clinical and laboratory synopsis. In: Vinken PJ, G.W. Bruyn GW, eds. Handbook of Clinical Neurology. Elsevier, Amsterdam, 1968, pp. 298–378.
20. Shoulson I. Huntington's disease. In: Asbury A, McKhann GM, McDonald WI, eds. Diseases of the Nervous System: Clinical Neurobiology. Ardmore Medical Books (WB Saunders), Philadelphia, 1986. pp. 85–99.
21. Graveland GA, Williams RS, DiFiglia M. Evidence for degenerative and regenerative changes in neostriatal spiny neurons in Huntington's disease. Science 1985;227:770–773.
22. Beal MF, Kowall NW, Ellison DW, Mazurek MF, Swartz KJ, Martin JB. Replication of the neurochemical characteristics of Huntington's disease by quinolinic acid. Nature 1986;321:168–171.
23. Beal MF, Brouillet E, Jenkins BG, Ferrante RJ, Kowall NW, Miller JM, Storey E, Srivastava R, Rosen BR, Bradley TH. Neurochemical and histologic characterization of striatal excitotoxic lesions produced by the mitochondrial toxin 3-nitropropionic acid. J Neurosci 1993;13:4181–4192.
24. Koutouzis TK, Emerich DF, Borlongan CV, Freeman TB, Cahill DW, Sanberg PR. Cell transplantation for central nervous system disorders. Critic Rev Neurobiol 1994;8:125–162.
25. Sanberg PR, Wictorin K, Isacson O. eds. Cell Transplantation for Huntington's Disease, RG Landes, CRC Press, Boca Raton, FL, 1994, p. 124.
26. Brouillet E, Jenkins BG, Hyman BT, Ferrante RJ, Kowall NW, Srivastava R, Roy DS, Rosen BR, Beal MF. Age-dependent vulnerability of the striatum to the mitochondrial toxin 3-nitropropionic acid. J Neurochem 1993;60:356–359.
27. Bossi SR, Simpson JR, Isacson O. Age dependence of striatal neuronal death caused by mitochondrial dysfunction. NeuroReport 1993;4:73–76.
28. Simpson JR, Isacson O. Mitochondrial impairment reduces the threshold for in vivo NMDA-mediated neuronal death in the striatum. Exp Neurol 1993;121:57–64.
29. Borlongan CV, Koutouzis TK, Freeman TB, Cahill DW, Sanberg PR. Behavioral pathology induced by repeated systemic injections of 3-nitropropionic acid mimics the motoric symptoms of Huntington's disease. Brain Res 1995;697:254–257.

30. Borlongan CV, Koutouzis TK, Randall TS, Freeman TB, Cahill DW, Sanberg PR. Systemic 3-nitropropionic acid: behavioral deficits and striatal damage in adult rats. Brain Res Bull 1995;36:549–556.
31. Koutouzis TK, Borlongan CV, Scorcia T, Creese I, Cahill DW, Freeman TB, Sanberg PR. Systemic 3-nitropropionic acid: long-term effects on locomotor behavior. Brain Res 1994;646:242–246.
32. Borlongan CV, Cahill DW, Freeman TB, Sanberg PR. Recent advances in neural transplantation: relevance to neurodegenerative disorders. J Florida Med Assoc 1994;10:689–694.
33. Hamilton BF, Gould DH. Nature and distribution of brain lesions in rats intoxicated with 3-nitropropionic acid: a type of hypoxic (energy deficient) brain damage. Acta Neuropathol 1987;72:286–297.
34. Lu SL, Giordano M, Norman AB, Shipley MT, Sanberg PR. Behavioral effects of neural transplants into the intact striatum. Pharmacol Biochem Behav 1990;37:135–148.
35. Sanberg PR, Calderon SF, Girodano M, Tew JM, Norman AB. The quinolinic acid model of Huntington's disease: locomotor abnormalities. Exp Neurol 1988;105:45–53.
36. Polgar S, Koutouzis TK, Cahill DW, Cahill DW, Sanberg PR. Fetal striatal transplantation and their relationship to transient hypoactivity. Soc Neurosci Abstr 1993;19:685.
37. Sanberg PR, Borlongan CV, Koutouzis TK, Norgren RB Jr, Cahill DW, Freeman TB. Human fetal striatal transplantation in an excitotoxic lesioned model of Huntington's disease. Ann NY Acad Sci 1997;831:452–60.
38. Pundt LL, Kondoh T, Conrad JA, Low WC. Transplantation of human striatal tissue into a rodent model of Huntington's disease: phenotypic expression of transplanted neurons and host-to-graft innervation. Brain Res Bull 1996;39:23–32.
39. Borlongan CV, Polgar S, Cahill DW, Freeman TB, Sanberg PR. Will fetal striatal transplantation correct the akinetic stage of Huntington's disease? Neurodegeneration 1996;5:189–195.
40. Quinn N, Brown R, Craufurd D, Goldman S, Hodges J, Kieburtz K, Lindvall O, MacMillan J, Roos R. Core assessment program for intracerebral transplantation in Huntington's disease (CAPIT-HD). Move Disord 1996;11:143–150.
41. Sanberg PR, Henault MA, Deckel AW. Locomotor hyperactivity: effects of multiple striatal transplants in an animal model of Huntington's disease. Pharamcol Biochem Behav 1986;25:297–300.
42. Deckel AW, Moran TH, Coyle JT, Sanberg PR, Robinson RG. Anatomical predictors of behavioral recovery following fetal striatal transplants. Brain Res 1986;365:249–258.
43. Burns LH, Pakzaban P, Deacon TW, Brownell AL, Tatter SB, Jenkins BG, Isacson O. Selective putaminal excitotoxic lesions in non-human primates model the movement disorder of Huntington's disease. Neuroscience 1995; 64:1007–1017.

44. Kanazawa I, Kimura M, Murata M, Tanaka Y, Cho F. Choreic movements in the macaque monkey induced by kainic acid lesions of the striatum combined with L-Dopa. Brain 1990;113:509–535.
45. Pakzaban P, Deacon TW, Burns LH, Isacson O. Increased proportion of AChE-rich zones and improved morphological integration in host striatum of fetal grafts derived from the lateral but not the medial ganglionic eminence. Exp Brain Res 1993;97:13–22.
46. Sanberg PR, Johnson DA, Moran TH, Coyle JT. Investigating Locomotion abnormalities in animal models of extrapyramidal disorders: a commentary. Physiol Psychol 1984;12:48–50.
47. Koutouzis TK, Borlongan CV, Freeman TB, Cahill DW, Sanberg PR. Intrastriatal 3-nitropropionic acid: a behavioral assessment. Neuroreport 1994; 5:2241–2245.
48. Borlongan CV, Randall TS, Cahill DW, Sanberg PR. Asymmetrical motor behavior in rats with unilateral striatal excitotoxic lesions as revealed by the elevated body swing test. Brain Res 1995;676:231–234.
49. Borlongan CV, Cahill DW, Sanberg PR. Asymmetrical behavior in rats following striatal lesions and fetal transplants: the elevated body swing test. Restor Neurol Neurosci 1995;9:15–19.
50. Sanberg PR, Henault MA, Hagenmeyer-Houser SH, Giordano M, Russell KH. Multiple transplants of fetal striatal tissue in the kainic acid model of Huntington's disease: behavioral recovery may not be related to acetylcholinesterase. In: Azmitia EC, Bjorklund A, eds. NY Acad Sci 1987;495: 781–785.
51. Fricker RA, Torres EM, Hume SP, Dunnett SB. Functional striatal grafts: correlations between positron emission tomography in vivo, graft survival, recovery of reaching behavior. Soc Neurosci Abstr 1994;20:473.
52. Brundin P, Fricker RA, Nakao N. Paucity of P-zones in striatal grafts prohibits commencement of clinical trials in Huntington's disease. Neuroscience 1996;71:895–897.
53. Freeman TB, Olanow CW, Hauser RA, Kordower JH, Holt DA, Borlongan CV, Sanberg PR. Human fetal tissue transplantation. In: Germano IM (ed.), Neurosurgical Treatment for Movement Disorders. NY: AANS, 1998, pp. 177–192.
54. Borlongan CV, Koutouzis TK, Freeman TB, Hauser RA, Cahill DW, Sanberg PR. Hyperactivity and hypoactivity in a rat model of Huntington's disease: the systemic 3-nitropropionic acid model. Brain Res Protoc 1997;1:253–237.
55. Borlongan CV, Nishino H, Sanberg PR. Systemic, but not intraparenchymal, administration of 3-nitropropionic acid mimics the neuropathology of Huntington's disease: a speculative explanation. Neurosci Res 1997;28:185–189.
56. Borlongan CV, Koutouzis TK, Sanberg PR. 3-Nitropropionic acid animal model and Huntington's disease. Neurosci Biobehav Rev 1997;21:289–293.
57. Borlongan CV, Kanning K, Poulos SG, Freeman TB, Cahill DW, Sanberg PR. Free radical damage and oxidative stress in Huntington's disease. J Fla Med Assoc 1996;83:335–341.

20

Neuroprotective Strategies in Parkinson's Disease and Huntington's Chorea

MPTP- and 3-NPA-Induced Neurodegeneration as Models

Moussa B. H. Youdim, Gopal Krishna, and Chuang C. Chiueh

INTRODUCTION

MPTP-Induced Animal Model of Parkinson's Disease

Parkinson's disease is a neurodegenerative disorder involving the progressive degeneration of dopamine neurons arising in the substantia nigra compacta area and terminating in the striatum. Dopamine replacement therapy by administration of L-dopa has been developed based on a specific loss of pigmented substantia nigra compacta (A9 nigral) neurons and striatal dopamine. However, dopamine replacement therapy has failed to stop the progression of the disease. The major objective is to develop a better therapeutic approach to the treatment and prevention of the disease. In the past few years much has been discovered about the chemical pathology of Parkinson's disease. This new information gives hope not only for finding the cause of the disease, but also for developing new preventive drugs that may either halt the progressive degeneration of the A9 nigral neurons or, perhaps, provide means to rescue these dopamine neurons. The current hypothesis concerning the pathogenesis of Parkinson's disease holds that there is an ongoing selective oxidative stress that cannot be thoroughly investigated using brain tissues obtained during autopsy. Much of what we have learned about oxidative neurodegeneration have come from studies with 6-hydroxydopamine *(1,2)* and 1-methyl-4-phenyl-1,2,3,6-tetrahydropyridine (MPTP)—a neurotoxin that produces animal model for investigating oxidative stress and Parkinsonian syndrome *(3,4,5)*.

MPTP is a manmade neurotoxin that produces a selective nigral loss and a Parkinsonian syndrome in humans *(6,7)*. In animal models, low doses of MPTP induce a selective destruction of the pigmented A9 nigral neurons of

From: *Mitochondrial Inhibitors and Neurodegenerative Disorders*
Edited by: P. R. Sanberg, H. Nishino, and C. V. Borlongan © Humana Press Inc., Totowa, NJ

primates *(3)*, while higher doses prompt a reversible dopamine depletion in rodents *(8)*. MPTP initiates its dopaminergic neurotoxicity via metabolism by monoamine oxidase to the reactive metabolites 1-methyl-4-phenyl dihydropyridine (MPDP) and 1-methyl-4-phenyl pyridinium ion (MPP^+) *(9,10)*. These toxic metabolites of MPTP induce sustained dopamine overflow, generation of hydroxyl radical, lipid peroxidation, and calcium overload, resulting in A9 nigral cell death and dopamine depletion *(11–15)*. It has also been shown that high concentrations of MPP^+ inhibit mitochondrial respiration and oxidative phosphorylation at complex I *(16,17)*.

Dopamine neuromelanin can be generated by mixing dopamine, iron, and oxygen; this free radical mediated polymerization of oxidized dopamine to neuromelanin is blocked by hydroxyl radical scavengers including selegiline *(5,18,19)*. These in vitro findings have led to a suggestion of a site-specific and age-dependent generation of cytotoxic hydroxyl radicals in these melanized A9 dopamine neurons. Among brain dopamine neurons, only the iron-containing A9 nigral neurons contain melanin which is subjected to a continuous oxidative stress, resulting in their demise. Therefore, neuromelanin may be a reliable biological marker for oxidative stress in vivo as dopamine melanin is a product of oxidant stress *(18)*.

The consequence of oxidative stress includes the initiation of reactive oxygen species generation (i.e., superoxide anion radical and hydroxyl radical), the propagation of lipid peroxidation (i.e., peroxyl lipid radical, hydroxyl radical), protein oxidation (i.e., thiyl radicals), and DNA damage (necrotic and apoptotic cell death). Moreover, oxidative stress can also deplete ATP and reduced glutathione (GSH). Prolonged oxidative stress may cause neuronal degeneration when cellular repair mechanisms and antioxidant defense systems are weakened by factors such as aging, brain injury, and/or neurotoxic insults, including mitochondrial poison. Early findings demonstrated MPTP (<1.5 mg/kg, i.v.) sparing of nonpigmented dopamine neurons while inducing a selective destruction of pigmented A9 dopamine neurons preceded by increased dopamine levels and turnover in monkey putamen *(20)*. In addition, suppression of exaggerated dopamine turnover or release by NSD-1015, a dopa decarboxylase inhibitor, completely protects nigrostriatal dopamine neurons from oxidative injury caused by MPTP in C57BL6 mice *(21)*. This in turn has stimulated current investigations of the putative role of free radicals in MPTP-induced selective nigral degeneration for supporting oxidant stress hypothesis in Parkinson's disease *(5,22,23)*. Furthermore, MPTP-induced Parkinsonian animal models are useful for evaluating new treatments, brain imaging, transplant, and pallidotomy procedures *(24)*.

3-Nitropropionic Acid: A Mycotoxin Producing Huntington's Disease-Like Striatal Lesions and Dystonia

Extrapyramidal lesions caused by mildewed sugarcane poisoning may be mediated by 3-nitropropionic acid (3-NPA), which is produced by the *Arthrinium* fungus *(25–28)*. In northern China, children who have ingested rotting sugarcane covered with this fungus have developed irreversible dystonia. The clinical syndrome of tardive dystonia (e.g., choreoathetosis, torsion spasms, paroxysmal spasm of extremities) differs from the progressive nature of Huntington's disease; this 3-NPA-induced brain disorder is also apparently associated with bilateral lenticular lesions and necrosis in the putamen *(28,29)*.

3-NPA, which irreversibly inhibits the Krebs cycle succinate dehydrogenase *(30)*, induces lactate accumulation, mitochondrial swelling, and striatal damage in experimental animals *(30,31)*. It also impairs the blood–brain barrier. These studies indicate that 3-NPA is an environmental neurotoxin that affects striatal neurons, astroglial cells, and endothelium in humans. Thus, this mycotoxin could be utilized to investigate selective neuronal degeneration in the striatum—the hallmark of Huntington's disease *(32)*.

In addition to murine models, 3-NPA-induced diffused striatal degeneration and choreiform movements have been demonstrated in nonhuman primates *(33,34)*. Similar to MPTP-induced animal models for the investigation of the pathogenesis of Parkinson's disease, the unique neurotoxicity of 3-NPA provides a means to develop an animal model that replicates many, if not all, of the clinical and pathological features of Huntington's disease.

Experimental results show that the neurotoxins MPTP and 3-NPA both cause striatal lesions; however, damage caused by low-dose MPTP is highly site specific and is limited to melanized A9 nigrostriatal neurons *(3,20)*. The striatal lesions caused by 3-NPA are more diffuse and extend to the blood–brain barrier, astrocytes, striatal interneurons, and nigrostriatal dopaminergic fibers. Significant attenuation of striatal neurotoxicity is reported in copper/zinc superoxide dismutase (Cu/Zn-SOD) transgenic mice treated with either MPTP or 3-NPA *(35,36)*. Therefore, reactive oxygen species may play a role in site-specific and diffuse neurotoxicity produced by MPTP and 3-NPA, respectively.

OXIDANT STRESS AND NEURODEGENERATION

MPTP

Generation of Hydroxyl Radicals by MPP^+

During the past three decades, the oxidant stress hypothesis in neurodegeneration has been questioned owing to lack of direct evidence in vivo.

We have developed a method for trapping of hydroxyl radicals by intracerebral microdialysis of salicylate. We designed this in vivo procedure to demonstrate the time course of increased generation of reactive hydroxyl radicals in the striatum *(12)*. Further, this trapping of hydroxyl radicals by salicylate, assayed as 2,3-dihydroxybenzoic acid, has been used to monitor the formation of hydroxyl radicals in the striatum following the administration of striatal neurotoxins, including MPTP analogs *(13)*, MPDP (14), MPP^+ *(12)*, 3-NPA *(15)*, and methamphetamine *(37)*.

A sustained generation of hydroxyl radicals, semiquinone radicals, and peroxyl lipid radicals is apparent during both enzymatic and nonenzymatic oxidation of striatal dopamine when it is released in large quantities by MPDP and MPP^+ *(12,14)*. High levels of iron in the basal ganglia *(22,38)* may exaggerate iron-catalyzed dopamine autooxidation and generation of cytotoxic hydroxyl radicals and peroxyl lipid radicals inside and/or near the A9 nigral terminals *(5)*, leading to retrograde degeneration of nigral neurons *(39)*. In addition, MPP^+ undergoes a redox cycling in the presence of oxygen resulting in generation of superoxide *(40,41)*.

MPTP and Accumulation of Iron in Nigral Neurons

Iron accumulates in an age-dependent fashion in the A9 nigral neurons and in the globus pallidum. Iron may also play a role in neurodegeneration caused by MPTP and 6-hydroxydopamine. MPTP increases nigral iron by twofold *(42,43)* while 6-hydroxydopamine induces a 1.25 fold increase *(44)*. An excess amount of iron complexes *(45)* but not manganese *(46)* in the A9 nigral neurons leads to increases in dopamine turnover, hydroxyl radical generation, lipid peroxidation, dopamine depletion, and nigral loss which are associated with significant motor dysfunction.

Most of the nigral iron is deposited as ferritin complexes that can be released as biologically active small molecular weight iron complexes (i.e., ferrous citrate) for regulating dopamine biosynthesis and other cellular functions. Progressive increases of the cellular pool of free iron may generate reactive hydroxyl radicals through the redox cycling of iron–oxygen complexes especially in the presence of dopamine *(5)*. Iron acts as a catalyst for activating tyrosine hydroxylase—the rate-limiting step of dopamine biosynthesis. Because of a relative high concentration of transferrin transporters in the A9 nigral neurons, intranigral infusion of ferrous citrate causes acute increases in dopamine turnover and oxidant stress *(45,46)* which results in nigral loss and chronic striatal dopamine depletion *(45,47–49)*. Iron-complex-induced hydroxyl radical generation, lipid peroxidation, rotational behavior, and associated dopamine depletion can be prevented by antioxidants, including manganese and *S*-nitrosoglutathione (GSNO)/nitric oxide *(46,50)*. Thus, a slow and progressive increase of free iron as ferrous citrate/isocitrate redox complexes may contribute to the progressive nigrostriatal degeneration during senescence and, perhaps, in Parkinson's disease. Because nigral uptake of iron is doubled by MPTP, this MPTP-induced site-specific dopamine autooxidation and oxidative stress may be augmented by iron. This ultimately leads to the MPTP-induced selective damage to melanized iron-rich A9 dopaminergic neurons in the substantia nigra compacta area.

Role of Dopamine in the Selective Neurotoxicity of MPTP

MPTP analogs such as 2'-methyl-MPTP, MPDP and MPP^+ produce a sustained dopamine efflux from the nigrostriatal nerve terminals that can last for hours in rat brain and days in primate brain. Most of the released dopamine is oxidized either enzymatically by monoamine oxidase or nonenzymatically via iron-catalyzed autooxidation *(5)*. Either case leads to the site-specific generation of cytotoxic species such as hydroxyl radicals, dopamine aldehyde, hydrogen peroxide, and semiquinone radicals inside and/or near the dopaminergic synapse. Moreover, the intranigral infusion of MPP^+ not only causes the generation of hydroxyl radicals but it also increases brain lipid peroxidation—a hallmark for free radical induced chain reactions *(51)*. It is also known that MPP^+ increases the release of iron from ferritin *(52)*. Eventually, MPP^+ causes calcium overload and the death of the A9 nigral neurons *(11,45)*. Therefore, the selective neurotoxicity of MPTP may be mediated by oxidative stress induced by iron and dopamine in the A9 nigrostriatal neurons.

Inhibition of Mitochondria Complex I by MPP^+

Free radicals are products of the reduction–oxidation (redox) cycling of oxygen and drugs. The inhibition of complex I in mitochondrial oxidative phosphorylation may donate unpaired electrons to oxygen, leading to the generation of reactive oxygen species. MPP^+ (500 μM) inhibits complex I as well as the synthesis of ATP in mitochondria isolated from cortex and striatum *(16,17,53)*. The inhibition of complex I caused by high concentrations of MPP^+ may be mediated by free radicals as antioxidants ameliorate its inhibition of oxidative phosphorylation *(54)*. However, tissue levels of MPP^+, in the intact locus ceruleus noradrenergic neurons are much higher than in the damaged A9 nigral dopaminergic neurons. It is not known whether the pigmented A9 nigral neurons can concentrate enough MPP^+ in vivo to inhibit complex I in the mitochondrial respiratory chain after administration of toxic doses of MPTP (<1.5 mg/kg) in primates. Moreover, MPP^+ inhibits complex I in all brain regions including areas and neurons that are not sensitive to MPTP *(16,17)*. Thus, the problem of complex I inhibition and its role in the dopamine neurotoxicity of MPTP needs further investigation.

Putative Role of Peroxynitrite

Peroxynitrite is the product of mixing equal moles of nitric oxide and superoxide anion at high pH *(55)*. This mixture is able to decompose to form hydroxyl and nitrogen dioxide radicals that can be blocked by carbon dioxide at neutral pH. Under normal physiological conditions, however, peroxynitrite is rapidly converted to nitrites (i.e., formation of nitrotyrosine) but not hydroxyl radicals (e.g., no lipid peroxidation or protein oxidation). Moreover, in vitro studies have shown that weak brain lipid peroxidation and methionine oxidation caused by peroxynitrite can be inhibited with nitric oxide and carbon dioxide *(50,56,57)*. Its proposed cytotoxic effect, hence, cannot be demonstrated in vivo because carbon dioxide, nitric oxide, and GSNO can effectively ameliorate peroxynitrite's weak prooxidant action *(50)*. In fact, nitric oxide and GSNO protect A9 nigral neurons against oxidative stress caused by iron complexes (i.e., ferrous citrate and hemoglobin). In addition, it is known that neuronal nitric oxide synthase is not located in the nigrostriatal dopamine neurons, and intranigral infusion of nitric oxide or peroxynitrite does not cause oxidative brain injury *(50,58)*. In neuronal nitric oxide synthase knockout mice, MPTP still causes significant striatal dopamine depletion, indicating that neurotoxic effects of MPTP/MPP^+ are not mediated by nitric oxide.

Role of Reactive Microglia

The first sign of possible involvement of an inflammatory process in MPP^+ induced striatal injury is revealed by a delayed increase in the cellular calcium and interleukin-2 at the lesion sites *(11,59)*. Further, oxidative brain damage may induce an inflammatory process that results in proliferation of reactive microglia near the Parkinsonian A9 nigral neurons *(60)*. The function of microglia activation is not clear. It is believed reactive microglia may generate free radicals and/or toxic agents *(61)*. However, other studies suggest that microglia can generate neurotrophins and cytokines which may provide protection against oxidative damages caused by free radicals generated by MPTP, 6-hydroxydopamine, and 3-NPA *(36,62–64)*. Thus the mechanisms by which microglia can alter their physiological function as a neuroprotective or a cytotoxic system needs clarification if we are to develop more effective neuroprotective agents.

Neurotoxic Mechanisms for 3-NPA in the Striatum

Augmentation of Dopamine-Mediated Oxidant Stress by Mitochondria Complex II Inhibition

3-NPA causes selective striatal lesions without inducing lesions in cortical neurons despite its ability to produce a uniform inhibition of succinate dehydrogenase in both the cortical and striatal brain regions *(30,33)*. These results indicate that impairment of mitochondrial complex II alone may not be responsible for bilateral striatal lesions caused by 3-NPA. In a similar situation, MPTP inhibits complex I in all brain regions. Although MPP^+ is taken up by all brain dopamine neurons, it causes a site-specific oxidative stress and tissue damage only in the melanized A9 nigral neurons that contain high levels of iron and dopamine.

Recently, it has also been shown that dopamine augments or mediates astrocyte and endothelial cell death caused by 3-NPA *(65)* and methyl malonate—a reversible succinate dehydrogenase inhibitor *(66)*. Thus, dopamine-induced oxidative damage to striatal neurons and other brain cells are greatly potentiated by minimal inhibition of mitochondrial complex II. Moreover, 3-NPA can also cause oxidative injury to the nigrostriatal dopamine terminals in the striatum. Because 3-NPA is not selectively taken up and accumulated by dopaminergic fibers, it is unable to produce MPP^+-like site-specific oxidative damage to the nigrostriatal dopaminergic nerve terminals.

3-NPA-Induced Oxidative Stress in the Striatum

It is intriguing to explain why 3-NPA generates hydroxyl radicals, increases lipid peroxidation, and causes diffused oxidative lesions only in the striatum but not in the cortex *(29,36)*. There are several clues to explain why 3-NPA produces global lesions only in the striatum. Similar to the effects of iron in the generation of dopamine melanin *(5,18)*, 3-NPA enhances adrenochrome formation *(29)*. Therefore, we propose that in the striatum, 3-NPA may increase ferrous citrate-like prooxidative iron complexes because 3-NPA increased acidosis, resulting in the accumulation of lactate and/or citrate *(31,67)*. These acids could enlarge the pool size of small molecular weight iron complexes in the striatum where high levels of nonheme iron and ferritin are located. Such increases have been shown to exaggerate dopamine turnover and cause greater oxidative stress through the generation of hydroxyl radicals and lipid peroxidation in the striatum. These acidic iron complexes may be able to diffuse freely in the striatum during 3-NPA-induced acidosis, resulting in diffuse oxidative lesions.

3-NPA-induced inhibition of ATP formation may also interfere with cellular antioxidative defense enzymes and thus prolong the half-life of cytotoxic free radicals. Citrate/isocitrate, ADP, and GSSG or thiyl radicals are known to chelate iron and generate more free radicals in the presence of ascorbate through the redox cycling. Metal chelating agents such as EDTA and deferoxamine at low concentrations inhibit iron-catalyzed formation of hydroxyl radicals but they may enter the redox cycle in the presence of ascorbate, resulting in detrimental side effects. Catechol dopamines may also chelate iron and thus promote oxidant stress by generating more free radicals. Further, cytotoxic free radicals can mediate 3-NPA-induced delayed apoptotic cell death via the caspase pathway *(68)*. Consequently, we propose that 3-NPA-induced striatal lesions are mediated by diffuse oxidative stress in the striatum where high levels of iron and dopamine are located.

PROTECTION OF BRAIN NEURONS FROM OXIDANT STRESS

MPTP

Prevention of MPP^+ uptake into nigral neurons may suppress the neurotoxicity of MPTP. Transgenic knockout of vesicular transports in conjunction with depletion of ATP may potentiate the dopaminergic toxicity of MPTP because intracellular dopamine can no longer be protected in the synaptic vesicles containing relatively high levels of ATP and ascorbate *(69)*.

MPP^+-induced oxidative injury in the midbrain dopamine neuronal cultures or PC12 cells is suppressed by antioxidants including pramipexole, apomorphine, lazaroid, trolox, selegiline, bromocryptine, estradiol, and melatonin *(70,71)*. The neuroprotective action of selegiline is similar to that of other atypical antioxidants that inhibit brain lipid peroxidation and polymerization of dopamine melanin and also protects nigral neurons from MPP^+-induced injury both in vitro and in vivo *(72,73)*. In addition to inhibition of monoamine oxidase B, selegiline also suppresses MPP^+-induced generation of hydroxyl radicals and lipid peroxidation. In PC12 cells, selegiline and estradiol increase Bcl2 and prevent p53-mediated apoptotic cell death *(74)*. Recent in vivo studies indicate that selegiline protects and/or rescues nigral neurons from mild to moderate but not severe oxidative injury induced by MPP^+ *(75)*.

Based on the peroxynitrite hypothesis, attempts have been made to protect nigral neurons from injury mediated by putative toxic nitric oxide derivatives through inhibiting neuronal nitric oxides such as 7-nitroindazole and L-N^G-nitroarginine methyl ester. However, only 7-nitroindazole but not L-N^G-nitroarginine methyl ester protects the striatum from oxidative stress caused by MPTP *(76,77)*. Other studies indicate that 7-nitroindazole can suppress the generation of hydroxyl radicals, the peroxidation of brain lipids, and the inhibition of the bioactivation of MPTP analogues *(15,78)*. Therefore, similar to broad pharmacological actions of selegiline, 7-nitroindazole may have atypical antioxidant properties. Recent reports indicate that based on their broad spectrum of actions selegiline and 7-nitroindazole also protect brain neurons from oxidative damage caused by ischemia-reperfusion injury *(79,80)*.

Protection Against the Neurotoxicity of 3-NPA

The inhibition of mitochondrial complex II in the cortical areas by 3-NPA fails to cause blood–brain barrier impairment; it produces neither slow necrotic cytotoxicity nor delayed apoptosis in these brain regions as they contain much less iron and dopamine. However, 3-NPA generates hydroxyl radicals and augments oxidative damages in the striatum, where high concentrations of iron and dopamine are located. It is interesting to note that neurons containing NADPH diaphorase (nitric oxide synthase) resist the oxidative injury of 3-NPA, which is consistent with a new finding that nitric oxide and GSNO are potent neuroprotective antioxidants in the brain *(49)*.

Most of the current neuroprotective studies focus on the possible use of a single therapeutic agent for the prevention of the cytotoxicity of 3-NPA. Treatments, however, that aim at antagonizing excitatory amino acids cannot fully protect striatal neurons and other brain cells from oxidant stress or delayed apoptotic cell death *(81)*. In fact, MK-801, a potent inhibitor of *N*-methyl-D-aspartate receptors, causes significant neurodegeneration in the brain *(82)*. Agents that improve oxidative phosphorylation such as creatine, coenzyme Q_{10} and nicotinamide may partially ameliorate cytotoxicity caused by 3-NPA and MPP^+ *(83)*. Agents that quench cytotoxic hydroxyl radicals (i.e., 5,5-dimethyl-1-pyrroline-*n*-oxide (DMPO), 7-nitroindazole) indeed provide significant protection against striatal lesions produced by MPTP *(84)*. Free radical spin trapping compounds such as α-phenyl-*tert*-butyl-nitron are effective in the MPTP but not the 3-NPA model *(85)*. Further, up-regulation of Cu/Zn-SOD also significantly decreases oxidative damages caused by MPTP and 3-NPA *(35,36)*. These results support the current notion that site-specific and diffused dopamine-induced oxidant stress may mediate different degrees of striatal lesions caused by MPTP and 3-NPA, respectively.

PROSPECTIVE NEUROPROTECTIVE STRATEGIES

Based on apparent neuroprotective effects of antioxidative cellular defense systems in animal models of MPTP and 3-NPA, we are developing

Table 1
Prospective Neuroprotective Strategies Against Oxidant-Induced Neurodegeneration

1. Hypothermia therapy
2. Antioxidative agents and free radical scavengers
3. Energizers/nutrients including coenzyme Q_{10}, creatine
4. Buffers, chelators, and antagonists against calcium and iron
5. Protease inhibitors against calpain, platelet activating factor, and caspases
6. Neurotrophins and cystokines
7. Up-regulation of antioxidative cellular defense systems including the SOD system and putative GSNO/GSH/NO pathway

Clinical application of neuroprotective agents: earlier intervention >> delayed treatment; combined therapeutics >> single therapy; gene activation >> gene therapy.

synergistic neuroprotective strategies against progressive degenerative brain disorders caused by reactive oxygen species (Table 1). Owing to the complex oxidative cascades and lipid peroxidation chain reactions, the administration of multiple therapeutics generally provides better and more efficient protection against oxidant-induced brain lesions. Interestingly, some of the atypical neuroprotective antioxidants (i.e., estrogen, selegiline, and melatonin), in fact, can up-regulate cellular defense enzymes such as bcl2, neurotrophins, and SOD. Additional studies are required to elucidate the beneficial molecular mechanisms novel gene activators that may provide similar clinical outcomes as complicated gene therapy. Moreover, our in vivo data reveal that neuroprotective agents may be able to protect and/or rescue mildly to moderately damaged neurons but not severely affected ones *(75)*. Similar observations have been reported in recent clinical trials with pramipexole, a dopamine agonist with antioxidative properties *(70)*; the beneficial effects of neuroprotectants are apparent only in the early clinical stage but not in the advanced stage. Moreover, there are numerous examples in drug development in which drugs with a single pharmaceutical action have failed to combat oxidative stress-induced progressive clinical problems. Therefore, there is an urgent need for the development of clinical/biological markers for identifying patients at their early preclinical stage. Clinical efficacy could be significantly improved by early treatment with neuroprotective agents to slow the progression of degenerative brain disorders. A small improvement in the cellular antioxidative defense system may retard progressive motor, cognitive, and/or mental disorders caused by brain atrophy, and thus improve quality of life.

SUMMARY

Based on the information collected during the investigation of drug-induced Parkinsonism in a Danish medicinal chemist and in drug abusers in the United States, MPTP has been used to develop animal models for studying Parkinson's disease. Similarly, animal models for investigating Huntington's disease have also been developed based on the discovery of pathogenic role of 3-NPA, a mycotoxin, in sugarcane poisoning cases in China. MPTP and 3-NPA models have provided relevant evidence for supporting the role of oxidant stress in the pathogenesis of degenerative brain disorders such as Parkinson's disease and Huntington's disease. Furthermore, elucidation of neurodegenerative mechanisms caused by synergistic prooxidative actions of iron and dopamine plus the impairment of the respiratory chain in oxidative phosphorylation (i.e., mitochondria complex I and II) could lead to the development of effective neuroprotective agents. Neuroprotective strategies could then be outlined for the treatment of progressive brain atrophy and its associated movement and mental disorders, including dementia.

ACKNOWLEDGMENTS

This chapter was prepared as part of the official duties of Moussa B. H. Youdim as Fogarty Scholar-in-Residence sponsored by the Fogarty International Center for Advanced Study in Health Sciences, NIH, Bethesda, MD. Gopal Krishna is an emeritus scientist. We appreciate the excellent editorial assistance provided by Ms. Margaret Nguyen.

REFERENCES

1. Cohen G, Heikkila RE. The generation of hydrogen peroxide superoxide radical and hydroxyl radical by 6-hydroxydopamine dialuric acid and related cytotoxic agents. J Biochem 1974;249:2447–2452.
2. Glinka Y, Tipton KF, Youdim MBH. Nature of inhibition of mitochondrial respiratory complex I by 6-hydroxydopamine. J Neurochem 1996;66:2004–2010.
3. Burns RS, Chiueh CC Markey SP, Ebert MH, Jacobowitz D, Kopin IJ. A primate model of Parkinson's disease: selective destruction of substantia nigra pars compacta dopaminergic neurons by *N*-methyl-4-phenyl-1,2,3,6-tetrahydropyridine. Proc Natl Acad Sci USA 1983;80:4546–4550.
4. Chiueh CC. Dopamine in the extrapyramidal motor function: a study based upon the MPTP-induced primate model of parkinsonism. Ann NY Acad Sci 1988;515:226–238.
5. Chiueh CC, Miyake H, Peng MT. Role of dopamine autoxidation hydroxyl radical generation and calcium overload in underlying mechanisms involved in MPTP-induced parkinsonism. Adv Neurol 1993;60:251–258.

6. Davis G, Williams A, Markey S, Ebert M, Calne E, Beichert C, Kopin IJ. Chronic parkinsonism secondary to intravenous injection of meperidine analogues. Pschiatry Res 1979;1:249–254.
7. Langston JW, Ballard P, Tetrud JW, Irwin I. Chronic parkinsonism in humans due to a product of meperidine-analog synthesis. Science 1983;4587:979–980.
8. Chiueh CC, Johannessen JN, Sun JL, Bacon JP, Markey SP. Reversible neurotoxicity of MPTP in the nigrostriatal dopaminergic system of mice. In: MPTP-Parkinsonian Syndrome Producing Neurotoxin. Markey SP, Trevor AJ, Castagnoli N, Kopin IJ, eds. Academic Press, New York, 1986, pp. 473–480.
9. Chiba K, Trevor A, Castagnoli N Jr. Metabolism of the neurotoxic tertiary amine MPTP, by brain monoamine oxidase. Biochem Biophys Res Commun 1984;120:574–578.
10. Markey SP, Johannessen JN, Chiueh CC, Burns RS, and Herkenham MA. Intraneuronal accumulation of a pyridinium metabolite may induce drug-induced parkinsonian syndrome in monkey. Nature (London) 1984;311:464–467.
11. Sun CJ, Johannessen JN, Gessner W, Namura I, Singhaniyom W, Brossi A, Chiueh CC. Neurotoxic damage to the nigrostriatal system in rats following intranigral administration of $MPDP^+$ and MPP^+. J Neural Transm 1988;74: 75–86.
12. Chiueh CC, Krishna G, Tulsi P, Obata T, Lang K, Huang S-J, and Murphy DL. Intracranial microdialysis of salicylic acid to detect hydroxyl radical generation through dopamine autoxidation in the caudate nucleus: effects of MPP^+. Free Radical Biol Med 1992a;13:581–583.
13. Chiueh CC, Huang S-J, Murphy DL. Enhanced hydroxyl radical generation by 2'-methyl analog of MPTP: reversal by clorgyline and deprenyl. Synapse 1992b;11:346–348.
14. Obata T, Chiueh CC. In vivo trapping of hydroxyl free radicals in the striatum utilizing intracranial microdialysis perfusion of salicylate: effects of MPTP, $MPDP^+$ and MPP^+ . J Neural Transm [Gen Sect] 1992;89:139–145.
15. Chiueh CC, Rauhala P. Free Radicals and MPTP-induced selective destruction of substantia nigra compacta neurons. Adv Pharmacol 1997;42:796–800.
16. Thakar JH, Hassan MN. Effects of 1-methyl-4-phenyl-1,2,3,6: tetrahydropyridine (MPTP) cyperquat (MPP^+) and paraquat on isolated mitochondria from rat striatum cortex and liver. Life Sci 1988;43:143–9.
17. Mizuno Y, Suzuki K, Sone N, Saitoh T. Inhibition of mitochondrial respiration by MPTP in mouse brain. Neurosci Lett 1988;3,349–353.
18. Chiueh CC, Murphy DL, Miyake H, Lang K, Tulsi PK, Huang SJ. Hydroxyl free radicals. (•OH) formation reflected by salicylate hydroxylation and neuromelanin: in vivo markers for oxidant injury of nigral neurons. Ann NY Acad Sci 1993;679,370–75.
19. Linert W, Herrlinger E, Jameson RF, Kienzl E, Jellinger K, Youdim MBH. Dopamine 6-hydroxydopamine iron and dioxygen: their mutual interactions and possible implication in the development of Parkinson's disease. Biochim Biophys Acta 1996;1316:160–168.
20. Chiueh CC, Markey SP, Burns SP, Johanneseen JN, Kopin IJ. Neurochemical and behaviorial effects of 1-methyl-4-phenyl-1,2,3,6-tetrahydropyridine in rat guinea pig and monkey. Psychopharmacol Bull 1984;20:548 553.

21. Chiueh CC, Johannessen JN, Sun JL, Bacon JP, Markey SP. Reversible neurotoxicity of MPTP in the nigrostriatal dopaminergic system of mice. In: MPTP-Parkinsonian Syndrome Producing Neurotoxin. Markey SP, Trevor AJ, Castagnoli N, Kopin IJ, eds. Academic Press, New York, 1986, pp. 473–480.
22. Riederer P, Sofic E, Rauch WD, Schmidt B, Reynolds GP, Jellinger K, Youdim MBH. Transition metals ferritin glutathione and ascorbic acid in Parkinsonian brains. J Neurochem 1989;52:512–520.
23. Fahn S, Cohen G. The oxidant stress hypothesis in Parkinson's disease: evidence supporting it. Ann Neurol 1992;32:804–812.
24. Chiueh CC. Dopamine in the extrapyramidal motor function: a study based upon the MPTP-induced primate model of parkinsonism. Ann NY Acad Sci 1988;515:226–238.
25. Hu WJ. Isolation and structure determination of *Arthrinium* toxin causing sugarcane poisoning. Chung Hua Yu Fang Hsueh Tsa Chih. 1986;20:321–323.
26. He FS. Extrapyrimidal lesions caused by mildewed cane poisoning (with a report of 3 cases). Chung Hua I Hsueh Tsa Chih. 1987;67:395–396.
27. Liu X, Luo X, Hu W. Studies on the epidemiology and etiology of moldy sugarcane poisoning in China. Biomed Environ Sci 1992;5:161–177.
28. He FS, Zhang S, Qian F, Zhang C. Delayed dystonia with striatal CT lucencies induced by a mycotoxin. 3-nitropropionic acid). Neurology 1995;45:2178–2183.
29. Fu YT, He FS, Zhang S, Jiao X. Consistent striatal damage in rats induced by 3-nitropropionic acid and cultures of arthrinium fungus. Neurotoxicol Teratol 1995;17:413–418.
30. Gould DH, Wilson MP, Hamar DW. Brain enzyme and clinical alterations induced in rats and mice by nitroalipathic toxicants. Toxicol Lett 1985;27:83–89.
31. Hamilton BF, Gould DH. Nature and distribution of brain lesions in rats intoxicated with 3-nitopropionic acid: a type of hypoxic. energy deficient;brain damage. Acta Neuropathol (Berl) 1987;72:286–297.
32. Ludolph AC, He F, Spencer PS, Hammerstad J, Sabri M. 3-Nitropropionic acid-exogenous animal neurotoxin and possible human striatal toxin. Can J Neurol Sci 1991;18:492–498.
33. Brouillet E, Hantraye P, Ferrante RJ , Dolan R, Leroy-Willig A, Kowall NW, Beal MF. Chronic mitochondrial energy impairment produces selective striatal degeneration and abnormal choreiform movements in primates. Proc Natl Acad Sci USA 1995;92:7105–7109.
34. Palfi S, Ferrante RJ, Brouillet E, Beal MF, Dolan R, Guyot MC, Peschanski M, Hantraye P. 1996;Chronic 3-nitropropionic acid treatment in baboons replicates the cognitive and motor deficits of Huntington's disease. J Neurosci 16:3019–3025.
35. Przedborski S, KosticV, Jacksonlewis V, Naini AB, Simonetti S, Fahn S, Carlson E, Epstein CJ, Cadet JL. Transgenic mice with increased Cu/Zn-superoxide dismutase activity are resistant to N-methyl-4-phenyl-1,2,3,6-tetrahydropyridine-induced neurotoxicity. J Neurosci 1992;12,1658–1667.
36. Beal MF, Ferrante RJ, Henshaw R, Matthews RT, Chan PH, Kowall NW, Epstein CJ, and Schulz JB. 3-Nitropropionic acid neurotoxicity is attenuated in copper/zinc superoxide dismutase transgenic mice. J Neurochem 1995; 65:919–922.

37. Fleckenstein AE, Wilkins DG, Gibb JW, Hanson GR. Interaction between hyperthermia and oxygen radical formation in the 5-hydroxytryptaminergic response to a single methamphetamine administration. J Pharmocol Exp Ther 1997;283:2681–285.
38. Youdim MBH, Ben-Shachar D, Riederer P. The possible role of iron in the etiopathology of Parkinson's disease. Move Disord 1993;8:1–12.
39. Johannessen JN, Chiueh CC, Bacon JP, Garrick NA, Burns RS, Weise VK, Kopin IJ, Parisi JE, Markey SP. Effects of 1-methyl-4-phenyl-1,2,3,6-tetrahydropyridine in the dog: effect of pargyline pretreatment. J Neurochem 1989;53:582–589.
40. Sinha BK, Singh Y, and Krishna G. Formation of superoxide and hydroxyl radicals from 1-methyl-4-phenylpyridinium ion. MPP^+); reductive activation by NADPH cytochrome P-450 reductase. Biochem Biophys Res Commun 1986;135:583–588.
41. Adams JD, Klaidman LK, Leung AC. MPP^+ and $MPDP^+$ induced oxygen radical formtion with mitochondrial-enzymes. Free Radical Biol Med 1993; 15:181–186.
42. Temlett JA, Landsberg JP, Watt F, Grime GW. Increased iron in the substantia nigra compacta of the MPTP-lesioned hemiparkinsonian african green monkey: evidence from proton microprobe element microanalysis. J Neurochem 1994;62:134–146.
43. Mochizuki H, Imai. H, Endo K, Yokomizo K, Murata Y, Hattori N, Mizuno Y. Iron accumulation in the substantia nigra of 1-methyl-4-phenyl-1,2,3,6-tetrahydropyridine. MPTP)-induced hemiparkinsonian monkeys. Neurosci Lett 1994;168:251–253.
44. Oestreicher E, Sengstock GJ, Riederer P, Olanow CW, Dunn AJ Arendash GW. Degeneration of nigrostriatal dopaminergic neurons increases iron within the substantia nigra: a histochemical and neurochemical study. Brain Res 1994;660:8–18.
45. Mohanakumar KP, de Bartolomeis A, Wu R-M, Yeh KJ, Sternberger LM, Peng S-Y, Murphy DL, Chiueh CC. Ferrous-citrate complex and nigral injury: evidence for free-radical formation and lipid peroxidation. Ann NY Acad Sci 1994;738:392–399.
46. Sziraki I, Mohanakumar KP, Rauhala P, Kim HG, Yeh JJ, Chiueh CC. Manganese: a transition metal protects nigrostriatal neurons from oxidative stress in the iron-induced animal model of parkinsonism. Neuroscience 1998; 85:1101–1011.
47. Ben-Shachar D, Eshei U, Finberg JPM, Youdim MBH. The iron chelator desferrioxamine (desferal) retards 6-hydroxydopamine-induced degeneration of nigrostriatal neurons. J Neurochem 1991;56:1441–1444.
48. Sengstock GJ, Olanow CW, Dunn AJ, Barone S Jr Arendash GW. Progressive changes in striatal dopaminergic markers nigral volume and rotational behavior following iron infusion into the rat substantia nigra. Exp Neurol 1994;130:82–94.
49. Rauhala P, Khaldi A, Mohanakumar KP, Chiueh CC. Apparent role of hydroxyl radicals in oxidative brain injury induced by sodium nitroprusside. Free Radical Biol Med 1998;24:1065–1073.

50. Rauhala P, Lin A M-Y, Chiueh CC. Neuroprotection by S-nitrosoglutathione of brain dopamine neurons from oxidative stress. FASEB J 1998;12:165–173.
51. Chiueh CC, Wu RM, Mohanakumar KP, Sternberger L, Krishna G, Obata T, Murphy DL. In vivo generation of hydroxyl radicals and MPTP-induced dopaminergic toxictiy in the basal ganglia. Ann NY Acad Sci 1994;738:25–36.
52. Youdim MBH, Riederer P. Understanding Parkinson's disease. Sci Am 1997;257:52–59.
53. Singer TP, Castagnoli N Jr Ramsay RR, Treor AJ. Biochemical events in the development of parkinsonism induced by 1-methyl-4-phenyl-1,2,3,6-tetrahydropyridine. J Neurochem 1987;49:1–8.
54. Seaton TA, Cooper JM, Schapira AHV. Free radical scavengers protect dopaminergic cell lines from apoptosis induced by complex I inhibitors. Brain Res 1997;777:110–118.
55. Beckman JS, Beckman TW, Chen J, Marshall PA, Freeman BA. Apparent hydroxyl radical production by peroxynitrite: implications for endothelial injury from nitric oxide and superoxide. Proc Natl Acad Sci USA 1990; 87:1620–1624.
56. Rauhala P, Sziraki I, Chiueh CC. Peroxidation of brain lipids in vitro: nitric oxide versus hydroxyl radicals. Free Radical Biol Med 1996;21:391–394.
57. Berlett BS, Levine RL, Stadtman E,R. Carbon dioxide stimulates peroxynitrite-mediated nitration of tyrosine residues and inhibits oxidation of methionine residues of glutamine synthetase: both modifications mimic effects of adenylation. Proc Natl Acad Sci USA 1998;95:2784–2789.
58. Rauhala P,Mohanakumar KP, Sziraki I, Lin AMY, Chiueh CC. *S*-nitrosothiols and nitric oxide but not sodium prusside protect nigrotstriatal dopamine nuerons against iron-induced oxidative stress in vivo. Synapse 1996;23:59–60.
59. Liang S-M, Liang C-M, Chiueh CC. 1989;Visualization of interleukin-2-like molecules in MPP^+-lesioned rat brain. Biochem Biophys Res Commun 165:1312–1318.
60. Jellinger K, Paulus W, and Grundke-Iqbal I, Riederer P, Youdim MBH. Brain iron and ferritin in Parkinson's and Alzheimer's diseases. J Neural Transm [PD Sect] 1990;2: 327–340.
61. Colton C, Gilbert D. Production of superoxide anions by a CNS macrophage the microglia. FEBS Lett 1987;223,284–288.
62. Hynes MA, Poulsen K Armanini M, Berkemeier L, Phillips H, Rosenthal A. Neutrophin-4/5 is a survival factor for embryonic midbrain dopaminergic-nuerons in enriched cultures. J Neurosci Res 1994;37:144–154.
63. Sautter J, Meyer M, Spenger C, Seiler RW, Widmer HR. Effects of combined BDNF and GDNF treatment on cultured dopaminergic midrain neurons. NeuroReport 1998;9:1093–1096.
64. Kirschner PB, Jenkins BG, Schulz JB, Finkelstein SP, Matthews RT, Rosen BR, Beal MF. NGF BDNF and NT–5 but not NT–3 protect against MPP^+ toxicity and oxidative stress in neonatal animals. Brain Res 1996;713:178–185.
65. Nishino H, Kumazaki M, Fukuda A, Fujimoto I, Shimano Y, Hida H, Sakurai T, Deshpande SB, Shimizu H, Morikawa S, Inubushi T. Acute 3-nitropropionic acid intoxication induces striatal astrocytic cell death and dys-

function of the blood brain barrier: involvement of dopamine toxicity. Neurosci Res 1997;27:343–355.
66. McLaughlin BA, Nelson D, Erecinska M, Chesselt M. Toxicity of dopamine to striatal neurons in vitro and potentiation of cell death by a mitochondrial inhibitor. J Neurochem 1998;70:2406–2415.
67. Tsai MJ, Goh CC, Wan YL, Chang , C. Metabolic alterations produced by 3-nitropropionic acid in rat striata and cultured astrocytes: quantitative in vitro 1H nuclear magnetic resonance spectroscopy and biochemical characterization. Neuroscience 1997;3:819–826.
68. Keller JN, Guo Q, Holtsberg FW, Bruce-Keller AJ, Mattson MP. Increased sensitivity to mitochondria toxin-induced apoptosis in neural cells expressing mutant presenilin-1 is linked to perturbed calcium homeostasis and enhanced oxyradical production. J Neurosci 1998;18:4439–4450.
69. Gainetdinov RR, Fumagalli F, Wang YM, Jones SR, Levey AI, Miller GW, Caron MG. Increased MPTP neurotoxcity in vesicular monoamine transporter 2 heterozygote knockout mice. J Neurochem 1998;70:73–78.
70. Cassarino DS, Fall CP, Smith TS, Bennet JP. Pramipexole reduces reactive oxygen species production in vivo and in vitro and inhibits the mitochondrial permeability transition produced by the Parkinsonian neurotoxin methylpyridinium ion. J Neurochem 1998;71:295–301.
71. Iacovitti L, Stull ND, Johnston K. Melatonin rescues dopamine neurons from cell death in tissue culture models of oxidative stress. Brain Research 1997;768,317–326.
72. Mytilineou C, Cohen G. Deprenyl protects dopamine neurons from the neurotoxic effect of 1-methyl-4-phenylpyridinium ion. J Neurochem 1985;6:1951–1953.
73. Wu R-M, Chiueh CC, Murphy DL. Apparent antioxidant effect of l-deprenyl. (selegiline) on suppression of hydroxyl radical formation and rescue of nigral injury elicited by 1-methyl-4-phenylpyridinium ion *in vivo*. Eur J Pharmacol 1993;243: 241–247.
74. Tatton WG. Selegline can mediate neuronal rescue rather than neuronal protection. Movement Disorders 1993;8(Suppl 1):S20–S30.
75. Wu RM, Murphy DL, Chiueh CC. Neuronal protective and rescue effects of deprenyl against MPP^+ dopaminergic toxicity. J Neural Transm 1995;100:53–61.
76. Schulz JB, Matthews RT, Muqit MMK, et al. Inhibition of neuronal ntiric oxide synthase by 7-nitroindazole protects against MPTP-induced neurotoxicity in mice. J Neurochem 1995;64:936–939.
77. Hantraye P, Brouillet E, Ferrante R, Palfi S, Dolan R, Matthews RT, Beal MF. Inhibition of neuronal nitric oxide synthase prevents MPTP-induced parkinsonism in baboons. Nature Med. 1996;2:1017–1021.
78. Castagnoli K, Palmer S Anderson A, Bueters T, Castagnoli N Jr. The neuronal nitric oxide synthase inhibitor 7-nitroindazole also inhibits the monoamine oxidase-B-catalyzed oxidation of 1-methyl-4-phenyl-1,2,3,6-tetrahydropyridine. Chem Res Toxicol 1997;10:364–368.

79. Nanri K, Montecot C, Springhetti V Seylaz J, Pinard E. The selective inhibitor of neuronal nitric oxide synthase 7-nitroindazole reduces the delayed neuronal damage due to forebrain ischemia in rats. Stroke 1998;29:1248–1253.
80. Lahtinen H, Kostinaho J, Kauppinen R, Haapalinna J, Keinanen R, Sivenius J. Selegiline treatment after transient global ischemia in gerbils enhances the survival of CA1 pyramidal cells in the hippocampus. Brain Res 1997; 757:260–267.
81. Pang Z, Geddes J W. Mechanisms of cell death induced by the mitochondrial toxin 3-nitropropionic acid: acute excitotoxic necrosis and delayed apoptosis. J Neurosci 1997;17,3064–3073.
82. Wang S, Yang Q, Moller CJ, Sharp FR, Haglid KG. Elirpodil prevents expression of the 70Kda heat shock protein in MK801-injured neurons. Pharmacol Toxicol 1996;79:166–168.
83. Matthews RT, Yang LC, Jenkins BG, Ferrante RJ, Rosen BR, Kaddurah-Daouk R, Beal MF. Neuroprotective effects of creatine and cyclocreatine in animal models of Huntington's disease. J Neurosci 1998;18:156–163.
84. Schulz JB, Henshaw DR, MacGarvey U, Beal MF. Involvement of oxidative stress in 3-nitropropionic acid neurotoxicity. Neurochem Int 1996;29:167–171.
85. Fredriksson A, Erikksson P Archer T. MPTP-induced deficits in motor activity: neuroprotective effects of spintrapping agent alpha-phenyl-*tert*-butylnitrone (PBN). J Neural Transm 1997;104:579–592.

Index